# Health in Late Prehistoric Thailand

Kathryn M. Domett

BAR International Series 946
2001

Published in 2016 by
BAR Publishing, Oxford

BAR International Series 946

*Health in Late Prehistoric Thailand*

ISBN 978 1 84171 238 3



BAR Publishing is the trading name of British Archaeological Reports (Oxford) Ltd. British Archaeological Reports was first incorporated in 1974 to publish the BAR Series, International and British. In 1992 Hadrian Books Ltd became part of the BAR group. This volume was originally published by Archaeopress in conjunction with British Archaeological Reports (Oxford) Ltd / Hadrian Books Ltd, the Series principal publisher, in 2001. This present volume is published by BAR Publishing, 2016.

Printed in England

BAR
PUBLISHING

BAR titles are available from:

BAR Publishing
122 Banbury Rd, Oxford, OX2 7BP, UK
EMAIL info@barpublishing.com
PHONE +44 (0)1865 310431
FAX +44 (0)1865 316916
www.barpublishing.com

# Abstract

The aim of this research was to provide a synthesis of the variation in health among prehistoric Thai communities and show that the health of these people was differentially affected by both their natural and cultural environment. Four skeletal samples comprising a total of 500 skeletons provided the material for this research. There were two coastal southeastern skeletal samples, Khok Phanom Di (2000-1500 BC, early agriculture) and Nong Nor (1100-700 BC, Bronze Age) and two inland northeastern samples, Ban Lum Khao (1400 BC, Bronze Age) and Ban Na Di (600-400 BC, Iron Age). It was hypothesised that the health profile of samples from within the same natural environment would be similar and, conversely, the health status of the northeast would be in contrast to the southeast. Additionally, it was hypothesised that changes in the cultural environment through time, including the intensification of rice agriculture, would see an improvement in general health.

Health, morbidity, and mortality were investigated through the analysis of a selection of parameters that included measures of mortality, growth, growth disturbances, joint disease, trauma, and dental health. This selection, although not exhaustive, enabled a *representation* of the health status of the four samples to be obtained and compared. Using a biocultural approach this information has been integrated with archaeological and ethnographic evidence for nutrition, pathogen load, and culture, to determine the effect each natural and cultural environment had on community health.

Within the southeast region the health profile of the Khok Phanom Di and Nong Nor samples were in contrast; Nong Nor had lower subadult mortality, taller adult statures and lower prevalences of dental pathologies. However, skeletal preservation was poor at Nong Nor and not all parameters were able to be observed. Within the northeastern samples health profiles were similar in overall prevalences of joint disease and dental health, but childhood morbidity and mortality were different. The latter, measured through observations of infant mortality, enamel hypoplasia and adult stature, were higher at Ban Lum Khao than Ban Na Di. The natural environment has had a significant influence on the health status of the people studied but not in the manner hypothesised. Comparisons within and between regions were complicated by time differences that may have affected the people's ability to cope with their environment. The comparatively poorer health suffered by the Khok Phanom Di and Ban Lum Khao communities may have been related to their settlement and adjustment of potentially new environments. In contrast, it is likely that both the Nong Nor and Ban Na Di communities were familiar with their natural environments as they were not the first settlers in their respective regions.

With respect to the cultural hypothesis, results indicated a general improvement in health had occurred through the time periods studied. The earliest sample, the people of Khok Phanom Di, were the least healthy. They had comparatively high prevalences of dental pathologies and joint disease, and high subadult mortality and morbidity, the latter reducing the attained adult stature. Health improved into the Bronze Age, particularly so in the Nong Nor sample, who had low subadult mortality and tall statures compared with the Ban Lum Khao sample. However, poor preservation of the Nong Nor skeletal material places a caveat over any interpretation. The people of the early Iron Age at Ban Na Di continued the trend for improving health. They had a moderate subadult mortality and morbidity, the latter was recovered from successfully as they reached tall statures. In addition, dental health improved and osteoarthritis decreased with time. Post-Iron Age, Thai people underwent an expansion of society led by the establishment of centralised political regime of the Angkorian civilisation. From this selective view of skeletal health it appears the people were healthy enough to withstand the effects of such a transition. Further integration with other prehistoric Southeast Asian skeletal samples is now required to support this statement

# Acknowledgements

Many people have helped me through the process of this research. Foremost in deserving of thanks is Dr Nancy Tayles. Nancy was the first to stimulate my interest in this field of research and, as my PhD supervisor, has provided me with endless guidance and support. Hallie Buckley, my fellow PhD student has also been a mainstay throughout. This research could not have gone ahead without the opportunity to work in Thailand provided by Professor Charles Higham, Dr Rachannie Thosarat and the Thai Fine Arts Department. The experience gained working on excavations in Thailand has been invaluable. Thanks go to all the people I have worked with in Thailand: Dougald, Paul, Nigel, Jeremy, Judy, Jill, Graeme, Sarah, Kirsten, Bec, and my sister Michelle. Thanks also to Dr Dougald O'Reilly for checking the facts in Chapter 2. Our Thai workers deserve thanks for always providing extra entertainment and being adept at getting rid of some unpleasant insects, spiders, and snakes!

Many other University of Otago colleagues have also helped along the way. Always a source of information, Drs Sue Mercer and George Dias were always happy to discuss aspects of my research. My HOD, Professor Gareth Jones, has also been unfailing in his support for our research. Statistical advice was most gratefully accepted from George Speirs. Without the very patient assistance of Robbie McPhee, I would still be dealing with computer graphics programmes. Thanks also to Mary Nolan who took some of the photos.

Unfailing support through this process has also come from my family and close friend Amanda, even if from a distance. My parents have continually emphasised the importance of education and have thus been supportive in all ways possible in any associated endeavours. I have also been fortunate enough to have my partner, Mark's, constant encouragement.

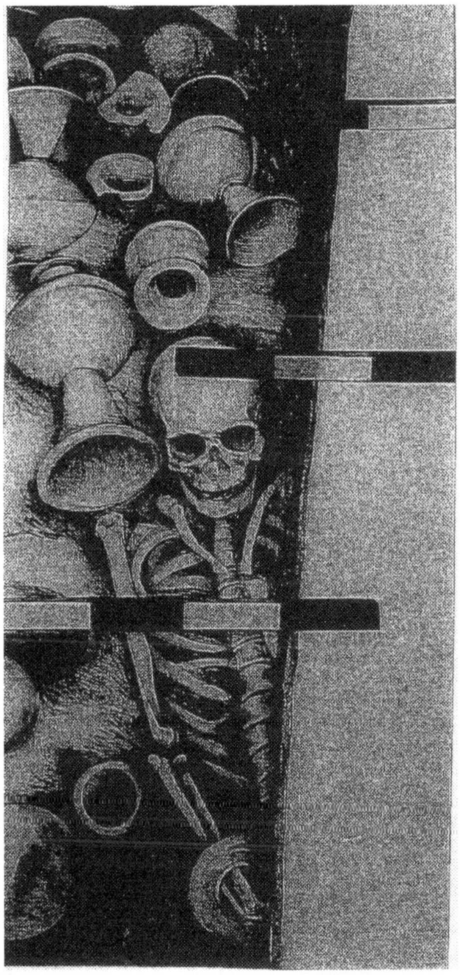

A northeastern Thai Bronze Age adult burial.
*Artist: Michelle Domett*

# Contents

# List of Tables

# List of Figures

# 1 Introduction

The study of prehistory in Southeast Asia has been substantially advanced over the last few decades through the activities of several large scale inter-disciplinary projects (for example Kijngam 1984; Higham and Bannanurag 1990; Higham and Thosarat 1997, 1998a). Insight into the settlement, culture, and technological and agricultural development in the prehistory of this region has particularly increased. Central to these projects, and to any study of the past, are the people. The human skeleton can provide information regarding health and disease and a skeletal collection, as a sample of individuals from a community, can provide information that suggests their general quality of life. This will be a reflection of both the environment they lived in and their ability to exploit it. Therefore, biological anthropology has a significant role to play in the understanding of Southeast Asian prehistory.

This thesis focuses on the health status of four prehistoric populations from within the modern state of Thailand, as examples of prehistoric people in Southeast Asia. In the last few decades archaeological work in Thailand has provided a wealth of human skeletal material and archaeological information, now enabling a synthesis of health during prehistory in the region. Particular advances in Thailand have been made in the understanding of the prehistory during the periods encompassing from approximately the end of the third millennium BC through to about 200 AD (Kijngam 1984; Higham and Bannanurag 1990; Higham and Thosarat 1997, 1998a). This time span includes the period of early rice agriculture through to the Bronze and Iron Ages that predate the development of state-level social structure (Higham 1996). The skeletal collections used in this thesis cover this range of time periods and are from coastal southeastern and inland northeastern Thailand.

As mentioned above, there have been many archaeological projects in Thailand over the last three decades. However, publication of the results from each project has not been as forthcoming as expected and has, therefore, limited the discussion of prehistory in Thailand. The results from other projects, such as the Thailand Archaeometallurgy Project led by V. Pigott and based in central Thailand, have not been published in full, as the projects of Higham and Thosarat have. The excavations of Non Nok Tha and Ban Chiang in northeastern Thailand in the 1960's and 1970's, respectively, have yet to be published either. Therefore, the information on these other projects is limited. This has resulted in a reliance on the work of Higham and Thosarat and their colleagues. While it would be preferable to include all available information, this is clearly impossible.

An impact on human health can be made by various factors changing through time. These factors include the development and intensification of agriculture, the development of technology, in particular metal working, and an increase in social complexity with advancing time. The environmental situation of a prehistoric site is also an important factor to be considered in an assessment of human health. It is with respect to these factors that periods within prehistory can be defined and they are summarised here for Southeast Asia based on archaeological evidence to date.

In Thailand, an early agricultural community, prior to approximately 1500 BC, was one reliant on small scale rice agriculture, probably cultivated in swamp margins (Higham 1989). Communities were initially small and autonomous (Higham 1996), but social hierarchy was increasing. An exchange network with other communities was also developing. Coastal settlements were particularly favoured for the richness of resources. During this time, settlements were becoming sedentary, which allowed for the domestication of some animals and plants. No knowledge of metal working had been established at this stage.

Post-1500 BC in Thailand, there was a settlement expansion along tributary streams of the Khorat Plateau to the northeast (Higham 1989). This period, the Bronze Age, saw the first use of copper alloy artefacts, and the trading of ingots and implements was evident. Subsistence was quite varied, but included an increasing reliance on rice agriculture (Higham 1989).The social structure of these communities was not greatly different from that during the early agricultural period.

Social structure became much more complex during the Iron Age, beginning around 500 BC. This period saw the establishment of hierarchical and centralised societies with a widening exchange network, including contact with Indian traders. Iron working techniques were adopted and assisted with a further intensification of rice agriculture (Higham 1996).

Four prehistoric populations from Thailand comprise the material for this thesis. These are Khok Phanom Di, Nong Nor, Ban Lum Khao, and Ban Na Di. Khok Phanom Di and Nong Nor are both from the southeastern region of Thailand and were close to the coast in prehistory, particularly Khok Phanom Di. Ban Lum Khao and Ban Na Di are located in inland northeastern Thailand, Ban Na Di further north than Ban Lum Khao. The basic information for each of these sites is provided in Table 1.1 and their location is indicated in Figure 1.1. A brief introduction to these sites follows, but a more detailed description is provided in Chapter 2.

**Table 1.1:** The dates, location and size of the skeletal cemetery samples from prehistoric Thailand analysed in this thesis

| Site | Dates | Period | Region | Environment | Sample Size |
|---|---|---|---|---|---|
| Khok Phanom Di | 2000-1500 BC | Early agriculture | southeast | coastal | 154 individuals |
| Nong Nor | 1100-700 BC | Bronze Age | southeast | coastal | 155 individuals |
| Ban Lum Khao | 1000-500 BC | Bronze Age | northeast | inland | 110 individuals |
| Ban Na Di | 600-400 BC | Late Bronze to early Iron Age | northeast | inland | 78 individuals |

Southeastern Thailand includes an extensive coastline, the expansive floodplain of the Bang Pakong River, and an upland region to the north, physically dividing it from the northeast (Higham 1996). The sea level was much higher in this region during particular periods of prehistory, but the same tropical monsoonal climate prevailed (Higham 1996). However, the sharp distinction in, for example, food availability, between the wet and dry season, may not have been felt so acutely as inland, because the availability of marine resources are largely unaffected by the seasons (Higham 1996).

A major, long-term archaeological project was begun in the early 1980's within the Bang Pakong Valley of southeastern Thailand, jointly led by the Archaeological Division of the Thai Fine Arts Department and the University of Otago, New Zealand (Tayles 1999). The aim of the Bang Pakong Valley research project was to examine human adaptation and settlement within coastal areas spanning prehistoric and historic periods. Previously research had focused on inland sites. The search for evidence of the origins of rice agriculture was also part of the project (Higham 1990). The area selected along the Bang Pakong Valley included a number of important sites that had been coastal in prehistory, including Khok Phanom Di (Higham 1990).

Khok Phanom Di was occupied by an early agricultural community for up to 500 years from approximately 2000 BC (Table 1.1.). During the initial settlement of Khok Phanom Di, the surrounding environment included mangal, estuarine, and freshwater conditions (Higham and Bannanurag 1990), and hence a rich and varied supply of food resources were available. A lowering of the sea level occurred during the occupation of this coastal site with subsequent environmental changes which are detailed in Chapter 2. The human skeletal remains from this site comprise the first sample used in this thesis. The skeletal material was very well preserved with many complete individuals excavated.

The excavation of nearby Nong Nor was initially undertaken to identify contemporary sites to Khok Phanom Di. However, the excavation provided evidence of an occupation period predating Khok Phanom Di and a later cemetery. The cemetery, in use during the Bronze Age, postdates that of Khok Phanom Di. The excavation uncovered a large number of human skeletons that comprise the second sample used in this thesis. Unfortunately, the skeletal material was poorly preserved. The environment surrounding Nong Nor was characterised by a large floodplain which provided an abundance of aquatic food resources (B.Boyd, pers comm). Coastal resources were also accessible.

The northeast of Thailand is a raised plateau. The area is bounded to the west, north, and south by mountain ranges, and to the east by the Mekong River. This region is particularly susceptible to a long dry season (Higham 1996). Temperatures in the dry season can fall much lower than in the southeast. Most prehistoric sites in the northeast have been found situated adjacent to rivers or large streams, most probably to ensure a constant supply of fresh water and sufficient and varied food resources.

The most recent multidisciplinary project undertaken in the northeast by the University of Otago and the Thai Fine Arts Department was the 'Origins of Angkor' archaeological project. This project involved the excavation of three prehistoric sites in the Mun River Valley with the aim of investigating the social, cultural, and technological developments that led to autonomous communities undergoing the transition to more centralised and hierarchical societies. Descendants of these communities later built large scale monuments and temples, some of which are still standing in the northeast today. They represent the northeastern extension of the Angkorian state. The main temple of Angkor is in Cambodia and was an ancient 'holy city' comprised of a "...complex of monuments, reservoirs and walls which served as the centre of political influence over much of mainland Southeast Asia..." (Higham 1989: 321) during the first millennium AD and later.

The sites excavated in the northeast under this project included the Bronze Age site, Ban Lum Khao. This excavation aimed to establish the baseline of the Iron Age. This cemetery site provided the third collection of human skeletal remains able to be included in this thesis. The other two sites excavated as part of the project, Noen U-Loke and Non Muang Gao, were both from the Iron Age and are presently under investigation. The former site was the most recently excavated and preliminary results from the human skeletal remains are incorporated into this thesis where appropriate. The skeletal material was excavated from 1997 to 1998, too late to be included in this thesis. The site of Non Muang Gao yielded no useful skeletal remains.

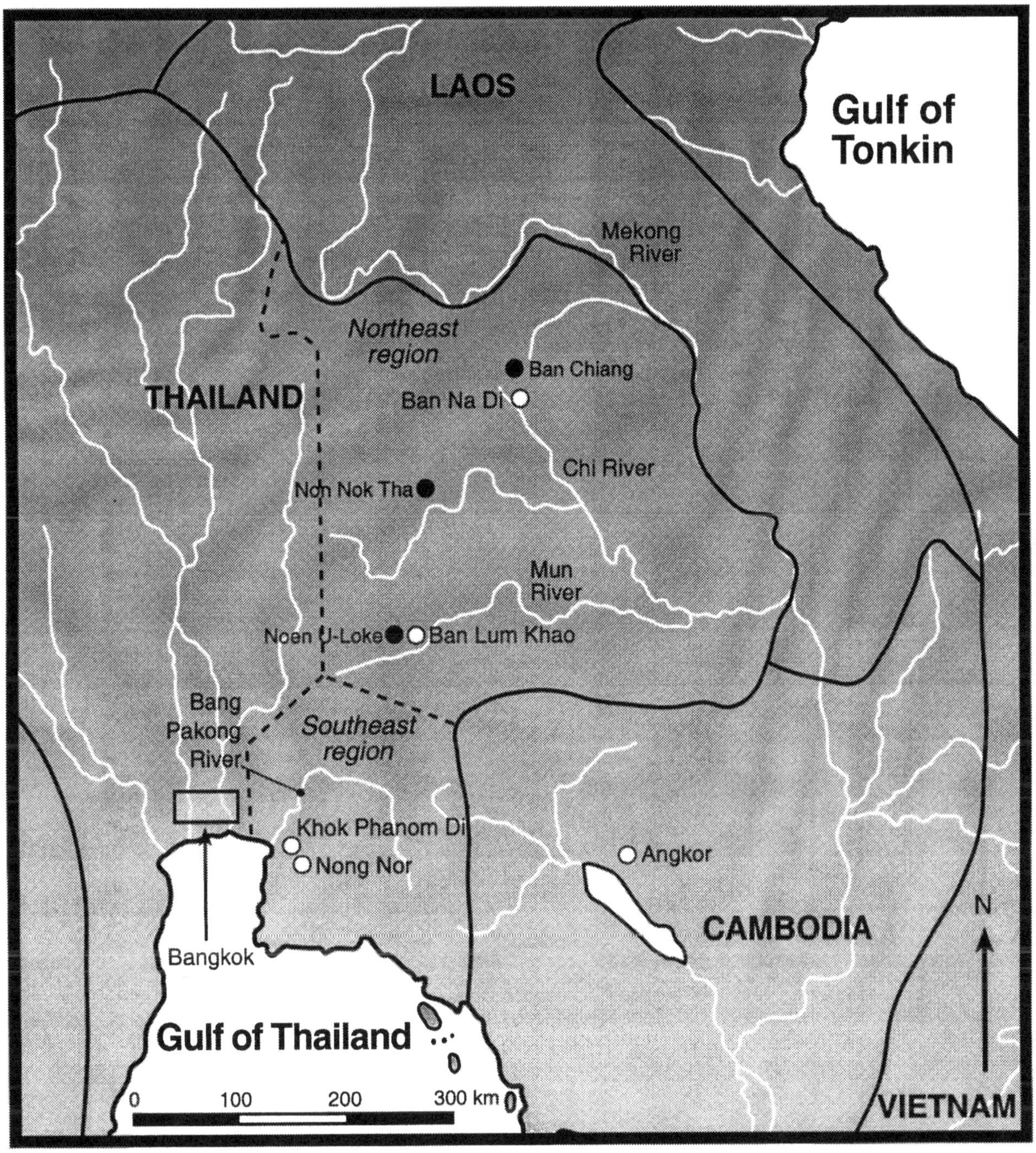


**Figure 1.1:** Map of Southeast Asia indicating the sites discussed in this research and the boundaries of the areas of northeast and southeast Thailand (dotted lines). Sites located with white circles are the primary sites discussed

Another project in the northeast, begun in 1980, was the Thai Fine Arts Department's 'Northeast Thailand Archaeological Project'. This was established to continue the research begun at Ban Chiang (discussed in Chapter 2) and encompassed the period of 2000 BC through to 300 AD. This incorporated the transition from autonomous to centralised societies (Higham and Kijngam 1984a). The aim was to test models so far based only on single sites such as Ban Chiang. The project undertook a site survey to establish the pattern of sites settled in the vicinity of Ban Chiang and also aimed to expand the knowledge of prehistoric people through an excavation of a prehistoric cemetery (Higham and Kijngam 1984b).
Ban Na Di was the cemetery excavated in the Middle Chi River valley of northeast Thailand. The site was situated near small streams, like Ban Lum Khao, and provided evidence of life during the late Bronze Age and early Iron Age. The excavation of parts of the cemetery provided the fourth collection of human skeletal remains for this thesis.

Overall, the skeletal collections come from sites which cover the period from approximately 2000 - 400 BC, with two from the southeastern area and two from inland northeastern Thailand.

# Research Aim

The four skeletal samples outlined above total nearly 500 individuals. Some of the skeletal material, that from Khok Phanom Di and Ban Na Di, has been analysed previously with the research emphasis on the health of each individual sample (Wiriyaromp 1984; Tayles 1992, 1999). The analysis of the Nong Nor sample was carried out by myself and N.Tayles (bone) and V.Hunt (dentition). The skeletal material from Ban Lum Khao was exclusively analysed by myself, as was a complete reworking of the Ban Na Di skeletal material. Together these skeletal collections provide sufficient material to enable extensive comparisons between the samples to be undertaken. This has only become possible in the last few years with the excavation of further skeletal samples.

The overall aim is to undertake a large scale synthesis of the variation in health among these samples and show that the health of the samples was differentially affected by both their natural and cultural environment. In order to fulfil this aim, the first objective is to characterise each skeletal sample based on an extensive analysis of the skeletal evidence for health, disease, and mortality, and integrate this with the archaeological evidence of the natural and cultural environments for each site. The second and primary objective is to compare the evidence for health and disease among the samples, also incorporating archaeological evidence for the natural and cultural environments. This will provide an integrated picture of life in late prehistoric Thailand over different natural environments and through changes in culture over time.

In taking this approach there are a number of factors to consider. First, it is necessary to consider the ways in which the natural and cultural environment of a community can affect health. These are briefly outlined below and are followed up further in Chapter 2.

### Natural Environment

A population's characteristics are intimately linked with their surroundings, and therefore an ecological approach, one that considers the relationship between a living organism and its environment, is appropriate in the study of human health. As Goodman *et al.* (1988: 177) point out, "...the environment is the source both of resources necessary for survival and the stressors that adversely affect adaptation". The relationship between the environment and the people who inhabit it, is further described by Kealhofer (1996):

> Environments and humans are intricately interrelated and codetermined. Change in one affects change in the other...Cultural changes are not always functionally related to the environment, nor are they necessarily adaptive or optimizing... Unintended consequences of human actions (e.g., deforestation) play a crucial role in changing environmental-human relationships. (Kealhofer 1996: 112)

One of the most important aspects of the environment for the people who inhabit it, is the resources, particularly the food, it can provide. An adequate diet can literally be the difference between life and death. Archaeological evidence of faunal and botanical remains, as well as palaeoenvironmental studies, can identify the resources available in a particular environment. This evidence can indicate the potential range and abundance of resources available to a prehistoric population and thus the adequacy of the diet. This will be discussed in further detail in Chapter 2.

### Cultural Environment

The cultural environment of a skeletal sample is defined here by its position in time. Factors associated with time include technological developments, including the level of development of metal working, agriculture and other subsistence activities, and population size and density changes, with a corresponding change in social complexity.

Studies relating the subsistence shift from a hunter-gatherer mode through to agricultural intensification (Larsen 1982), have analysed how such cultural changes have affected human health. Morphological changes in human skeletal remains associated with age, mechanical function, diet and nutrition, and disease have particularly been focused on. This transition has been extensively analysed and reviewed in the literature for many areas of the world (for example Turner 1979; Cohen and Armelagos 1984a; Goodman *et al.* 1984a; Johansson and Horowitz 1986; Bridges 1989; Larsen 1995). Notably absent from these studies are the events occurring in Southeast Asia during this transition and how they affected humans. Only one study (Douglas 1996) has focused on this transition in Thailand. This research was based on two prehistoric skeletal samples from the northeast, but the controversial dating of the samples is a disadvantage. However, this work provides a useful comparison to this study and more details are included in Chapter 2.

The effects of the transition to agriculture in the Americas and in Europe, including the nature of the resource base, its dependability, its diversity, and its composition, can be different in the tropics from in more temperate or subtropical environments (Jackes *et al.* 1997). Therefore, while these earlier studies in other parts of the world can be useful in the establishment of hypotheses to be tested in prehistoric Thailand, there are likely to be other factors, unique to this region's transition to agriculture, that may influence human health differently from other regions of the world. Therefore, the focus here is on a comparison of the four Thai sites, with only limited comparison of them with other studies from around the world.

# Hypotheses

Following on from the research aim of this thesis and taking into consideration similar previous studies in other parts of the world, a number of hypotheses have been developed. Although here they are presented under the separate headings of the natural and cultural environment, at times it is difficult to differentiate between the effects of place and time on human health.

### 1. Natural Environment

There are distinct differences in the natural environment between the northeast and southeast of Thailand. It is expected that this would have had a differential impact on the health of the communities being compared.

The prehistoric sites of Khok Phanom Di and Nong Nor are both located in the southeast of Thailand. However, they were occupied during different periods of prehistory and this invokes caution against comparing these sites solely on the basis of their natural environment. The natural environments within the immediate vicinity of each site were different in some important aspects. The Khok Phanom Di people were predominantly reliant on coastal and marine resources, while the Nong Nor people probably exploited riverine resources of the Bang Pakong River floodplains, although they did have access to distant coastal resources. It is possible that the cemetery at Nong Nor was used by people who were living elsewhere, possibly closer to the coast, however it is logical to assume that they occupied an environment close to the cemetery. Little faunal and botanical evidence was associated with the cemetery at Nong Nor, thereby limiting the archaeological evidence concerning the natural environment. However, based on the available archaeological evidence, the diet of the people from these two samples was generally not similar and may have affected their health differentially.

The skeletal sample from Ban Lum Khao predates that from Ban Na Di but both are from the northeastern region of Thailand. The natural environments of these two sites are thought to have been similar in prehistory. Both sites are, and were in prehistory, located adjacent to a river, providing access to the fish, shellfish, and other resources a river can provide. The people living at these sites in prehistory were involved in rice agriculture, possibly more intensely at Ban Na Di and with more advanced tools than at Ban Lum Khao. However, in general the natural environment was a constant factor through time between these two samples. Therefore, the health of these two populations may have been differentially affected by other factors, such as their cultural environment.

Summary

1. It is hypothesised that there will be *similarities* in the general level of health of samples from the same natural environment. That is Khok Phanom Di and Nong Nor will have a similar level of health to each other and Ban Lum Khao and Ban Na Di will be similar to each other. Thus, health is hypothesised to follow the pattern:
'Khok Phanom Di = Nong Nor'
'Ban Lum Khao = Ban Na Di'

2. It is also hypothesised that there will be *differences* in the general health between the samples from different natural environments. That is, differences will be evident in health between the samples from the southeast (Khok Phanom Di and Nong Nor) compared with the northeast (Ban Lum Khao and Ban Na Di). Thus, health is hypothesised to follow the pattern:
'Southeast ≠ Northeast'

### 2. Cultural Environment

The cultural environment has changed through time in prehistoric Thailand. Changes that occurred among the early agricultural period, Bronze, and Iron Ages have been significant, including changes in the predominant subsistence mode, as technology advanced and rice agriculture intensified. Changes are also evident in terms of social structure, complexity, and population size.

The cemetery sites of Khok Phanom Di and Ban Na Di were used at different times, up to 1500 years apart. These populations used different technologies in their exploitation of their environment. Their social structure was also different. The samples from Ban Lum Khao and Nong Nor provide the intermediary step between Khok Phanom Di and Ban Na Di. The former samples may also be similar in their general health to each other given their contemporaneity.

Changes in health are expected over time. In order for complex societies to develop, such as that of the Angkorian civilisation, the preceding populations must have had an adequate level of health. Societies must also have had the appropriate structure in place to support some specialist members who did not contribute directly to the survival of the group. It has been suggested that this state was reached during the Iron Age from the mid first millennium BC in Thailand, and enabled communities to become more centralised politically and able to support and organise larger numbers of people (Higham 1996).

Through time the health of the prehistoric people may be seen to have improved as their ability to exploit their environment developed. The development of agriculture, knowledge of metal working and tools to make subsistence activities more efficient, are all related to the cultural development of the community, and can improve the general quality of life. However, in contrast, increasing social complexity can lead to hierarchical groups within each community, with some subgroups gaining access to better nutrition and other resources. This kind of differentiation in health within the Thai skeletal samples may be increased through time, with Khok Phanom Di showing the least and Ban Na Di the most evidence for inequalities within their samples.

Summary

It is hypothesised that there will be an improvement in health with time in prehistoric Thailand from 2500 BC through to 400 BC. The Khok Phanom Di sample will show the lowest level of health and the Ban Na Di sample will be the healthiest. The Nong Nor and Ban Lum Khao samples will be in between. The two latter samples will have a similar level of health. Along with the improvement in health, an increase in the variability of health within the samples may increase. Thus, health is hypothesised to follow the pattern:
'Khok Phanom Di < Nong Nor/Ban Lum Khao < Ban Na Di'

# Research Approach

### Theoretical basis and background literature

> Comparative studies of regional samples of human remains provide an important perspective on community health in relation to ecological and biocultural variability. (Larsen 1997: 91)

This comparative environmental and cultural approach presented has been undertaken in a number of previous studies which address similar questions for other parts of the world (for example Larsen 1982; Goodman *et al.* 1984a; Kennedy 1984). In the last two decades or more, the move has been towards population-scale analysis rather than individual accounts of such aspects as specific diseases or trauma. In 1980, Buikstra and Cook outlined the development of population-based strategies with an emphasis on the benefits of the integration of archaeological information in the assessment of population morbidity and mortality. This is termed the biocultural approach.

The biocultural approach in the analysis of human skeletal remains includes more than just incorporating cultural aspects from the archaeological record, but also considers environmental and non-human biological evidence, such as faunal and botanical remains. It provides a broader view of health in prehistory than observations of single diseases can provide. While the presence of diseases are, obviously, important, the significance of them also needs to be addressed. However, oversimplified explanations can be misleading when in reality the interaction of cultural, environmental, and population factors are dynamic and complex (Bush and Zvelebil 1991).

The biocultural approach has been well used over the last three decades. For example, the implications of the development of agriculture on health in various parts of the world were put together by Cohen and Armelagos (1984a). Kennedy (1984), for example, presented the skeletal evidence of health from sites spanning the change from hunting and gathering to farming communities in South Asia. Powell's (1988) study on the prehistoric Moundville chiefdom in Alabama, U.S.A, integrated the human biological observations, such as infectious disease and growth disturbances, with the social dimensions established from extensive archaeological work. A further example of the use of the biocultural approach is the work of Bridges (1989, 1991, 1992) in attempting to link the skeletal evidence of arthritis with cultural and ethnographic evidence for activity.

One of the most well-known adaptations of this biocultural theory was the development of an ecological model for interpreting the relationship between health, stress, and adaptation in skeletal populations. The model was first presented by Alan Goodman and later published in conjunction with colleagues (Huss-Ashmore *et al.* 1982). The Goodman model (Figure 1.2) is used at a population level to determine the impact of chronic and acute stress through the assessment of specific and non-specific skeletal indicators.

The Goodman model emphasises the importance of considering the effects of not only the natural environment, but also the cultural environment and human biological factors in interpreting a population's health status (Martin *et al.* 1985). For example, the natural environment is the source of food for a population but may be limiting in some way. Food shortages may be overcome by technological advances that can increase the subsistence base (Martin *et al.* 1985) and buffer the environmental effects. On the other hand, the cultural organisation of the population may also limit access to certain food resources for some members of the community, perhaps because of a hierarchical social structure or food taboos. The ability of each individual to cope with nutritional inadequacy and respond to disease, or host resistance, is also incorporated into Goodman's model. This will determine how severely affected the individual is by the environment and overall how well the community adapts.

> Health is a function of the organism's ability to adjust to environmental constraints and stressors. It is the result of the adaptive process. Poor health constrains adaptation. It is a condition that needs to be alleviated or adjusted to. In this sense, health, classically defined in the terms of morbidity and mortality, is a measure of the success of adaptation. (Goodman 1991: 35).

Further modifications of this model appeared in Goodman *et al.* (1988) and Goodman and Armelagos (1989). The changes included assessing the impact of 'stress' on the population indicating the circularity of the model: "The process by which health is related to environment and

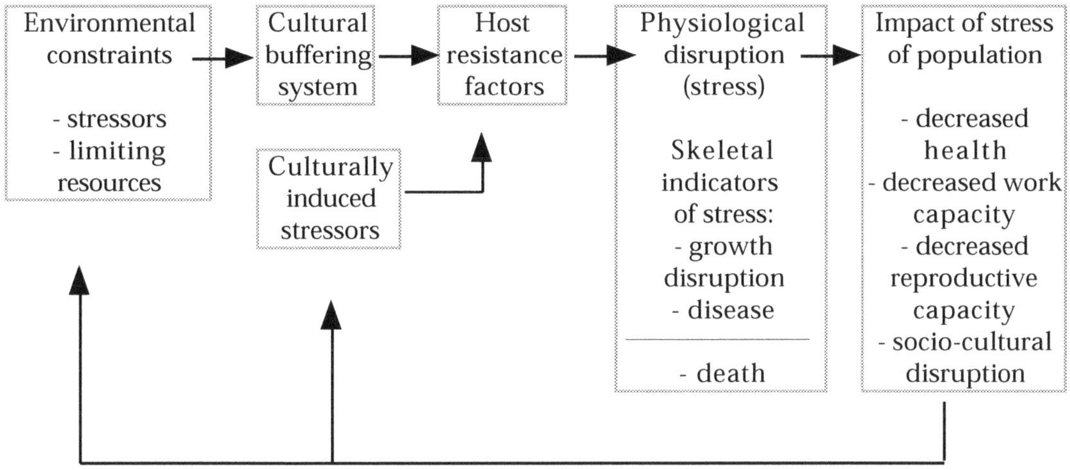


**Figure 1.2:** A biocultural model for studying the relationship between health, stress, and adaptation in skeletal populations (adapted from Goodman and Armelagos 1989).

adaptation is nothing less than intrinsic, interactive, and cyclical" (Goodman 1991: 35). Thus, increasing levels of stress have a negative affect on the ability to exploit environmental resources (work) and on the cultural environment (indicated by feedback arrows in Figure 1.2). The cultural environment was also acknowledged to be a further source of 'stress'.

Huss-Ashmore *et al.* (1982: 396, 395) defined stress "...as the physiological disruption of an organism resulting from environmental perturbation" and adaptation as "...a dynamic interaction between human populations and the environment...". This ecological model outlines useful concepts in health assessment, however the word 'stress', and associated discrepancies among its multiple definitions, can be avoided. Perhaps the terminology 'indicators of morbidity and mortality' implies the same thing (Wood *et al.* 1992).

Buikstra and Cook's (1980) review acknowledged advances in particular areas of palaeopathology including the, then recent, development of the investigation of 'non-specific indicators of stress' that were central to the work of Goodman and others. However, Buikstra and Cook (1980) rightly acknowledged the lack of critical review with respect to such indicators as Harris lines, enamel hypoplasia, and growth attainment. Following Selye's concept, the term 'stress' was defined as "...any extrinsic variable or combination of variables which causes the organism to react" (Buikstra and Cook 1980: 444). This concept has been accepted by many biological anthropologists, long after the clinical literature has ignored its value (Bush 1991). Stress has been defined in many ways, only some of which have described here. Stress is a generic term and perhaps should only be used as a way of tying "...together the threatening or taxing demands of the environment on living organisms..." (Mason 1975b cited in Bush 1991: 17).

Goodman *et al.* (1988) define stress using the Selye concept, mentioned above, which relies on the premise that the stress response is non-specific (Bush 1991). However, the Selye concept of stress had been criticised by Cox *et al.* (1983) as it has been shown that different stressors may produce different reaction patterns (Cox 1978; Cox and Mackay 1981 cited in Cox *et al.* 1983). A more recent review by Higgins (1992), also outlines further criticism of Selye's theories, particularly focusing on experimental design. Selye carried out his experiments in the 1930's when only non-specific and indirect measurements of hormones were possible (Mason *et al.* 1968 cited in Higgins 1992). In a search of the scientific literature concerning the concept of stress "...no discipline other than palaeopathology, continues to use the non-specific model as devised by Selye" (Higgins 1992: 152). Bush's (1991) initiative in stating "...that anthropological studies would benefit from the striking of the word "stress" from the vocabulary, given the problems of definition" (Bush 1991: 17) and the criticism above, is also taken here.

The Goodman model is used in the following analysis of skeletal health, *purely* as a means of combining environmental, cultural, and biological factors in interpreting the morbidity and mortality of a population. However, the use of the term 'stress' is avoided because of the problems with definition and its generic nature in defining the problems affecting a population.

In 1991, Ortner and Aufderheide recognised the need to address the theoretical questions of palaeopathology. The question of whether the evidence for disease in archaeological human bone is actually indicative of poor health in the individual was first raised by Ortner (1979). This was further addressed by Ortner and Aufderheide (1991) and, although not named as such, basically outlined the 'osteological paradox' concept presented in more detail later by Wood *et al.* (1992). Ortner (1991) states that it is theoretically possible that skeletal evidence of infectious disease, for example, may, in some cases, be evidence of an effective immune system and therefore of relatively good health, as a healed lesion indicates the individual was healthy enough to overcome the infection. However, "the easy assumption is that a relatively high prevalence of cases of infectious skeletal disease in a skeletal sample is, indeed, indicative of decreased population health" (Ortner 1991: 9). Ortner (1991) accepts that this may be true, but not in every case. It is known that the skeleton is commonly the last part of the body to be affected in many diseases (Ortner 1991). Therefore, an individual skeleton with evidence of an infectious disease would have had to survive the initial acute phase for the disease to become chronic and for skeletal lesions to be apparent. This implies a relatively good immune system, which implies a relatively good level of health, especially in contrast with an individual who suffered the same disease but died before the skeleton was affected. Those individuals with the best immune system will of course overcome the infection in the early stages (Ortner 1991).

This leads on to two important aspects in the interpretation of morbidity and mortality in prehistoric skeletal populations. Firstly, there is an individual response to disease. Therefore, within any one population, there will probably be "...an unknown mixture of individuals who varied in their underlying *frailty* or susceptibility to disease and death" (Wood *et al.* 1992: 345) (their emphasis). This is termed *hidden heterogeneity* in the landmark paper 'The Osteological Paradox' (Wood *et al.* 1992). The heterogeneous response to disease may not only be a consequence of a genetic difference in immunity, but also a differential distribution of food (the nutrition-infection synergism is discussed later) and exposure to disease. Not forgetting that many human archaeological samples represent information from a wide span of time, therefore chronological trends in health may be present which will in turn produce a heterogeneous frailty in a sample (Wood *et al.* 1992).

The second aspect, following Ortner's (1991) ideas, is concerned with how to interpret an individual with no specific or non-specific health indicators. They may represent an individual with good health, with an effective immune system, able to recover from acute diseases and periods of malnutrition without compromising their skeletal system. Alternatively, they may represent a person who died after suffering acutely without time to allow the problem to affect the skeleton. More information would be required to interpret this. For example, a person with no evidence of chronic disease but died at a young age may indicate a certain level of acute ill health (or an accident).

Further problems outlined in Wood *et al.* (1992) are discussed in a later section which addresses the limitations and biases of archaeological skeletal populations.

# Measures of Health

Human bone is a dynamic living tissue, constantly remodelling under the processes of resorption and deposition by specialist cells, osteoclasts and osteoblasts, respectively. This remodelling allows for the general maintenance of bone and renewal of the tissue, growth, and as a response to biomechanical forces and lesion repair. The outcome of these responses may be either a loss of bone, no change, or excess formation. The difference between the rates of resorption and deposition will determine the outcome. Because of the limited ways in which bone can respond, the final outcome can be very similar for different diseases and other conditions. Reference only to a broad category of disease or other variable is frequently all that can be stated.

Most diseases are frequently under-represented or totally absent in skeletal populations as the associated pathology, especially those of an acute nature, do not affect bone. However, as discussed above, the absence of disease in a skeleton is not necessarily an indication of good health. As a result, skeletal populations should only be compared with other skeletal populations and not with modern populations in terms of disease prevalence (Cohen 1989), as the nature of the samples are very different.

Some diseases produce characteristic skeletal lesions but often a specific diagnosis cannot be made. Some diseases can be recognised by the distribution of lesions within the skeleton (the pattern of bones affected). This emphasises the importance of analysing all available skeletal elements rather than single bones. However, relatively rarely can specific diagnoses be made in palaeopathological analysis:

> Researchers may have to be satisfied with defining a disease cluster rather than naming a specific pathogen; yet this may be entirely appropriate to research designs that emphasize the evolution of disease patterns or general measures of the quality of life in extinct human groups. (Buikstra and Cook 1980: 435-6)

There are other so called non-specific signs of morbidity and mortality that may be evident in a skeletal population. These provide a more general indication of the health level of a population. The specific and non-specific parameters that are used to measure health, morbidity and mortality in this thesis are outlined below. By taking this multifactorial approach a better understanding of a community's quality of life can be obtained.

> Whereas data from a single source are likely to be open to a number of equally consistent interpretations, the use of multiple types of data greatly reduces the number of likely interpretations and consequently the chances of error in interpretation. (Goodman 1993: 285)

When assessing a sample's health status, signs of both good and ill health are included. The term 'health' frequently used in this thesis is often used as an abbreviation of all aspects of 'health status'; that is, good health, ill health (morbidity/disease), and mortality. The following provides a brief outline of each parameter to be observed. This study does not consider every possible indicator of health and disease. For example, evidence of non-specific infectious lesions such as periostitis were not able to be consistently recorded because of the concretions covering the bone in the Ban Lum Khao skeletal material and some of the Ban Na Di and Khok Phanom Di material. However, the range of parameters were chosen, firstly, to provide a *representation* of health and disease in prehistory and secondly, they had to be able to be compared among skeletal samples. Therefore, universal measures of health rather than specific skeletal pathologies were more suited to this type of study. A more detailed introduction to these measures of health are provided within the relevant chapters.

### Age at death and the sex ratio

Ironically, the age at which an individual died is an important indicator of health in life. The presence of a skeleton represents a person who experienced the ultimate level of ill health, that is death. The death of an older adult is to be expected. However, the death of an infant, child, or young adult is considered premature, and may be the result of detrimental environmental pressures. The consideration of the age structure of a population is also essential in the analysis of age progressive pathologies, such as joint disease and dental pathology. Age at death estimates of all individuals when possible will be undertaken and used in the above ways.

The adult sex ratio will also be determined for each sample. This can reflect the level of differentiation between the sexes in terms of survival of each sex. An uneven ratio may indicate a particular sex was coping better in the same environment than the other, and further investigation into the cause may be found in other observations of health prevalences within each sex.

### Growth and disturbances of growth

Childhood growth is assessed by comparing dental developmental age with skeletal size (long bone lengths). This is because dental development is relatively unmodified by the environment, in contrast to skeletal development (Moorrees *et al.* 1963a; Ubelaker 1987). Therefore, a comparison of genetically similar populations can indicate if growth has been retarded due to environmental effects.

Stature is estimated from adult long bone lengths and is a useful comparative measure to make among populations of similar genetic backgrounds. With homogeneity, differences in stature will no doubt reflect environmental conditions, particularly nutrition. A good diet and the absence of disease will enable a population to reach their genetic potential for height.

Growth disturbances are to be analysed in the skeletons and dentitions of both adults and children. Harris lines are lines of arrested growth and recovery. They are seen as transverse lines of radio-opacity in long bones (Roberts and Manchester 1995). In palaeopathology, Harris lines are considered to be indicators of generally non-specific problems in childhood (Martin *et al.* 1985; Grolleau-Raoux *et al.* 1997) induced by periods of malnutrition, infection, and other diseases. However, there are problems with this observation that are discussed later.

Linear enamel hypoplasia is seen macroscopically as a transverse line or row of pitting across the external surface of the tooth crown. It is a nonspecific indicator of a disturbance in growth during the formation of the tooth crown. For permanent teeth this relates to the first seven years of life except for the third molar (Lewis and Roberts 1997). As with Harris lines, the disturbance in growth can be caused by a variety of health problems during childhood, including malnutrition and infectious illness.

Bone mass is a valuable indicator of, firstly, how well bone has been maintained throughout life, secondly, of nutritional status and, thirdly, as an index of the strength of bone (Martin *et al.* 1985; Aloia 1994; Resnick and Niwayama 1995c). Bone mass measurements are used as a measure of adult health, particularly reflecting the quality of diet. Low levels, for example, can be indicative of osteoporosis and leave an individual with an increased risk of fracture (Mazess 1982). There is also a general decline in bone mass with age. In this study bone mass is observed radiographically and measured in the second metacarpal.

### Joint disease

Evidence of joint disease is recorded in the major appendicular joints and the intervertebral body joints of all adults. Osteoarthritis is the most common form of joint degeneration and is seen as a non-inflammatory disorder of freely moveable joints. It is characterised by deterioration of articular cartilage finally resulting in bone rubbing on bone, as well as the formation of new bone around the joint margins. This condition has been associated with numerous aetiological factors including advancing age, joint use (physical activity), trauma, obesity, and possibly even has a genetic predisposition (Resnick and Niwayama 1981; Rogers *et al.* 1987).

Vertebral osteophytosis (or spondylosis) is the degeneration of the fibrocartilaginous joints of the vertebral discs with a characteristic formation of new bone around the joint margin (Rogers *et al.* 1987). Vertebral osteophytosis can increase with age, and it can also be increased by hard physical labour (Kennedy 1989). The presence of osteophytes can therefore be used as an indicator of physical activity involving strain on the back.

### Trauma

Trauma indicates an extrinsic influence on the skeleton (Lovell 1997) that includes the environment (type of terrain for example), and physical activity associated with subsistence patterns, or as the result of interpersonal violence. Also the nutritional status and genetic composition of an individual can predispose them to fracture as mentioned in connection with loss of bone mass in osteoporosis (Mazess 1982). Both non-vertebral acute fractures and trauma to the vertebrae will be considered.

### Dental Pathology

Teeth are continuously subjected to mechanical, chemical, and pathogenic stressors that are all strongly influenced by dietary factors. Therefore, the assessment of dental pathologies can provide information on diet, techniques of food preparation, and food consumption (Powell 1985). The dentition is observed for caries, advanced attrition, secondary infection in the form of periapical cavities, and antemortem tooth loss.

Caries is a pathological condition caused by bacteria in plaque that produce acid while digesting carbohydrates (sugars and starches). The acid can destroy the enamel and eventually may create a cavity (Powell 1985; Hillson 1996).

The wear or attrition of the occlusal tooth surface is not a pathological condition but the natural result of masticatory stress on that tooth. Attrition can become pathological when wear has advanced so far that the pulp cavity is exposed and infection may result with the possibility of tooth loss (Powell 1985).

Periapical cavities can be seen in the alveolar bone around the tooth's root apex. They indicate an infection has passed from the dental pulp of a tooth, from such problems as caries and severe attrition, into the periapical tissues. The resulting inflammation that occurs can destroy the surrounding bone leaving a periapical cavity (Dias and Tayles 1997).

Antemortem tooth loss is the premature loss of a tooth and can be the result of a number of pathological problems, such as caries, calculus accumulation, periodontal disease, or advanced attrition. However, ritual ablation is also practiced by some cultures. The cause of tooth loss is not always going to be obvious as the evidence is lost when the tooth is lost.

**Nutrition**

Many of the above parameters can be useful in determining the quality of nutrition in a population and these will be brought together, with other archaeological evidence, to specifically assess the quality of each population's diet.

Throughout prehistory the prevalence of specific and non-specific health indicators has changed. It is interesting to note how the changes are related to such things as subsistence, sedentism, population growth, living conditions, and social changes. For example, with the shift from hunter-gathering to agricultural subsistence economies, the level of health of a population is said to have decreased (Ribot and Roberts 1996). However, these trends are not entirely universal. This emphasises the importance of treating each study separately as environments differ throughout the world. This is especially relevant in relatively uninvestigated regions such as Southeast Asia.

Statistical tests used in this analysis of the skeletal and dental data are provided in Appendix B. The tests are described and the level of significance chosen for each type of test is discussed.

# Limitations and Biases

There are often a number of unavoidable assumptions that are made about an excavated skeletal sample that can limit and bias the results obtained from an analysis of such a sample. Some have already been touched on, but are included with others below.

Each of the skeletal samples under investigation in this thesis are from cemeteries. This raises the issue of how related this subsample of the dead was to the living population from which they came. In epidemiology, random samples are drawn from large populations, however this is virtually impossible in palaeoepidemiology (Waldron 1994). Therefore, "...sample representivity is often difficult to assess" (Roberts and Manchester 1995: 9).

The factors that will have an effect on whether the excavated sample is representative of the living population are outlined in Waldron (1994). These factors all result in reducing the sample size. The factors include the proportion of dead buried at the site, the proportion lost due to disturbance and poor preservation, the proportion discovered, and the proportion excavated (however, the latter is often the same as the proportion discovered).

It has been observed that some cemeteries are reserved for the wealthy, the sick, or the young. These factors relate to the proportion of dead buried at the site and would become apparent after an investigation into the health or demographics of the sample and can be taken into account in any interpretations of the population. Archaeological analysis of grave wealth may also be of assistance in this regard.

Postdepositional processes can result in the movement and subsequent loss or disarticulation of skeletal elements. Studies generally concur that such taphonomic processes often affect the small bones and these are the most likely to disappear first (Waldron 1994). The articular ends of long bones are also prone to fragmentation due to the reduced amount of cortical (compact) bone compared to the shafts.

Preservation of human remains can vary significantly between samples and is often related to the soil composition and level of the water table. The rising and falling of the water table, which happens seasonally in monsoonal Thailand, can also result in the movement of part or parts of the skeleton away from the rest creating a disarticulated burial or at very worst resulting in only a scatter of bones being recovered. Disturbance can also be the result of human activity in prehistory. For example, an earlier burial may have been discovered during the digging of a grave for a later burial. This may be the result of descendants being unaware of where the dead are buried or of demands for space. Modern day farmers may also disturb burial sites when developing land for cultivation or through the digging of pits. Looting of prehistoric sites is also an unfortunate occurrence in both modern and prehistoric times, as was the case at Nong Nor.

The proportion of the cemetery excavated determines the sample size a skeletal biologist has available and this is usually further influenced by financial and time constraints of an excavation. The sample size has implications when it comes to the statistical analysis of the observations made. The number of burials excavated is also often limited by the aims of the archaeological excavation. Archaeologists often work within set dimensions that can mean only part of a skeleton is uncovered. Also, as implied by previous comments on preservation and taphonomy, parts of a skeleton will often have disappeared or be moved away by soil and water movement. This results in a certain amount of missing data and consequently, results of various disease prevalences, for example, only represent the lower limit of the condition (Waldron 1994). It is possible to allow for this in the calculation of some prevalences (Waldron 1994) and this has been applied in this thesis where it is thought applicable.

# Book Structure

Chapter 2 will further outline the prehistory of Thailand, setting the scene in which human health will be assessed. An outline of all relevant archaeological evidence for the four sites is provided, particularly in terms of the environment and available natural resources, the material culture and technology, as well as the mortuary ritual. The excavations are summarised including dates and a description of the human skeletal remains and their degree of preservation. This chapter also expands on the influence of the natural and cultural environment on a community's health.

The third chapter details the age at death and sex of all individuals from each of the four samples. It also provides an indication of the level of fertility. This chapter aims to establish differences in age and sex structures of the four skeletal samples which is essential for further analyses of age related pathologies. Age at death also provides a measure of health for each sample.

Chapter 4, entitled 'Growth and disturbances of growth' considers aspects that can reflect growth, such as stature. It also considers alterations in the normal pattern of growth during childhood, as indicated by enamel hypoplasia and Harris lines. These parameters provide information concerning health and ill health during childhood. Bone mass in adults is also included as a measure of bone maintenance in adulthood.

Chapter 5 presents evidence for the assessment of joint disease. This is divided into osteoarthritis of the appendicular skeleton and vertebral osteophytosis of the spine. These two pathologies are compared among the samples to gain an impression of the degree of physical activity and how the workload may have differed among the samples.

Chapter 6 deals with the evidence for trauma and while it is a continuation of skeletal pathology it is considered separately as it has different implications for a community. Trauma analysis is restricted here to fractures, with both non-vertebral fractures and spondylolysis assessed. A comparison of the distribution of such fractures among communities can provide an indication of the relative risk of trauma through interactions with the environment or interpersonal violence.

Chapter 7 details dental health and pathology including caries, attrition, periapical cavities, and antemortem tooth loss. This evidence is compared among the samples to obtain information concerning their differences and similarities and how they relate to the diet of each sample.

Chapter 8 discusses all the above skeletal evidence integrated with the archaeological and modern evidence in a biocultural approach to fulfil the overall aim of this thesis. Details of the natural environment, specifically the pathogen load and nutrition, are discussed, and the hypotheses based on the natural environment variable are specifically addressed. The archaeological evidence relevant to the cultural environment is integrated with the skeletal evidence to address the hypothesis of health changes through time. This will provide an integrated picture of health in late prehistoric Thailand over different natural environments and through changes in culture over time.

# 2 Bioarchaeology and Thai Prehistory

## Introduction

In order to address the questions raised in the introduction, archaeological evidence is considered in conjunction with the assessment of the skeletal remains. This chapter firstly outlines the archaeological evidence relevant to this approach. It particularly focuses on the environment and the food resources it can provide. This chapter also addresses the evidence of technological advances, particularly rice agriculture and metallurgy, and the subsequent increase in population size and social complexity which includes settlement patterns and population density issues.

Background details of the excavation for each prehistoric site from which human skeletal remains were recovered are provided. Descriptions of the human skeletal remains, including the completeness of individuals, the preservation of bone and their treatment on site, are provided.

How archaeological evidence can help with the analysis of population health is described first, as introduced in Chapter 1. This is divided into material relating to the natural environment and the cultural environment. However, it is recognised there can be considerable overlap in evidence for the two aspects.

## Natural Environment

The natural environment provides all of the resources for a prehistoric population. This includes animal products, vegetation, and fresh water.

The analysis of faunal remains from an archaeological site includes the identification of species present in prehistory. It may also be possible to determine whether the animal identified was domesticated or wild. This information can provide an indication of, firstly, the subsistence mode of the community and, secondly, how much control the population had over their food supply. The domestication of animals would allow for the control of protein supply, among other dietary requirements, in the community's diet. This supply may be distributed to individuals or families in order of rank, with some societies placing women and girls last in line (Carloni 1981; Gittelsohn 1991 cited in Ortner 1998), although this is not always the case.

The interaction of humans with animals is also important from another angle. The care of domesticated animals requires close contact with them which can result in the transmission of zoonoses, including parasites to humans. Brothwell (1991) presents several different definitions of the term zoonosis, but the main point, particularly for palaeopathologists, is the sharing of infectious diseases with other taxa. An increase in human population density shared with other vertebrates as well as cultural factors, such as trade of animals, may also affect the transmission of zoonoses (Brothwell 1991).

The analysis of faunal remains can also assist those interested in describing a prehistoric environment. Each species has a favoured environment. Therefore, the presence of a particular species in an archaeological context may help indicate the type of environment that surrounded the site in prehistory. The analysis of botanical remains will also add information in the same way.

The analysis of botanical remains can also provide information about plant varieties, including whether they were wild or domesticated. The presence of domesticated varieties indicates some knowledge of agriculture. Plants provide the main carbohydrate source in a population's diet and through the development of agriculture a population can become reliant on a single source of carbohydrate which can have repercussions on their nutrition. However, many botanical remains leave no archaeological traces and the range of vegetation is likely to be more varied than archaeological evidence infers. The effects of the development of agriculture on the health of a community have been considered extensively in the literature and will be reviewed later. The presence of wild varieties of plants indicate that at least part of the subsistence mode is taken up by gathering. This type of subsistence base is more likely to provide a greater variety and therefore a more nutritionally adequate diet. Most plants, including seeds or fruit, are seasonal, so that is there is a period when they are unavailable for human consumption. There is, therefore, the possibility of a seasonally deficient diet and a potential for seasonal malnutrition unless knowledge of storage methods or other resources counter this.

Fresh water is essential for survival, and most prehistoric sites were situated near a reliable supply. Many prehistoric sites in Thailand were located near rivers or a confluence of streams (Higham 1989). Where there is no obvious source of water supply, geoarchaeological analysis of surrounding environments can provide evidence of ancient rivers. The presence of freshwater fish and shellfish also provide an additional advantage of living near a river and, if salt is also nearby or obtained through an exchange network, fish can be preserved when the supply is plentiful. If crops fail this can be a more reliable food source throughout the year.

Rivers also provide a method of transport to enable a population to participate in exchange networks. An exchange network can, for example, provide essential elements missing in the local diet, or pass on knowledge of technology, such as metal working. However, a negative aspect of this contact can be the transmission of disease between populations.

The tropical monsoon environment of Southeast Asia makes rivers prone to flooding in the wet season, which can be advantageous to human populations. The inundation of floodplains provides excellent opportunities for rice cultivation. It also provides an increase in other riverine food resources which would make an advantageous contribution to the diet. However, the same rivers are often much reduced in capacity in the dry season, which is especially severe in the northeast of Thailand which suffers from a long dry season. Riverine resources may become scarce and the reduced availability of potable water could have serious consequences for a population. Knowledge of storage methods and salt preserving, as mentioned above, would be very useful during the dry season. The coastal environment, in comparison, feels the effects of the monsoon less than the dry northeast as the sea is a rich source of food all year round. Not only does the sea provide fish and aquatic mammals, the coastline can provide a variety of shellfish species.

# Cultural Environment

### Technology and subsistence

A discussion of the technology available to those in prehistory is generally biased as the recovery of such evidence is usually limited to non-biodegradable materials, such as metal and pottery. This is generally the case in the following descriptions of technology at each Thai site. However, there is the possibility of considering ethnographic evidence. Bamboo, for example, is extensively used in modern Thailand as fishtraps, basketry (Thompson 1996), and for temporary construction and ladders. It is likely that bamboo was used in prehistory in Thailand and has been found at prehistoric sites, such as Spirit Cave in the north (Thompson 1996).

The development and knowledge of metal working would have provided a community with certain advantages. Tools to assist with clearing land of trees and a plough to cultivate the land would have made the transition to intensive agricultural practices possible. Knowledge of metal working may also have had an effect on the efficiency of a population's subsistence mode. The development of the plough, and the domestication of the water buffalo to pull it, would have made cultivation easier. Populations that could provide metal ingots and/or were particularly skilled at metal working would have profited from an exchange network.

One of the most important advances in prehistory was the ability to domesticate plants, providing a population with a reliable food source. Knowledge of rice agriculture has been present in Thailand for at least 4000 years. The cultivation of domesticate varieties of rice was evident at Khok Phanom Di from 2000 BC (Thompson 1996). Such early agricultural communities did have some knowledge of the exploitation of rice, but probably combined it with hunting and gathering. There was a gradual increase on reliance on rice through to its intensive use in the Iron Age. A hunter-gatherer lifestyle may possibly have provided a more varied, and therefore more nutritionally adequate diet in comparison to rice agriculture. On the other hand it has been proposed that a hunter-gatherer lifestyle was more stressful in terms of labour, whereas an agricultural community was of a more sedentary nature. However, Larsen (1997) cautions researchers not to over-simplify their interpretation of a population's subsistence mode. In reality, many prehistoric populations were probably reliant on a mix of the hunter-gatherer and agricultural subsistence bases, even after intensive agricultural development. Human communities are much more complex in their ability to adapt to changes than is often inferred from the available archaeological evidence.

The domestication of animals ensured a more reliable source of food and, as mentioned, assisted with some labour, such as water buffalo for pulling ploughs.

### Population Size

An agriculturally based subsistence provides a more reliable and abundant food source for a community than a hunter-gatherer lifestyle. The increased reliability of a food supply and the sedentary nature of agricultural communities would enable a population to increase in size. However, population expansion carries with it certain negative aspects.

By increasing the size, and presumably the density of a population, the number of potential hosts for infectious agents is increased and can create a relationship that forms the basis of endemic infections (Larsen 1997). Population crowding often leads to a reduction in the level of hygiene and contamination of the water supply, promoting the existence of diarrhoea-causing enteric bacteria (Larsen 1997). Therefore, an increase in population size can be concomitant with an increase in morbidity.

Specialisation is often a further consequence of a population increase. It enables a population to be part of an exchange network, for example. Once a reliable source of food is organised, not all individuals need to be involved in food production and are able to support the specialists. Specialists would include those individuals with skills to work metal, make pottery or jewellery. These could be exchanged for the population's deficiencies, such as dietary needs, or salt for food preservation. The degree of specialisation can also be considered an aspect of social complexity.

### Social Complexity

Larger populations require more organisation of resources than smaller ones, especially in the control of water and food, as well as exotic or valued goods. This results in the creation of an authority; a hierarchical society. Division of labour may be established based on sex, age, wealth, or familial relationship, for example. In terms of the health of a community, it may result in inequality among groups within a community, with some being provided with less access to nutritional food and water, and being required to carry out more work than others. Reflecting the inequality in access to food, higher status individuals have been observed to receive more animal protein than lower ranked individuals in caries infections. In contrast to this, higher ranked groups in other societies may receive softer and more refined foods which are easier to eat but promote more caries (Larsen 1997). This type of social structure makes for a complex set of interactions, and as a consequence, makes interpretations more difficult.

Even before highly ordered, developed societies existed, it is likely that there was a sexual division of labour with consequences for the diet of these people. This has been seen in ethnographic studies which have observed men hunting for large game and eating more protein, while women gathered and ate more plant material, small animals, and shellfish (Larsen 1997). This has repercussions for the health of males and females. It may be possible to see this in skeletal material, for example, through observations of dental health.

While identifying the types of grave goods may indicate technological knowledge, the actual distribution and quantity of grave goods may help indicate socioeconomic divisions based on wealth or status (Tainter 1978). They may also indicate potential sex and age divisions within a community.

Based on the above theories concerning the effects the natural and cultural environment on health, the relevant information from each Thai sample is now presented. This provides a background to both the natural and cultural environment that each group of people were exposed to. Subsequently, differences and similarities among these variables may provide explanations concerning the differences and/or similarities among the health patterns of the skeletal samples.

# Southeast Thailand

As mentioned in Chapter 1, a major archaeological project was carried out in the early 1980's in the southeastern region of Thailand. The project was restricted to the area of the Bang Pakong River Valley in order to examine human settlement and adaptation of what were coastal areas in prehistory. The excavation of Khok Phanom Di was the major undertaking and a collection of extremely well-preserved human skeletons were uncovered. The involvement of a multidisciplinary group of researchers has led to considerable information published regarding many aspects of the environment, social structure, and health of this prehistoric group (see Higham and Bannanurag 1990, 1991a; Higham and Thosarat 1993; Thompson 1996; Tayles 1999). The second site to be excavated within this area was initially chosen to provide a contemporary comparison to prehistoric Khok Phanom Di. However, the chosen site, Nong Nor, instead provided an occupation phase predating Khok Phanom Di and a cemetery postdating it. A single volume covering the results of research into the natural and cultural environment, as well as the preliminary analysis of the skeletons has been published (Higham and Thosarat 1998a). Despite the lack of evidence for occupation during the cemetery phase, the people were no doubt living very close. prehistoric groups and are consequently observed to have less The skeletal sample, although providing over 150 individuals, was very poorly preserved and fragmented, which has consequently made comparisons with other skeletal samples difficult.

## Khok Phanom Di

### The Excavation

Khok Phanom Di is located in the lower Bang Pakong River Valley, Chonburi Province, 22 kilometres east of the coast of the Gulf of Thailand and approximately 50 kilometres east of Bangkok. The site of Khok Phanom Di is a 12 m mound rising above the surrounding rice fields and covers an area of approximately five hectares. Over a period of seven months in 1985, a 10 m x 10 m square was excavated near the centre of the mound to a depth of 6.8 m, where natural substrate was reached.

The excavation was under the direction of Professor Charles Higham and Dr. Rachanie Bannanurag (Bannanurag 1989; Higham 1989; Higham and Bannanurag 1990), and was a combined project between the University of Otago, New Zealand, and the Fine Arts Department of Thailand.

This early agricultural site is thought to have been occupied over a period of about 500 years from 2000 BC - 1500 BC from radiocarbon dating of *in situ* charcoal (Higham and Bannanurag 1990). Seven mortuary phases (MP 1 - 7) were identified based on the presence of apparent hiatuses between burial groups (Higham and Bannanurag 1990). They were confined to the lower 5.6 m of deposit, with the upper 1.2 m containing no burials (Tayles 1999). The majority of burials were associated with evidence of occupation, including ash, shell middens, biological remains, and pottery (Higham 1989). Figure 2.1 presents the layout of the burials excavated from the Khok Phanom Di cemetery.

### The Human Skeletal Remains

An extensive analysis of the human skeletal remains has previously been carried out. The raw data from this analysis by Tayles (1992, 1999) has been used with the permission of the author. The following is a summary from Tayles (1992) regarding the excavation and cleaning of the human skeletal remains.

Burials identified during the excavation were allocated burial numbers from 1 to 154 with numbers 148 and 149 not allocated. Two burials contained two skeletons each, bringing the total number of skeletons to 154. Some infant and adult remains were found during the faunal analysis, but as these could not confidently be assigned to a particular burial they were not included in further analysis. There was very little disturbance or mixing of burials and the majority of skeletons were complete and articulated. Less than 10 individuals had major bones missing. The bone preservation was generally very good. However, it was necessary for many bones to be reconstructed. This was not a significant problem as the edges of the postmortem breaks were not fragmented.

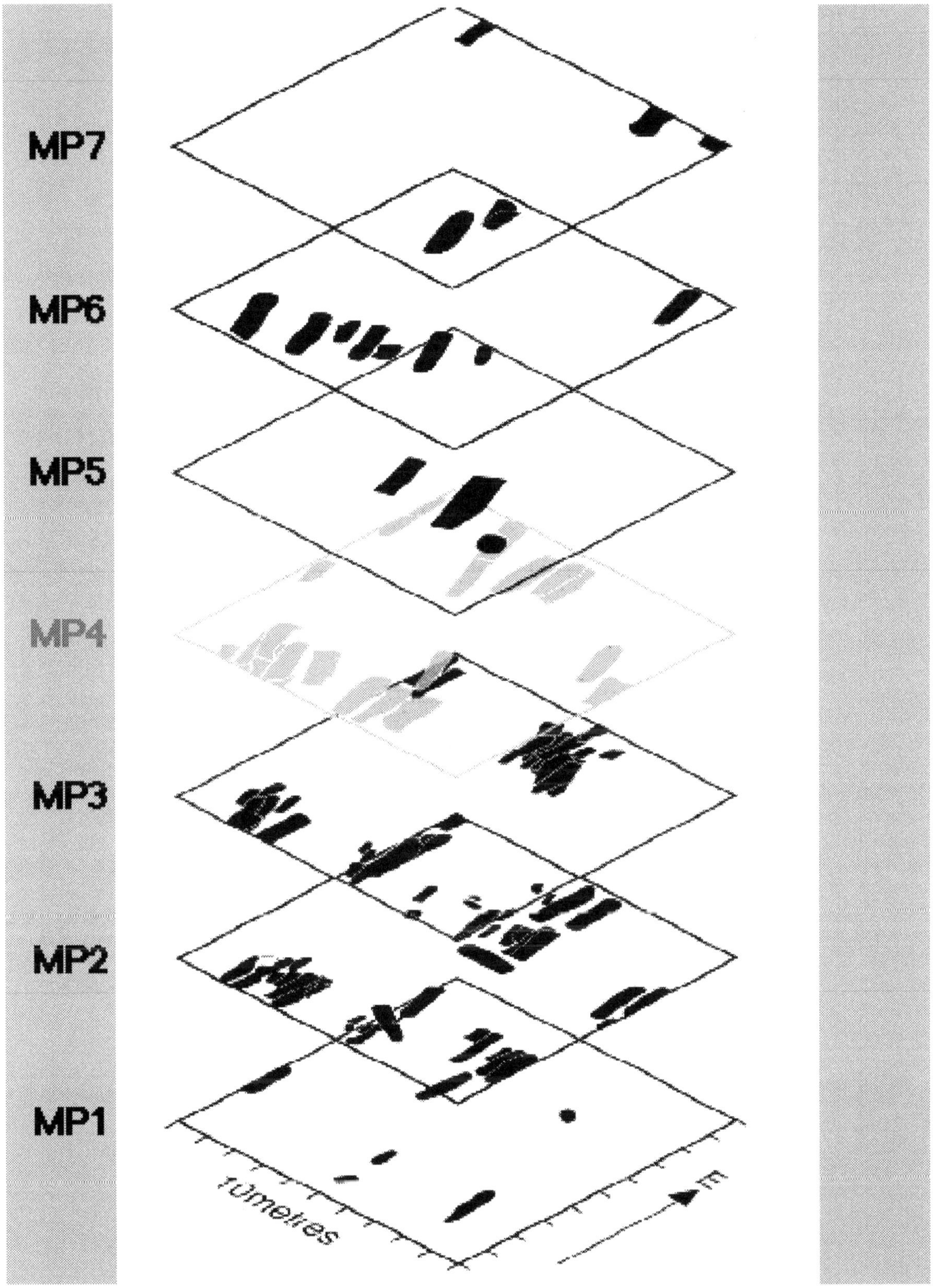


**Figure 2.1:** Khok Phanom Di burial plan indicating the seven mortuary phases and the spatial distribution of many burials (from Higham and Thosarat 1998e). MP = mortuary phase.

The presence of a calcareous deposit over many of the bones was initially a problem. Not only did it cement many bones together, it also obscured the periosteal bone surface. Attempts to remove this deposit were generally successful. A combination of a chemical treatment and the use of small tools such as dental probes and scalpels were used. The chemical treatment involved the soaking of bones in 7% acetic acid with 2% calcium acetate for between 15 minutes and overnight when possible. This softened the deposits so they could be removed more easily. The bones were then cleaned in water and dried before analysis.

An analysis of the genetic material in the Khok Phanom Di bone has also been carried out (Pierson 1998). However, this proved difficult and no conclusive results were obtained.

### The Environment and Natural Resources

There was a considerable amount of faunal material found during the excavation of Khok Phanom Di. However, mammalian species, in particular, were not in abundance (Higham and Bannanurag 1991b). There were a few pigs, macaque, chickens, deer, cattle, and water buffalo, but the only evidence of domesticated stock was the dog (Higham and Bannanurag 1991b). The majority of biological remains were those associated with aquatic ecosystems, particularly the mangroves, but also fish and shellfish from the open sea, estuarine, and freshwater habitats (Higham and Bannanurag 1991b).

Archaeobotanical and ethnographic evidence concluded that the mangrove environment in which the Khok Phanom Di site was immediately surrounded did not provide abundant plant remains (Thompson 1996). One example, is the hypocotyl (germinating embryo) of *Brugiera*, a mangrove fruit, which has high caloric and protein values, although involved time consuming processing to make it edible (Thompson 1996). The adjacent freshwater ecosystems may have provided many wild vegetables that grow along streams or in freshwater swamps (Thompson 1996). Evidence of coconuts were identified and as well as their flesh, may also have provided drinking fluid when freshwater supplies were depleted in the dry season (Thompson 1996).

There is evidence of domesticated rice (Thompson 1996) but the saline conditions of the site probably forced the rice cultivation further inland (Maloney 1991). This site represents one of the earliest known sites of rice cultivation in southeast Asia at 2000 BC (Higham 1996) although some dispute this (for example White 1995). The presence of a beetle known to thrive in rice storage facilities is indirect evidence that might suggest rice was being harvested and stored (Higham and Bannanurag 1991b). There is further indirect evidence of rice cultivation discussed with reference to the technology available to the people of Khok Phanom Di.

Core samples of the neighbouring environment around Khok Phanom Di were taken and these were used for the analysis of sediment and pollen remains to obtain information about the palaeoenvironment (Higham 1989). It has been deduced that at the initial settlement of Khok Phanom Di, the site was situated in the general vicinity of mangal, estuarine, and freshwater conditions with the open sea also accessible (Thompson 1996).

A tropical estuary may provide abundant resources, but this type of environment can carry a much higher pathogen load than drier areas (discussed in further detail in Chapter 8). However, the estuarine habitat did not remain the same throughout the occupation of Khok Phanom Di. There appears to have been a dramatic change in the environment during MP 4 which resulted in the sea being less accessible and the estuary drying up, perhaps due to a change in course of the nearby Bang Pakong River (Higham 1996). The evidence for drier land has been indicated by the presence of land snails during MP 3-4 and the decline in the exploitation of marine resources after this time also suggests environmental change (Mason 1991).

### Technology

Perhaps the most marked skill of the people at Khok Phanom Di was their ability to manufacture pottery vessels, some of which are considered masterpieces. Local marine clays were one of the major resources available to these people and were used from the time of initial settlement of the site for pottery manufacturing. There was an abundance of pottery vessels excavated from all layers. There was also evidence of burnishing pebbles and clay anvils used to shape pots from the earliest depths of occupation and clay cylinders were found which represent the formative stage of a pottery vessel (Higham 1989; 1996).

The level of subsistence technology present at this site includes some knowledge of rice agriculture. Not only does the presence of domesticated rice species imply this (Thompson 1996), the presence of certain artefacts such as shell knives supports it. Shell knives were the single most abundant artefact found and, in a blind test, the type of shell used, cut grasses and rice most easily (Higham 1993b).

The domestication of animals was not undertaken on any large scale, instead reliance of a marine diet was most likely. The presence of fishhooks, clay weights (net sinkers), and harpoons support this, as does the abundance of aquatic species discussed above. Harpoons may also have been able to be used for the hunting of terrestrial animals (Higham 1993a).

Another of the major resources available to these people was marine shell. This was used in a variety of ways including jewellery and more utilitarian tools, such as the shell knives mentioned above. This, and their ability to make pots, would have enabled the people of Khok Phanom Di to participate in an exchange network in order for this population to acquire items the site was deficient in, such as good quality stone for items such as adzes. Their position on an estuary, with a major river nearby, would have also placed them in an advantageous location to participate in such a network (Bannanurag 1989). The change in the environment may have led to the inaccessibility of some valuable marine resources and the river; hence they would have been less able to participate in the exchange network. It is interesting to note that the exotic stone adzes were always sharpened to a small size indicating maximum usage (Higham 1989) implying they were expensive or difficult items to obtain.

### Social Structure

There were seven mortuary phases (MP) identified, as mentioned earlier (Figure 2.1). The six earliest burials in MP

1 were not related to each other, but the remaining burials were placed in rows or interred above each other in discrete clusters. The clusters of burials began in MP 2 and most remained evident into MP 4. They are identified from A to G and comprise burials interred closely together, but not disturbed; a space between clusters distinguishes one cluster from another (Higham and Bannanurag 1990). It has been suggested on the basis of spatial contiguity that these clusters that continue through time represent family and/or social groups (Higham and Bannanurag 1990), but there is no conclusive biological evidence for this (Choosiri 1988).

In general the process of burying the dead involved a shallow grave with the body placed so that the head was to the east. The body was more often than not covered in red ochre. Then, from the observation of fragments of unwoven fabric or *tapa*, individuals were wrapped in a white cloth of the unwoven variety, like bark cloth (Bannanurag 1989). Grave goods were placed with the body and included pots, shell beads, burnishing pebbles, clay cylinders, and a polished stone adze. In later burials there was more variety of grave goods and included, in addition to those above, clay anvils, large shell bracelets, shell discs, shell earrings, a fish hook, and turtle carapace ornaments. Throughout the cemetery infant burials were also interred with grave goods, with the exception of 21 infants buried in scoops with no characteristic ritual (Higham 1989).

Pottery manufacturing implements were found only within the graves of women and children. This implies it was the women who were the potters. This indicates a division of labour and also a level of specialisation within the community. Only men were found with turtle carapaces. Based on these divisions in grave goods and possible divisions of labour, this group of people had some degree of social organisation with some inequality among members.

### Summary

Khok Phanom Di is in the southeast of Thailand, along the lower Bang Pakong River valley, however the site was coastal during prehistory. Khok Phanom Di was inhabited in prehistory by early agriculturalists. They occupied the site for approximately 500 years from 2000 - 1500 BC, and represent the oldest skeletal sample in this thesis. A total of 154 skeletons, of excellent preservation, were excavated from a portion of the prehistoric cemetery. The results of an analysis of this human material by Tayles (1992, 1999) will be used. The people of Khok Phanom Di undertook incipient rice agriculture, relying more predominantly on coastal, marine, and mangal resources for food. This diet did not include large quantities of red meat. This site was premetal, but the people were very skilled at pottery manufacturing. The tools for this technology were distributed predominantly among female adults. These people appeared to have had a not entirely egalitarian community with some divisions of labour present. However, both men and women were able to attain high status as indicated by the distribution of grave goods.

## Nong Nor

### The Excavation

Nong Nor is also located in the lower Bang Pakong River Valley, some 27 kilometres east of the coast. However, during prehistory it was much closer to the sea. This site is 14 kilometres south of Khok Phanom Di.

Nong Nor was excavated over three seasons from 1991 to 1993 under the direction of Dr Rachanie Thosarat and Professor Charles Higham. The excavation uncovered an area of 404 $m^2$ including a number of test squares (Higham and Thosarat 1998c). The site had been on a low mound, but half had been recently levelled for rice production. The other half, owned by a different individual, remained unmodified. Both areas had been subject to recent looting. Access to only the levelled section was granted, the owner of the unlevelled portion refusing to allow the excavation of his land. The excavation uncovered shallow cultural deposits that extended up to two metres in depth but generally were much shallower, not exceeding one metre in most places (Higham and Thosarat 1998c). Figure 2.2 presents a schematic plan of the burial layout in the cemetery.

There were two prehistoric phases identified at Nong Nor. The first was an occupation level which included an extensive shell midden. This was dated at about 2450 BC, prior to the occupation of Khok Phanom Di, using carbon 14 dating of charcoal from hearths underlying a shell midden, as well as shell from the midden itself (Higham and Hogg 1998). The second phase at Nong Nor was denoted by a Bronze Age cemetery. Burials were dug into the shell midden over 1000 years later than the occupation phase. This period has been dated at 1100 - 700 BC, from carbon 14 content in charcoal, less well-provenanced than above, and also from rice chaff, found as a tempering agent in pots (Higham and Hogg 1998). This period is considerably after the occupation of Khok Phanom Di, therefore neither period at Nong Nor was contemporaneous with Khok Phanom Di. Because there were no human burials found definitively associated with the earlier occupation at Nong Nor, only the information from the latter cemetery phase is of interest here. Unfortunately, little is known about the environment and subsistence of this phase.

### The Human Skeletal Remains

During excavation, the burials were photographed and drawn, the skeletal elements present recorded and preliminary estimates made of age and sex where possible. Complete or almost complete long bones were measured *in situ* and any evidence of pathology was recorded before the bones were lifted. However, as mentioned the preservation of the human bone was extremely fragile and there were very few complete long bones and no undamaged crania in the sample.

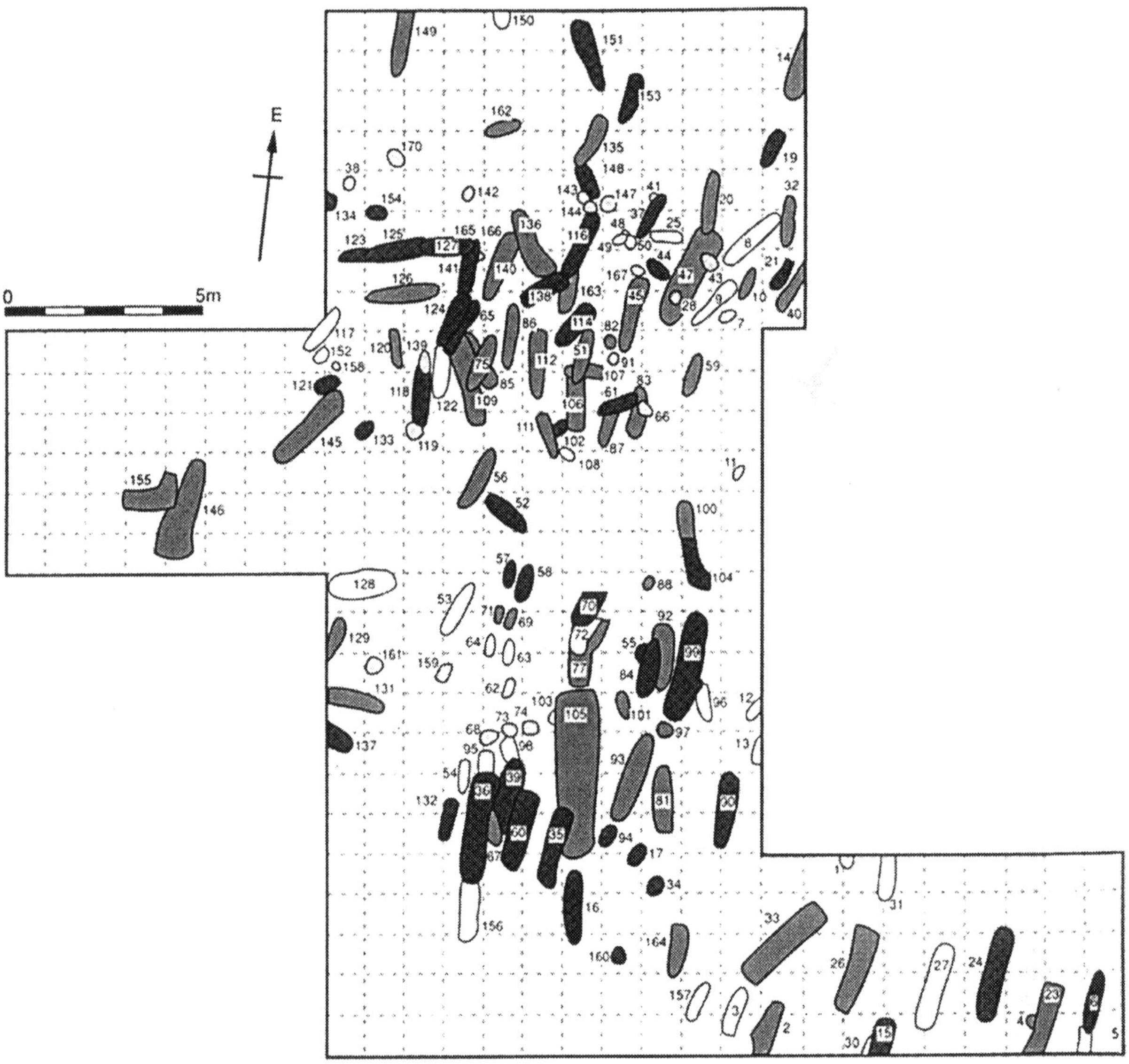


**Figure 2.2:** Nong Nor burial plan (from Higham and Thosarat 1998e)

The bones were washed, dried and packed in plastic bags in the field, with individual bones or groups of bones separately packed. Crania were lifted whole and cleaned without washing where appropriate. The material was shipped to the University of Otago, New Zealand. The reconstruction and recording was done in the laboratories of the Anthropology Department by myself and Nancy Tayles; the dentition was recorded by Veronica Hunt. Following this analysis, this material was returned to Thailand.

A total of 172 burials were identified during the three excavation seasons and subsequent analysis. A total of 15 burials, including nine infant burial jars, did not contain any human bone. The six adult burials included in this number were identified from grave cuts, some with pots and other grave goods. From the total of 172, the number of individuals represented by skeletal remains, after allowing for some amalgamation of incomplete and scattered skeletons, and the addition and deletion of others, totalled 155.

Some skeletons are incomplete because they extended beyond the excavation area. Many burials were damaged by human activity, including the removal of the upper layers of the mound during construction of the paddy field, subsequent agricultural activities, and looting of graves. This had apparently occurred in prehistory along with disturbance through superpositioning of graves, as well as in very recent history. In addition to this human activity, there is evidence of disturbance of the bones by natural processes, such as the annual inundation of the field during the wet season.

Bone tissue was particularly poorly preserved in burials at the northern edge of the mound, where the composition of the surrounding soil matrix resulted in chemical dissolution of the bone. The state of preservation of the bone tissue is generally better in skeletons buried in graves cut into the natural deposits than in those buried in the midden. Most bones had suffered some damage for one or more of the reasons noted above. Very few long bones were intact with complete articular surfaces and in most cases only shafts were recovered. However, many had shadows of complete bones identified during excavation, enabling an estimate of bone length to be made *in situ*. No crania were complete, with very few facial bones surviving and most calvaria broken and/or crushed into many pieces. No crania could be fully reconstructed, largely because of the crushed, broken edges of the bone fragments.

### The Environment and Natural Resources

There is not a lot of information regarding the resources available to the people burying their dead during the Bronze Age at Nong Nor. The earlier occupation period provided more information. Extensive palaeogeographical work was carried out in the area surrounding the site. Boyd *et al.* (1996) have described three palaeogeographical phases and four geoarchaeological periods. Phase 3 of the former and Period 3 of the latter are those that occurred during the Bronze Age when the site of Nong Nor was used as a cemetery.

The landscape is thought to have been a seasonally inundated floodplain with some rice cultivation possible. The people had access to coastal and marine resources, but at a greater distance than for those people at Khok Phanom Di. The presence of a large river and the floodplain associated with it would have provided a partially aquatic environment. An abundance of aquatic resources would therefore have been available during the period the site was used as a cemetery (B. Boyd, pers comm).

The faunal remains present as grave goods may indicate the types of resources available. Remains of the domestic dog were particularly abundant. Other animal remains found include pig, cattle, and chickens (Higham 1998).

### Technology

Local clay was used to make pottery vessels (Higham and Bannanurag 1992) and these were tempered with rice chaff (Debreceny 1998). These may have been used as part of the exchange network. Bronze artefacts were in abundance and were commonly in the form of jewellery. They were mostly made of copper alloys, although some artefacts were made of tin, which is unusual (Reay and Chang 1998). Throughout the cemetery, however, there is no evidence of the production of metal artefacts (Reay and Chang 1998). There is however, evidence for the manufacturing of shell jewellery on site as was indicated by the presence of raw shell material and manufacturing debris (Chang 1998). The people of Nong Nor therefore, may have imported items already manufactured, as well as importing copper, tin, sandstone, and an unusually wide range of exotic stone, such as serpentine and talc (Higham and Thosarat 1998e). It appears exotic materials came from a wider area than those at Khok Phanom Di (Higham 1996).

Subsistence for the people of Nong Nor was based on rice agriculture and some domesticated stock. Hunting and fishing are evidenced by the presence of points and fishhooks.

### Social Structure

The mortuary customs of prehistoric Nong Nor are consistent with those of early agricultural and Bronze Age cemetery sites in Southeast Asia (Higham and Thosarat 1998c). There is a relatively uniform pattern of mortuary ritual, especially of the adults, throughout the cemetery. The body was frequently interred with the head oriented in an easterly direction and with a number of grave goods. Such grave goods included pottery vessels, animal remains (often dog), grinding and polishing stones. Many individuals were also wearing jewellery made of bone, stone, shell, bronze, and a few of tin (Higham and Thosarat 1998b). The burial plan shows some degree of organisation spatially (Figure 2.2), but certainly not as clear as at other sites, for example as seen at Khok Phanom Di (Figure 2.1) (Higham and Thosarat 1998b).

The type and number of grave goods were analysed using a multivariate analysis for 49 of the most intact burials by Debreceny (1998). This method was used to identify possible groupings, such as social groups based on rank or wealth. However, no particularly rich group was identified, with no age or sex bias in mortuary wealth(Debreceny *et al.* 1998). Children were treated the same in death as adults. Overall, no major dissimilarities in grave goods were found among the burials (Higham and Thosarat 1998b) . This indicates a community not particularly socially complex with no strict hierarchical structure evident in the burial ritual.

Summary

The prehistoric mound of Nong Nor is situated in the southeast of Thailand, only 14 km from the site of Khok Phanom Di. Therefore, Nong Nor was also close to the coast in prehistory. Bronze Age people used the mound as a cemetery from 1100 - 700 BC. A total of 155 individuals have been excavated from part of the cemetery, however most were poorly preserved. The surrounding floodplains favoured rice agriculture and this was undertaken at a more intensive level than at Khok Phanom Di, as was the raising of domestic stock, such as cattle and pig. This environment also provided a good source of freshwater aquatic foods. Marine resources were also exploited but were not as predominant as at Khok Phanom Di. Pottery vessels were still being made by the people of Nong Nor, but bronze was also present, mostly in the form of jewellery. A wide range of stone artefacts, some exotic, was also present. However, no marked hierarchy were evident in the distribution of such grave goods. Bronze Age Nong Nor was an autonomous community with an increasing exchange network, but no external authority (Higham and Thosarat 1998e) .

# Northeast Thailand

As mentioned in Chapter 1, a number of archaeological projects have been undertaken in the last few decades in the northeast of Thailand. The most recent is the Origins of Angkor archaeological project based in the Mun River Valley. This project has excavated three prehistoric sites with the aim of uncovering the baseline for the Iron Age in this region. Of these three sites, Ban Lum Khao is the only one set totally within the Bronze Age. This excavation uncovered a cemetery totalling 110 human skeletons of relatively good preservation. Research on this site is still ongoing, although unpublished and/or preliminary reports have been written. This includes a general summary of the site (Higham and Thosarat 1997) and a thesis examining the social aspects of Ban Lum Khao and other Mun River prehistoric sites (O'Reilly 1999).

An earlier project further north in the Chi River Valley aimed to extend the research began at the archaeological excavation of Ban Chiang. Ban Na Di was excavated and provided insight into the early period of Iron Age in this region. The cemetery excavated uncovered 78 human skeletons of varying quality and completeness. There is a considerable amount of research from this excavation that has been published in three volumes (Higham and Kijngam 1984c).

### Ban Lum Khao

The Excavation

Ban Lum Khao is a modern village situated in the Northeast of Thailand on the Khorat Plateau, within the upper Mun River Valley. Under the direction of Dr Rachanie Thosarat and Professor Charles Higham, this prehistoric site was excavated from November 1995 to February 1996. A 10 m x 14.5 m square was excavated to a maximum depth of 1.7 m. Stratigraphically, little could be detected because of extensive bioturbation, however, three 'layers' were identified. Layer 1 was an occupation layer with an initial settlement date of between 1400 - 1000 BC. Layer 2 was a Bronze Age cemetery probably used between 1000 - 500 BC, and the third layer was of a late Bronze age occupation (Higham and Thosarat 1998e) . Dating based on radiocarbon analysis was only available for the initial occupation of the site and was from charcoal in pits (Higham and Thosarat 1997) . Dates of the cemetery have been inferred from the initial dates and other archaeological evidence. Figure 2.3 provides a plan of the burial layout of the cemetery.

The Human Skeletal Remains

As each burial was exposed, it was photographed and drawn before being removed from the ground. During removal, the skeletal elements present were recorded and preliminary estimates of age at death and sex were made where possible. Complete long bones were measured with sliding calipers. Notes were made on any pathology present. The bones were washed and dried. All were packed into bags and boxes for transportation and storage at the Archaeological Laboratory in the grounds of the Ninth Office of Archaeology and National Museum, Phimai. Further analysis of the skeletal remains was carried out there by myself in June 1996. The skeletal material is now stored at the Phimai National Museum, Thailand.

Approximately 110 burials were identified during excavation. However, post-excavation analysis revealed two burials contained no human bone, two individuals each were found within two burials, and one burial was amalgamated with the one above it. During further laboratory analysis of unprovenanced human bone, two more individuals were identified. Overall the skeletal remains represent 110 individuals.

The bone tissue is at least partly mineralised with concretions on the surface, but the deeper burials are in good condition. The concretion was a calcium carbonate deposit on the bone that needed to be removed as it obscured the periosteal surface. After washing, bones were put through a simple acid treatment protocol involving a dilute solution of acetic acid with a phosphate buffer ($NaH_2PO_4$, sodium dihydrogen phosphate) overnight. The bones were then washed again. However, this method produced variable results. This concretion is consequently still a limiting factor in the observation of some periosteal surfaces.

Nine burials contained extra bone belonging to infants or children in adult or child burials. Burial 99, a 13 year old child, also contained adult bone. It was decided not to give these additional burial numbers because only one or two extra bones were found within these burials. They could not be confidently assigned to other burials.

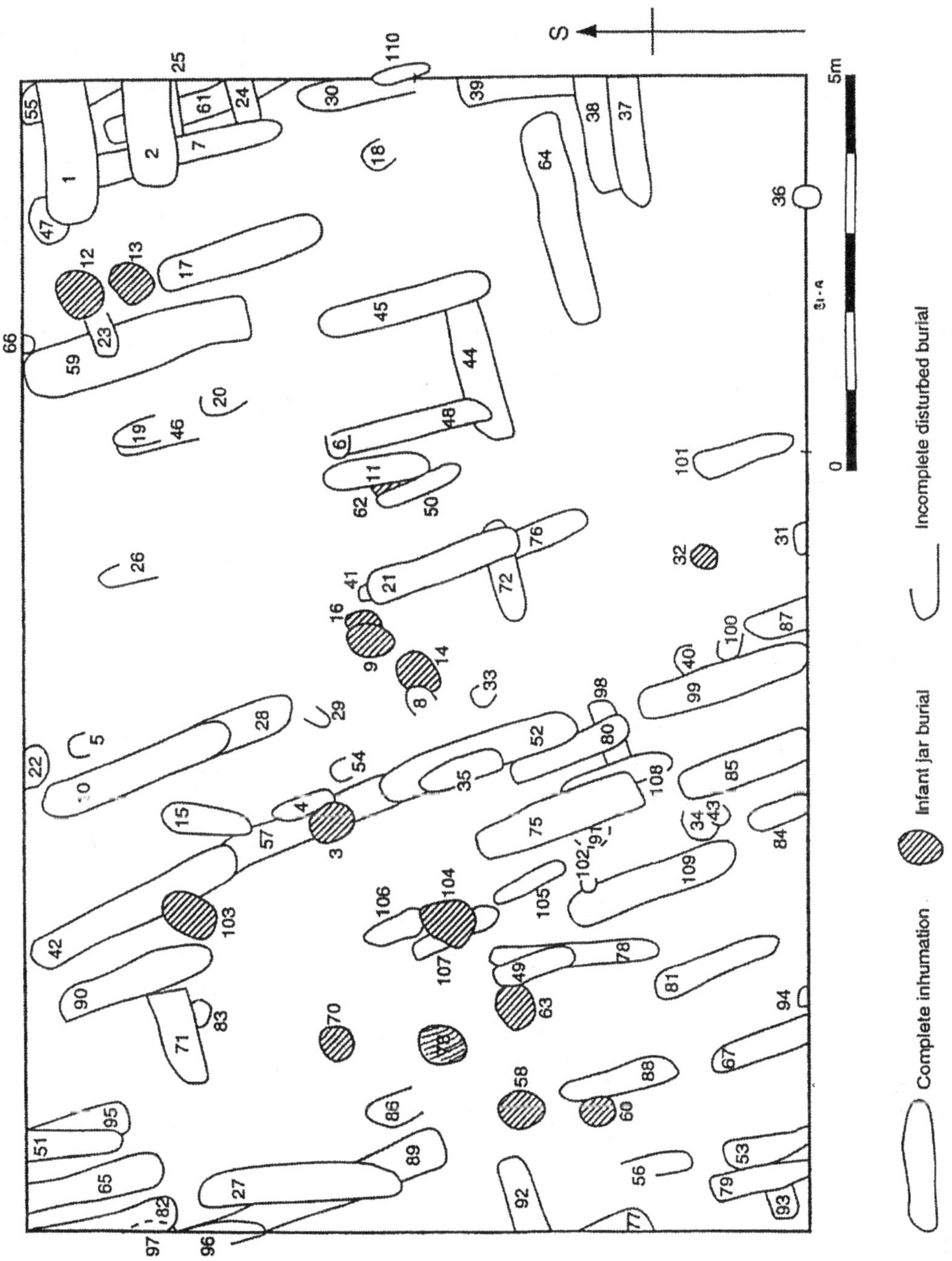


**Figure 2.3:** Ban Lum Khao burial plan (C.F.W. Higham, unpublished)

### The Environment and Natural Resources

The initial occupation layer included a series of pits containing rich organic remains, including shellfish, fish, turtles, and mammalian bones such as water buffalo, deer, pig, and the domesticated dog (Higham and Thosarat 1997). This indicates the type of biological resources available nearby. The size of the fish and water buffalo remains indicated that this area had probably not been previously exploited and the prehistoric people of Ban Lum Khao had arrived to an uninhabited area (O'Reilly 1999).

The surrounding environment would have been suitable for intensive irrigated wet rice agriculture (Welch 1985) but there is, as yet, no direct evidence for the presence of rice. The results are not yet available from research on the botanical material collected and recovered from flotation (G. Thompson, pers comm). However, preliminary analysis of the temper from a small sample of potsherds show that rice was abundant in the upper levels of the site. The evidence from potsherds from the lower levels, which include most of the burials, is as yet unclear, but does not exclude the possibility of the use of rice chaff as temper (J. Voelker, pers comm). This suggests that rice was part of the diet of the people buried in the cemetery.

The nearby Mun River forms part of the primary drainage system for the Khorat Plateau (Moore 1988). On the Khorat Plateau the monsoon has a very specific dry period (November to February) with no rain.

### Technology

Pots were plentiful, often burnished, red slipped, and found as grave goods. No graves contained actual bronze artefacts, unlike at other Bronze Age sites such as Nong Nor, but the remains of crucibles and moulds were found in associated layers. This implies the people of Ban Lum Khao were familiar with this technology.

The varied resources of the upper Mun Valley may have made trade possible (Welch 1985) and the presence of marine shell jewellery also implies exchange was occurring (Higham and Thosarat 1998c). Stone adzes and marble bracelets (Higham and Thosarat 1998c) may also have been imported.

The people of Ban Lum Khao had domesticated pigs, cattle, and dogs. Water buffalo remains were probably hunted and their presence indicates low-lying wetlands were nearby. This may have been the area in which these people would have harvested shellfish and fish. Deer from the forest fringe would also have been hunted (Higham and Thosarat 1998c).

### Social Structure

The cemetery was laid out with many of the adults and/or child burials in rows. Many of their graves faced a southeasterly direction, although a few were directly at right angles to this. The majority of infants were interred in very large burial jars, up to 60 cm across (Higham and Thosarat 1998e). These were frequently placed at the head of an adult. Inside these pots, along with the infant, were smaller pots. Grave goods of adults included pots at the head and feet, and also personal jewellery of stone and shell.

An analysis of the grave goods by age and sex revealed very few patterns. In general it was concluded that there was no evidence that Ban Lum Khao was a hierarchical society (O'Reilly 1999).

### Summary

The village of Ban Lum Khao is in the northeast of Thailand along the upper Mun River Valley. The Bronze Age cemetery at Ban Lum Khao is dated to between 1000 - 500 BC. A total of 110 individuals, most in a reasonable condition, have been excavated from a portion of this cemetery. Indirect evidence implies rice agriculture was probably a subsistence activity. However, this would have been supplemented with freshwater resources from the nearby river, as well as meat, both hunted and domestic. Pottery manufacturing appeared to be a common activity, given the number of pots present. These may have been used to participate in an exchange network. The people of Ban Lum Khao were also familiar with Bronze technology, although bronze artefacts were not present as they were at Nong Nor. Mortuary ritual indicates the social structure of this group was egalitarian and autonomous, like Nong Nor.

## Ban Na Di

### The Excavation

Ban Na Di is located in the Upper Songkhram area of the Khorat Plateau in Northeast Thailand. A well-known neighbour is the site of Ban Chiang, 23 kilometres to the southwest (Higham 1989). The excavations at Ban Na Di were carried out from 1980 to 1981 under the direction of Professor Charles Higham and Dr Amphan Kijngam. An area of 65 $m^2$ was uncovered over two separate areas; a depth of 4.2 m was the maximum reached (Higham 1996). The two areas, A and F, were about two meters apart. A test excavation had already been carried out at Ban Na Di in 1980 and the human skeletal remains from this test square are included in this thesis.

The prehistoric site of Ban Na Di was in use as a cemetery during the period of about 600 - 400 BC covering the transition from the Bronze to the Iron Ages. These are the latest revised dates published in Higham (1996). A much wider period of time was postulated previously (Higham 1996). Dating was initially based on radiocarbon analysis of charcoal from a variety of sources, including hearths (Higham and Kijngam 1984d). However, based purely on subsequent experience at other prehistoric Thai sites, Higham (1996) narrowed the range of occupation to the dates above. There were three distinct mortuary phases, each burial assigned to a phase based on the level from which it was cut (Higham and Kijngam 1984c). The first mortuary phase was the most significant as it contained the most burials, approximately 63 individuals (Houghton and Wiriyaromp 1984). The second phase comprised some infant jar burials located only in squares A1-2 and the third phase was late prehistoric where only three disturbed burials were found

(Higham and Kijngam 1984c). Figure 2.4 provides a plan of the layout of some of the burials.

### The Human Skeletal Remains

An analysis of the human skeletal remains from Ban Na Di has been carried out previously (Wiriyaromp 1984; Houghton and Wiriyaromp 1984). However the skeletal material, stored at the University of Otago, has been reworked for this thesis. The previous research stated that there were 73 individuals represented by the skeletal remains, but my reworking of the material identified a total of 78 individuals. Four further multiple burials, with enough material to be classed as individuals, were found, one containing an extra three individuals, the others, one extra.

One burial, T.22, a seven year old child, is now in a museum in Thailand and consequently not included in the present analysis. The 1984 analysis identifies multiple burials with an alphabetical suffix, while the present study uses numbers (e.g. 17.1). Skeletal material from the test square (F5) are identified by a prefix of 'T', and total 23 in number (Burial T. 1 - 21), including three multiple burials. Those from the main squares (A1-4 and F6) are identified by a 'M'. Of the 55 burials from the main area, 30 were in A1-4, and 25 were in F6.

The bone preservation is moderate in comparison to the other samples and much reconstruction was required. Many bones were incomplete and most skulls fragmented beyond repair. Many individuals were represented by incomplete skeletons. There was considerable disturbance, both prehistoric and modern and therefore a generous amount of extra bone with burials was present.

The material was transported to the Anatomy and Structural Biology Department, University of Otago, where the bone was cleaned and reconstructed. No water or chemicals were used to clean the bone, but small implements such as dental probes were used instead.

### The Environment and Natural Resources

Biological remains were quite numerous and included the presence of the domesticated pig, dog, cattle, and deer. These domesticated species were preferred for sacrificial purposes (Higham and Kijngam 1984c). Butchering marks have been seen on some wild water buffalo remains, indicating their consumption. These animals favour low lying, well watered terrain (Higham and Kijngam 1984c). A number of different species of deer were also consumed, including some that favoured woodland habitats with access to open grassy areas (Higham and Kijngam 1984c). The presence of these animal remains implies that these types of environments were accessible to the site. Many different species of fish were also found. This includes many which were from different types of aquatic habitat such as lakes, ponds, swamps, and streams. Shellfish were also abundant but only during Level 8 and may indicate a change in the environment. Many types of smaller animals and rodents were represented at Ban Na Di, including frogs, hares, and rats (Higham and Kijngam 1984c).

This site is located near three small streams which would have provided a source of fresh water. However, smaller streams are prone to drying up in the dry season. The area around the site also provided access to low terrace soils thought to be best suited for rice cultivation (Higham 1989). Rice remains were determined to be of the cultivated type (Higham and Kijngam 1984c) and at this stage rice agriculture would have been a predominant subsistence activity.

### Technology

Pottery was probably manufactured on site as the presence of clay anvils implies. Bronze metal working was carried out at the site from its beginnings, and almost all stages of the casting process have been found. Axes and arrowpoints were the predominant bronze products (Higham 1984), the latter indicating that hunting was one of the subsistence activities. Iron artefacts were first identified in the late stage of the earliest mortuary phase, at about Layer 6. Iron slag was then found in Layer 5 indicating that the manufacturing of iron goods was occurring locally (Higham 1989).

Worked bone, red and blue glass beads (Kijngam 1984), shell and clay beads, bracelets of trochus shell, bronze, and stone, dog tooth necklaces, ivory ornaments, pottery vessels, and bronze rings, anklets, arrowheads, beads, and bowls (Higham and Kijngam 1984c) were present at some stage throughout the site. Some were able to be made at the site, while many were made of exotic materials that would have had to be obtained from some distance away. Such materials as trochus shell, marble, limestone, travertine, and slate are not found locally and indicate a substantial trade network. For example, the closest supply of trochus in the present day is in the clear waters off peninsular Thailand, some 1000 kilometres south of Ban Na Di (Higham 1984).

### Social Structure

Most of the dead were interred in graves with a north-south orientation (Higham 1996). Grave goods included pottery vessels often containing food, such as rice and fish, and sometimes the leg of an animal was placed beside the body (Higham 1984, 1989).

In the consideration of the major burial phase, there seems to be no clear association of sex or age with grave goods although there were some differences between the two excavation squares, A (A1-4) and B (F5-6) (Higham and Kijngam 1984c). The burials in Area B were somewhat richer in terms of imported objects as grave goods, and some objects were entirely restricted to Area B, including clay figurines and iron objects. Over time, the burials in Area A did become wealthy compared to earlier phases, but Area B was always richer. It would seem that there are differences in the material culture of these separate areas. The dating indicates the areas are contemporary, but whether they are part of one large cemetery or perhaps two distinct lobes based on wealth is unknown. It was concluded that the people of Ban Na Di did not live in an egalitarian society. There was some degree of hierarchy and possible hereditary inequality (Higham and Kijngam 1984c) but the community would have been autonomous (Higham and Thosarat 1998e).

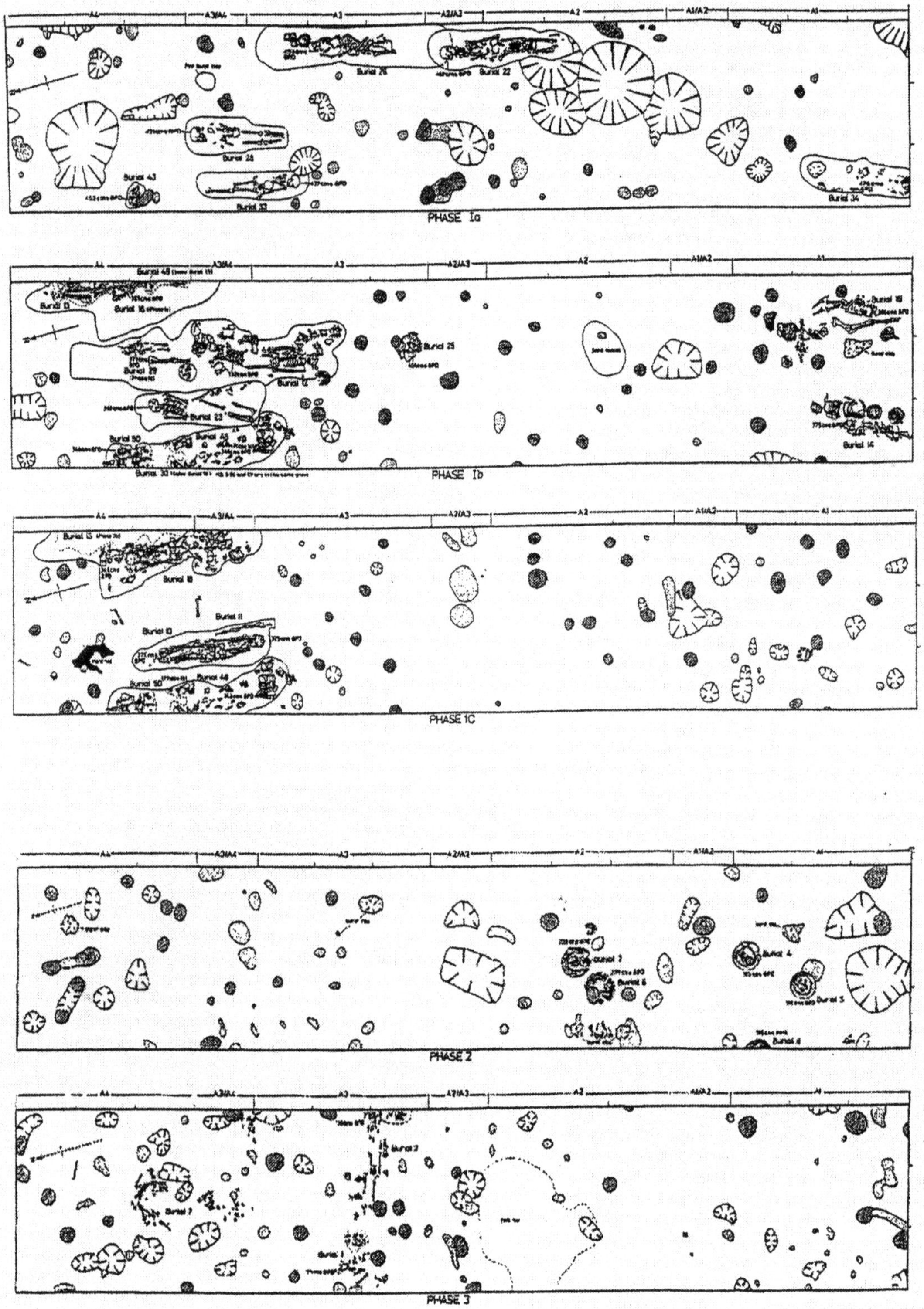


**Figure 2.4:** Ban Na Di burial plans (from Higham and Kijngam 1984e)

Summary

The site of Ban Na Di is located in the Upper Songkhram area of northeast Thailand. It was settled and occupied during the late Bronze Age to the early Iron Age, between 600 - 400 BC. Two areas of excavation of this prehistoric mound have yielded 78 individuals, with moderately preserved bone. This skeletal sample is the most recent of the four compared in this thesis. The people of Ban Na Di were familiar with rice agriculture and the raising of domestic stock, more so than at any of the other sites mentioned above. However, they also would have supplemented their diet with hunting, fishing, trapping animals, and gathering wild plant remains. Pottery was manufactured at this site and evidence of bronze metallurgy was abundant. Knowledge of iron forging was evident only during the later stages of the occupation of this site. An exchange network provided the raw materials for the bronze working, as well as providing exotic stone and shell. Mortuary analysis indicates the one group of individuals within this community had preferential access to these goods. This implies a community with a hierarchical structure.

# Other relevant prehistoric Thai skeletal samples

As mentioned in Chapter 1, there are a number of other prehistoric sites previously excavated in Thailand that contained human skeletal material. For reasons outlined in Chapter 1, such as accessibility and timing, only the four detailed so far in this chapter have been included for extensive analysis in this thesis. However, a number of the other skeletal samples will be compared with these, where appropriate, although not in any considerable detail.

The health of two skeletal samples from northeastern Thailand, Ban Chiang and Non Nok Tha, has recently been presented in a PhD thesis (Douglas 1996). The skeletal material and relevant archaeological information from these two sites is presented below. Additionally, recent excavations at Noen U-Loke, as part of the Origins of Angkor project in the northeast, has also provided a skeletal collection. Preliminary results from this sample are also compared to the four collections. The dentition from this sample has been included as an unpublished Master of Science thesis (Nelsen 1999) and further data collection and interpretation is currently underway (N.Tayles, pers comm). This sample is not included in this thesis in any more detail as the excavation was only recently completed. Other skeletal samples from Non Pai Wai and Nil Kham Haeng in central Thailand have been analysed recently (Agelarakis 1996). However, the published data on skeletal health was presented only in a very general way making these samples incomparable to the samples included in this thesis (Agelarakis 1996).

**Ban Chiang**

Ban Chiang is located in the northern part of the northeastern Khorat Plateau in the vicinity of the Songkhram River (Figure 1.1). The Thai Fine Arts Department began excavations in the modern village of Ban Chiang in 1967. A joint project with the University of Pennsylvania Museum continued the excavations in 1974 and 1975 (Douglas 1996).

A total of 140 individuals were excavated during this project and Douglas (1996) has reworked preliminary analyses to present an extensive picture of their health, particularly focusing on a comparison with the Non Nok Tha skeletal sample. However, the dating of the Ban Chiang material has not been straightforward, although recent advances have been made. AMS dates now indicate the earliest burial was dated to about 2100 BC and other conventional dates show the site continues through to the Bronze Age, approximately 1300 - 1000 BC, and iron artefacts were found in the latest burials (Higham and Thosarat 1998e). Ban Chiang is, therefore, characterised by a very long prehistoric occupation. The results from the latest analysis of the skeletal remains (Douglas 1996) will be compared with the results of the four skeletal samples analysed in this thesis where appropriate.

**Non Nok Tha**

Non Nok Tha is a prehistoric mound in the upper Chi Valley on the Khorat Plateau in the northeast of Thailand (Figure 1.1). The site was excavated over two periods, first in 1966 and later in 1968. The excavation was part of a joint venture between the Thai Fine Arts Department and the University of Hawaii, under the direction of Dr Wilhelm Solheim and Mr Donn Bayard (Douglas 1996). The stratigraphy of the site only reached 1.4 m but it incorporated many burials intercut and overlying one another (Bayard 1984; Higham and Thosarat 1998e). AMS dates indicate that the Bronze Age graves (approximately 136 burials of 180 (Douglas 1996)) from within Non Nok Tha are dated to 1500 - 1000 BC (Higham and Thosarat 1998e) although there are earlier and later burials.

Douglas (1996) presents a comprehensive reworking of the Non Nok Tha skeletal material. The research of Douglas (1996) provides extensive information able to be compared with the skeletal samples presented in this thesis as many of the methods are similar. A total of 180 individuals from both the 1966 and 1968 excavations were included in Douglas" (1996) analysis.

**Noen U-Loke**

Noen U-Loke is a recently excavated Iron Age mound on the Khorat Plateau in northeast Thailand (Figure 1.1). It was excavated in 1997-1998 as part of the 'Origins of Angkor' archaeological project described earlier (Chapter 1). Radiocarbon dates the mortuary phase of the mound from approximately 300 BC to 300 AD (Higham and Thosarat 1998d). A total of 128 individuals were identified from within the mound. Grave goods were abundant with adornments of gold beads, silver rings, bronze bracelets and rings, as well as glass beads (Higham and Thosarat 1998d). Iron tools, such as sickles and knives, were also present and may have been smelted locally. Mortuary ritual included, in some cases, burying the dead in beds of rice and although this had a detrimental effect on the bone, it signifies the abundance of rice at this site (Tayles *et al.* 1998b).

I was involved in the collection of information concerning the human remains from the first field season. This was carried out in a similar manner to the Ban Lum Khao burials. The second season of excavation involved the same process, although I could not be present. An analysis of the dentition has been carried out by Nelsen (1999). The analysis of the remainder of the skeletal material is now under investigation by Dr Nancy Tayles. As a comparison with the main skeletal samples presented in this thesis, the prehistoric people of Noen U-Loke provided the next step in time. Appropriate preliminary comparisons will be made to see if trends in health established in earlier times continue through to the late Iron Age.

# 3 Census

## Age at death estimation

## Introduction

> Age structure is an essential component of any analysis, since many pathological conditions are age-cumulative (e.g., degenerative joint disease,...osteoporosis); the longer someone is alive, the greater the exposure to factors leading to these conditions. Therefore, knowledge of age structure contributes to a more comprehensive understanding of health, disease, and growth from birth through adulthood in earlier societies. (Larsen 1987 cited in Larsen 1997: 340)

**Infants**

The development of the dentition is the most useful estimator of age at death as it provides a close approximation to chronological age (Saunders 1992). The development of the dentition begins in utero and calcification and eruption continue throughout childhood (Moorrees *et al.* 1963b; Costa 1986). These processes follow a progressive development and allow an individual from foetal age through to late adolescence to be aged more accurately than adults (Saunders 1992). However, as with the estimation of age of any individual, adult or child, the estimate obtained represents their biological or developmental age, which does not necessarily equate with their chronological age (Moorrees *et al.* 1963a).

Calcification of the deciduous dentition begins early in the foetal period and is useful in identifying infants up until three years of age. Calcification of the permanent teeth begins at birth and continues through to approximately 15 years. However, the third molar, the last to calcify, has a variable completion age (Moorrees *et al.* 1963b; Costa 1986). Teeth calcify from the crown to the root and, in the case of the deciduous dentition, then resorb from the root towards the crown until not enough root is left to hold the tooth in place and it is lost. This allows room for the permanent tooth to erupt into place. Eruption of the deciduous dentition begins at approximately six to nine months with the eruption of the central incisor. The permanent dentition begins erupting at around six years.

Moorrees *et al.* (1963a, 1963b) present charts for the age range of calcification of a sample of deciduous and permanent teeth. Radiographs are usually required to analyse the state of formation or resorption of the teeth. Separate charts are presented for males and females in the Moorrees *et al.* (1963a, 1963b) studies. As it is rarely possible to estimate the sex of the skeletal remains of children with any confidence, they suggest that the average of the male and female values are used when the sex of the individual is unknown (Moorrees *et al.* 1963a).

Charts documenting the sequence of eruption are also available. Ubelaker (1989) provides a summation of information based on American Indians and other 'non-white' populations regarding both the calcification and eruption of the permanent dentition. The information concerning the deciduous dentition is based on United States white populations from the Moorrees *et al.* (1963a, 1963b) studies mentioned above. The Ubelaker (1989) chart is in a useful form and all teeth can be considered in establishing an age range for each individual. It combines male and female data, which is more suitable for skeletal studies given the difficulty in estimating the sex of children. However, this does introduce a wider age range for each step. The eruption sequence represents the eruption through the gum (Ubelaker 1989) which is not actually able to be determined in skeletal material. Depending on the definition of alveolar emergence (through the bone) compared with gingival emergence (through the gum), the difference can be a matter of months or possibly up to a year (Haavikko 1970 cited in Hillson 1996). This may have a significant effect on age estimates unless an appropriate allowance is made.

Population differences in the sequence and timing of calcification and eruption are apparent. The subjects of the Moorrees *et al.* (1963a) study of permanent teeth were modern white North American children from Boston and Ohio. These two groups showed a difference in the calcification of some permanent teeth by up to six months (Moorrees *et al.* 1963a). What this indicates is that population differences in dental development are apparent and that this study may not be directly applicable to other populations. Therefore, using population-specific charts would provide the most reliable age estimation. However, not all populations have such information available. The limited information available for modern Thai children regarding tooth eruption (Kamalanathan *et al.* 1960b) will be discussed in the methods section.

Dental calcification is more highly correlated with chronological age and less influenced by environmental factors in comparison to dental eruption (Moorrees *et al.* 1963a; Ubelaker 1987). The entire dental developmental process is under much stronger genetic control than long bone growth (Ubelaker 1987; Saunders 1992), which is another variable used in estimating young children discussed below. The degree of preservation of a skeletal collection will determine the methods possible for aging infants. Infant skeletal material, including the dentition, is more frequently lost due to its poorer mineralisation and development compared with older individuals. When infant material is available it is preferable to use the dentition to obtain an age estimate.

Measures of diaphyseal length are commonly used for skeletal aging in conjunction with, or in the absence of, the dentition. There are many standards published for different populations that give age ranges for diaphyseal lengths (Merchant and Ubelaker 1977; Sundick 1978; Hoffman 1979; Angel *et al.* 1986; Ubelaker 1989). It is possible to establish population-specific standards within each skeletal population if the sample size is large enough. To obtain population-specific standards, the diaphyseal length measurements of those individuals aged from dental development are used to estimate the age of those individuals with only diaphyseal length measurements.

### Children/Adolescents

In children older than approximately 12 years, the fusion of epiphyses is the most accurate method for estimating their age at death. There are charts available for both males and females. However, as mentioned earlier, sexing of child skeletal remains is generally not possible. The complete pattern of fusion, as well as the *stage* of fusion for each epiphysis, over all available skeletal material for an individual is used in estimating age at death (McKern 1970; Krogman and Iscan 1986; Ubelaker 1987). The analysis of some epiphyses provide a higher level of accuracy compared with others. They include the proximal and distal humerus, the distal radius, and the femur head (Ubelaker 1987). Therefore these epiphyses are given more emphasis when estimating age.

Different biological populations can differ in their rate of maturity and this should be taken into account if no specific population standards are available for the population in question. These differences can be up to two years (Stewart 1979) and with the differences in male and female maturation rates as well, the best option would be to present the average age together with a range.

Dental development continues until the age of approximately 15 years when the second permanent molar finishes erupting. The variability of eruption age of the third molar precludes it from accurately estimating age, other than providing a general indication that the individual was probably older than 15 years.

### Adults

> Adult ages are much harder to determine because they are calculated using rates of degeneration, which are far more sensitive to individual (and group) differences in activities than biologically programmed patterns of development in children. (Cohen 1989: 110)

A complex, combined method of aging was presented by Acsádi and Nemeskéri (1970) and involved the use of four aging methods with an accuracy of 80-85% overall. Age at death indicators are not all of equal value and Acsádi and Nemeskéri (1970) applied a weighting factor to each method. Lovejoy *et al.* (1985) took this method further and it is now termed a multifactorial approach. Previously, age estimation was carried out on an individual basis rather than at the population level. The use of seriation over the entire population provides a systematic method of aging and enables researchers to establish a summary age based on as many aging methods as possible and taking into account their reliability. This type of approach is now considered to be the most reliable as it has been tested and found to be valid by later studies (for example Bedford *et al.* 1993). The following discussion outlines many of the methods available for estimating adult age at death from skeletal remains. However, the preservation of the skeletal material, which can often be incomplete, will determine which of the methods are possible to estimate adult age at death in any one sample.

Up until the age of 30 years there are epiphyses still fusing. These late fusing epiphyses include the sternal clavicle, iliac crest, and the sacral bodies. These should be given first priority in aging individuals thought to be in their twenties (McKern 1970; Webb and Suchey 1985; Krogman and Iscan 1986).

The fibrocartilaginous joint between the two pubic bones undergoes a pattern of development and degeneration that is age related and this has been used widely in estimating age at death. One of the first studies to present the progression of the morphology changes of the pubic symphysis was Todd (1921). Subsequently, these methods have been modified by other researchers (McKern 1970; Gilbert 1973; Gilbert and McKern 1973; Meindl *et al.* 1985; Katz and Suchey 1989). Pubic metamorphosis begins at around 16 years of age, but in males the changes develop at a faster rate than in females (Gilbert 1973) and standards used reflect this. As this is a sexually dimorphic structure, there are separate casts available for comparison for males and females. McKern's (1970) method only provides estimates for males up until about 36 years, while the female version (Gilbert and McKern 1973) estimates age at death up until 59 years of age.

The pubic symphysis is also affected by pregnancy and can produce changes in females that are not consistent with recorded patterns of age change. This may affect their reliability in estimating age. However, Gilbert and McKern (1973) state this only rarely inhibited aging of the pubis in their analysis. The traumatic changes due to pregnancy did not involve the ventral aspect required to estimate age, although the dorsal aspect was sometimes damaged. A further problem is the rate of survival of a pubic symphysis in skeletal populations. This is estimated to be only around 30% (Waldron 1987 cited in Jackes 1992), therefore its use is limited in many skeletal collections.

Aging adults from dental attrition has been shown to be highly reliable as long as seriation procedures are followed (Nowell 1978; Kieser *et al.* 1983; Lovejoy 1985; Lovejoy *et al.* 1985). The initial method was proposed by Miles (1963) and involved the establishment of a baseline using immature individuals aged from dental development. Their level of molar wear sets the baseline against which adult specimens of the skeletal sample are seriated. However, this method assumes that diet, dental wear rates, and dental pathology within the skeletal sample are uniform over all individuals. This is not always the case, particularly in socially complex societies where personal status can affect diet and, therefore, dental wear and pathology (Nowell 1978; Kieser *et al.* 1983). Despite this, the population specificity of this method is an advantage (Nowell 1978; Lovejoy *et al.* 1985) as it requires no comparison with biologically unrelated populations that can introduce their own biases into the assessment.

As previously emphasised, the use of multifactorial methods are most reliable. However, the best single indicator for estimating age at death in skeletal populations is considered to be the analysis of dental wear (Meindl *et al.* 1983; Lovejoy *et al.* 1985). Rather than seriating dental wear purely by observation it is useful to assign numerical grades to each tooth. There are a number of scoring techniques available (for example Molnar 1971; Scott 1979) and the selection of one depends on the aim of the investigation. The Scott (1979) method involves assigning a grade to each quadrant of each tooth and would perhaps be more useful in the analysis of types of attrition. The Molnar (1971) method is simpler with a chart outlining the progression of wear for each tooth type from grade one through to eight.

Other techniques involving the assessment of the level of degeneration of the skeleton can also be used. Bone mass is lost with age in both compact and cancellous bone. Within the cancellous bone trabecular involution progresses with age and is commonly assessed in radiographs of the proximal femur, proximal humerus, or the clavicle (Jowsey 1960; Hall 1961; Singh *et al.* 1970; Walker and Lovejoy 1985; Macchiarelli and Bondilioli 1994). Walker and Lovejoy (1985) present eight descriptive phases for the proximal femur, humerus, clavicle, and calcaneus for which radiographs can be seriated against. Physical activity, nutrition, and health can also have an affect on the amount and appearance of the trabeculae and this may affect results (Walker and Lovejoy 1985; Macchiarelli and Bondilioli 1994). Sex differences are also apparent with an acceleration of bone loss in postmenopausal women (Macchiarelli and Bondilioli 1994). As prehistoric women are assumed not to have lived this long, the effect of menopause on the trabeculae may not have been a significant problem for them.

The costochondral junction is another region which undergoes age progressive changes. These are identifiable at the sternal (medial) end of the ribs (Kerley 1970; Krogman and Iscan 1986; Iscan and Loth 1989). The main disadvantage with this method is that the descriptions presented in Iscan and Loth (1989) and their previous papers, are restricted to the fourth rib. This can be a difficult bone to identify in excavated human skeletons as emphasised in a test of Iscan's method by Russell *et al.* (1993).

The assessment of the degree of cranial suture closure is generally now regarded as a last resort criterion in estimating age at death given the inconsistencies of its progression with age (Meindl and Lovejoy 1985). However, it may be useful in a multifactorial approach if necessary. There is some debate about which surface to use, whether ectocranial (Meindl and Lovejoy 1985) or endocranial (Krogman and Iscan 1986). Whichever surface is chosen a distinction must be made between the data collected from the two surfaces (Acsádi and Nemeskéri 1970).

Degenerative changes of other joints, including osteoarthritis and vertebral spondylosis, can serve as a further indicator of age (Ubelaker 1989). However, they can provide only a very general estimate, as their aetiology is not always associated with age. For example, trauma can induce joint disease. The use of joint disease in age estimation can also limit the use of the joint disease data in further analysis. For example, a comparison of the prevalence of osteoarthritis between young and old adults to see if there is an association with age or not, cannot be carried out if it is used as an age criterion.

# Methods

All four samples, Khok Phanom Di, Nong Nor, Ban Lum Khao, and Ban Na Di, were aged using the same approach where possible. However, preservation differences among the samples limited the choice of method in some cases. As the population of Khok Phanom Di had been aged in a previous analysis by Tayles (1992), it was this approach that provided the basis of the methods used for the other samples. The data from Tayles (1992) was used once it was established that there was no significant interobserver error between Tayles and myself (Appendix A). The descriptions of the estimation of age at death in the subadults and adults from Khok Phanom Di are from Tayles (1992). The age at death estimation methods and results for the Nong Nor people have also previously been reported by myself and colleagues (Tayles *et al.* 1998a). The methods are first described in general, outlining the approach considered for all four samples. The actual methods used for each sample are then further explained.

### Subadults

For each skeletal sample the following were recorded for subadults where possible:

1. Dental calcification stages (Symington and Rankin 1908; Moorrees *et al.* 1963a, 1963b; van der Linden and Duterloo 1976; Ubelaker 1989);
2. Dental eruption stages (Kamalanathan *et al.* 1960b; Ubelaker 1989);
3. Diaphyseal length, population specific or others (Johnston 1962; Merchant and Ubelaker 1977; Sundick 1978);
4. Epiphyseal fusion (McKern 1970; Webb and Suchey 1985; Krogman and Iscan 1986);
4. Skeletal maturity and size - comparative aging of incomplete bones with those already aged.

With respect to dental eruption, the chart from Ubelaker (1989) is used in this study to estimate the age at death. However, some additions were made for the age at death estimation. As stated, dental development is population-specific and information from modern Thai children from a study by Kamalanathan *et al.* (1960b) has been added to the Ubelaker chart to obtain the most accurate results possible. A number of Thai teeth showed later eruption times compared with those presented in the Ubelaker chart.

**Table 3.1:** Comparison of dental eruption age ranges for the permanent teeth that showed differences between the multiple populations presented in Ubelaker (1989) and modern Thai in Kamalanathan *et al.* (1960b)

| Tooth | Jaw | Ubelaker (1989) Age range | Kamalanathan (1960b) Combined* | Male | Female |
|---|---|---|---|---|---|
| First molar | Mandible and Maxilla | 6y +/- 24 months | 7.0 y | 7.0 y | 7.0 y |
| Central incisor | Maxillary | 7y +/- 24 months | ~8 y | 8.1 y | 7.8 y |
| Lateral incisor | Mandibular | 7y +/- 24 months | ~8 y | 8.2 y | 7.6 y |
| Lateral incisor | Maxillary | 8y +/- 24 months | ~ 9 y | 9.1 y | 8.8 y |
| First premolar | Mandibular | 10y +/- 30 months | ~ 10-11 y | 11.1 y | 10.4 y |
| Second premolar | Mandible and Maxilla | 11y +/- 30 months | ~11-12 y | 11.8-11.9 y | 11.5-11.6 y |
| Second molar | Mandibular | 11y +/- 30 months | ~ 11.5 y | 11.7 y | 11.6 y |

*This data was not provided in the Kamalanathan *et al.* (1960b) study but is an estimate of the combined male and female age.

Table 3.1 indicates the teeth that were different in eruption age between the two methods and represents the changes incorporated into the chart for this study. Both studies define eruption as emergence through the gum.

The dental development differences from the Thai study were not consistently used in the estimation of age of the Khok Phanom Di sample but they were for the other samples. No differences between the two methods amounted to more than one year and were always within the range presented by Ubelaker (1989). Additionally, all comparisons of subadults among the samples involved using age groups rather than single ages.

The non-dental aging methods frequently required comparisons with those already aged. In some cases there were no population-specific data available. In these cases the Khok Phanom Di sample provided the standards for comparisons to be made against. This sample was chosen as it provided the most reliable data given the large number of subadults and their excellent preservation.

Khok Phanom Di

The age estimates for the Khok Phanom Di people were obtained from a previous study by Tayles (1992). The following is an outline of the methods used in that study. Infants at Khok Phanom Di were primarily aged using dental development, which included the analysis of both calcification, (radiographically determined) and eruption stages. Those without dentition were aged using the diaphyseal length standards based on those with dentition. However, there was no comparable material for some individuals in some size ranges, therefore, standards for diaphyseal lengths of prehistoric American Indians had to be used (Johnston 1962; Merchant and Ubelaker 1977; Sundick 1978). Some infants were aged by only a general indication of skeletal size, by comparing a fragment of a bone with other aged infants. Children aged between approximately five and twelve years were aged using dental calcification based on Moorrees *et al.* (1963a) and van der Linden and Duterloo (1976). However, one child from this group was aged using comparative diaphyseal lengths.

Nong Nor

The details of the aging of the Nong Nor subadults have been published as a report by a small research group, including myself (Tayles *et al.* 1998a). Nong Nor subadults were primarily aged using tooth eruption (Kamalanathan *et al.* 1960b; Ubelaker 1978), the development of epiphyses (McKern 1970), and diaphyseal lengths. Not all factors were able to be considered in all individuals. Subadults with diaphyseal lengths were aged by comparison of long bone lengths with those subadults aged from more reliable indicators or from the long bone lengths from children and infants of the Khok Phanom Di sample. Those individuals with none of these factors were aged comparatively with subadults from Khok Phanom Di using the following method. The most complete long bone, preferably including a metaphysis, from each individual unable to be aged from Nong Nor was compared, visually, in general size, to a range of subadults from Khok Phanom Di.

Ban Lum Khao and Ban Na Di

Ban Lum Khao and Ban Na Di infants and children were aged in a similar manner to those at Khok Phanom Di. However, no radiographs were taken of children's mandibles, therefore estimates of dental calcification stage were only able to be obtained from a comparison to Ubelaker's chart (1989) when roots were visible (loose teeth). Dental eruption was used more frequently to estimate age. Diaphyseal length standards, developed from those individuals aged from their dentition, were used for those without dentition. However, as no population specific standards were available for those infants less than one year of age, diaphyseal length standards established from the Khok Phanom Di subadult were used.

**Adults**

The following information was collected from all adult individuals where possible:

1. Epiphyseal union (McKern 1970; Webb and Suchey 1985; Krogman and Iscan 1986);
2. Pubic symphysis morphology score (Todd 1921; Brooks 1955; McKern and Stewart 1957; Gilbert and McKern 1973);

**Table 3.2:** Average wear grade ranges used for age at death determination in the Nong Nor sample. (Grading scheme following Molnar (1971))

| Age Category | First molars | | Second molars | | Third molars | |
|---|---|---|---|---|---|---|
| | Grades | N | Grades | N | Grades | N |
| Young adult | 2 - 4.5 | 25 | 1 - 5 | 24 | 1 - 3 | 18 |
| Middle aged adult | 5 - 5.75 | 23 | 2.5 - 5.5 | 24 | 1 - 4 | 23 |
| Old adult | 6 - 8 | 12 | 2 - 8 | 6 | 1.5 - 7 | 8 |

3. Dental wear (Molnar 1971);
4. Dental pathology;
5. Joint degeneration;
6. Comparative aging involving the comparison of the development of muscle attachments with those already aged (only in cases were no other method was possible).

This is the order of priority used, with epiphyseal union and pubic symphysis scores taking first precedent in the aging of the younger adults. However, dental wear was taken as the first priority, followed by pubic symphysis score, in estimating the age at death of older adults.

The procedure for using dental wear as an age estimator is based on an adaptation of the Miles (1963) method. It involves seriating individuals according to their degree of molar wear (assumes a constant rate of wear within each sample). Each molar in all individuals possible was assessed for its level of wear using the scale presented by Molnar (1971). Within each sample, each individual was seriated according to their average grade of molar wear. For each individual the average grade of wear for the first molars was calculated using all the available scores for first molars from that individual. This is the first level of seriation. The next level of seriation uses the average wear of the second molars, then the average wear of the third molars. Those young adults aged from other factors, such as epiphyseal union and pubic symphyses, provide a baseline for delimiting the range of molar wear for the young adults. This provides an indication of the rate of wear and older adults can be subdivided into categories of five or ten year ranges. When anomalies arose within individuals, dental pathology, such as antemortem tooth loss, was also taken into consideration. The wear of anterior teeth was also used in some cases to confirm an age estimate.

Khok Phanom Di

The age at death estimates for the people of Khok Phanom Di were obtained from Tayles (1992). The following summarises the methodology used in that study. As indicated above, epiphyseal union was given first priority for aging those up until the late twenties. Pubic symphysis scores and dental wear were used as second and third priority, respectively. The calcification of the third molar was also used following the stages from Moorrees *et al.* (1963a). After the age of about 28 years, the order of priority changed with dental wear being first, then the pubic symphysis score and, finally, joint degeneration and dental pathology. The multifactorial seriation approach was used and individuals were assigned an age in single years. However, this does not indicate that accuracy was this high, only infants can be aged to within one year with any confidence. The assignment of age to a single year, being based on multifactorial seriation, is a reflection of each individual's position in the series, relative to others in the group (Tayles 1999).

Nong Nor

Because of the poor standard of preservation of the Nong Nor material, adults were classified in relative age groups of young, middle, and old age ranges, rather than any more precise chronological estimates being attempted. They are assumed to correlate with the age ranges of 15-29 years, 30-39 years, and 40+ years, respectively, in order for comparisons to be made with the other samples. The same order of priority of aging methods used for the Khok Phanom Di sample were applied to this sample. However, due to preservation problems, the primary criterion used for age estimation was molar wear. Seriation was used in the manner described above. Table 3.2 provides the range of average molar wear grades for each age group. The range of average molar wear for each age range are obtained by considering individuals aged from other factors, such as epiphyseal union, as providing an indication of the rate of wear in this sample. Using this information an estimate of the range can be made specific for this sample.

Once all age allocations were made, each individual was considered separately to determine if all other teeth fitted into the general pattern and that other, more subjective age indicators, such as antemortem tooth loss, joint degeneration and muscle markings, also correlated with the assigned age.

Those adults without dentition were aged comparatively with those that had already been aged using dental wear on the basis of the development of muscle and ligament attachment sites (entheses) on the long bones. Six burials (two young, two middle aged and two older adults) were used as standards for the basis of comparison. Other criteria for aging included joint condition, i.e. level of joint degeneration (although little evidence was available) and bone size as an indicator of skeletal maturity. Unfortunately, many of the adults could not be aged at all, and are simply assigned as adults.

Ban Lum Khao

Epiphyseal union was given first priority for aging those individuals up until the late twenties, as in the Khok Phanom Di sample. Those who could be aged primarily by epiphyseal union were used as a baseline for seriation of the average molar wear for the first, second, and third molars as described above for the Nong Nor sample. The limits for

**Table 3.3:** Average wear grade ranges used for age at death determination in the Ban Lum Khao sample. (Grading scheme according to Molnar (1971))

| Age Category | First molars | | Second molars | | Third molars | |
|---|---|---|---|---|---|---|
| | Grades | N | Grades | N | Grades | N |
| 17 - 29 years | 2.0-4.5 | 16 | 2.0-3.75 | 16 | 1.0-3.5 | 13 |
| 30-39 years | 3.75-5.5 | 10 | 2.5-5.0 | 12 | 1.0-4.0 | 5 |
| 40 + years | 5.75-7.50 | 3 | 4.0-6.0 | 3 | 2.0-6.0 | 3 |

**Table 3.4:** Average wear grade ranges used for age at death determination in the Ban Na Di sample. (Grading scheme according to Molnar (1971))

| Age Category | First molars | | Second molars | | Third molars | |
|---|---|---|---|---|---|---|
| | Grades | N | Grades | N | Grades | N |
| 17 - 29 years | 2.0-4.0 | 13 | 2.0-3.0 | 12 | 1.0-2.5 | 9 |
| 30-39 years | 4.0-4.5 | 11 | 3.0-4.0 | 10 | 2.0-3.3 | 9 |
| 40 + years | 5.0-7.0* | 6 | 4.5-6.7 | 8 | 2.0-6.3 | 5 |

*one individual had an average of 3.0.

each age group are provided in Table 3.3. A few pubic symphyses were also available for estimating age at death in the younger adults and these individuals also provided an indication of the rate of wear in this sample and where the 'cut-off' points should be for the molar wear averages. The ranges presented in Table 3.3 are different from those for Nong Nor and Ban Na Di as this method is population specific.

When anomalies arose adult aging also incorporated the wear of the other teeth and the presence and degree of dental pathologies, such as antemortem tooth loss. The condition of the joints, such as the degree of osteoarthritis, and the degree of marking of muscle attachment sites, were also taken in to consideration in a few cases where little other information was available.

Ban Na Di

Ban Na Di adults were aged using the seriation of molar wear in a similar manner to that described above, again taking into account the presence of dental pathology when anomalies arose. Pubic symphysis scores were taken into account when present and they, along with the late fusing epiphyses, helped in delimiting the age groups assigned to the ranges of molar wear, indicated in Table 3.4.

# Sex estimation

# Introduction

Most parameters used to estimate the sex of adult skeletal remains are based on either morphological or metric analyses. The morphology of the pelvis can provide a high level of accuracy in determining sex (Krogman and Iscan 1986; Saul and Saul 1989). The sexually dimorphic features observed in the pelvis are based on the differences in reproductive function of males and females. The pelvic cavity (birth passage) is the major area from which bony differences stem between the sexes. The wider and shorter pelvic cavity necessary for females to give birth is reflected in a number of different areas of the pelvis. The sacrum is wider and shorter in females compared with males, the angle of the greater sciatic notch of the pelvic bone is more open, and the pubic arch angle is wider anteriorly in females than in males (Krogman and Iscan 1986; Saul and Saul 1989) .

The pubic bone has a number of sexually dimorphic features, as detailed by Phenice (1969), that relate to the attachment of the genitalia. The presence of a ventral arc on the anterior surface of the pubic symphysis is a feminine characteristic, while the absence of this arc is seen more frequently in males. The second feature is the subpubic concavity, which is present in females more often than in males. Lastly, the observation of a ridge on the medial aspect of the ischio-pubic ramus is more likely to be found in a female (Phenice 1969) . There is some overlap between males and females, particularly in the latter feature, but used together they can be very reliable. Phenice (1969) obtained a 96% accuracy level on a skeletal sample of known sex.

A multifactorial approach in the assessment of the pelvis is most reliable as traits presenting intermediary results can be weighted against more definitive results. However, a test of accuracy and reliability of the many traits of the pelvis used in sex estimation by Rogers and Saunders (1994) , found that a combination of all the morphological sex traits of the pelvis did not provide the highest accuracy. There are some features that should not be included. They recommended that the least reliance should be placed on features such as the ridge on the ischio-pubic ramus, and features of the acetabulum and auricular surface which are not discussed here.

While priority is given to features of the pelvis, other skeletal characteristics may assist in identifying sex when necessary. These following traits are based on the universal size differences between males and females. Males, on average, are larger and more robust than females. Observations of certain features of the cranium and mandible provide the highest level of accuracy in sex estimation after the pelvis.

Many of the sexually dimorphic features observable in the skull stem from the more robust musculature of males. The supraorbital ridge, temporal lines, nuchal lines, external occipital protuberance, and the mastoid of the cranium, and the gonial angle of the mandible are more robust and prominent features in males. The shape of the orbit and the form of the orbital rim are also used, with males characteristically having squarer orbits and rounder orbital rims than females (Buikstra and Mielke 1985; Krogman and Iscan 1986) .

Human adult male and female skeletons also differ in size as well as shape, and some of these features can be metrically assessed. Sexual dimorphism in size arises from differences in growth during childhood. Females begin and cease the adolescent growth spurt earlier than males, and it occurs at a slower rate. Males have a longer prepubertal growth period as well as a faster rate of growth during their later growth spurt (Preece 1998) . These factors contribute to the sexual dimorphism seen in adults. In particular, long bone lengths and joint widths, for example, the diameter of the femur head, have been used to estimate sex using discriminant function analysis (Giles 1970; Houghton and de Souza 1975; DiBennardo and Taylor 1979; Kieser *et al.* 1992) . Dental metrics may also be used to estimate sex (Rosing 1983) .

Discriminant function analysis involves taking a number of different measurements, for example the femur head diameter and mid shaft circumferences, from individuals whose sex has been estimated from reliable features, preferably the pelvis. These measurements are used to obtain a discriminant function and a section point. The latter is the average of the male and female mean scores obtained from the discriminant function. The discriminant function equation is then applied to each individual of unknown sex that have had the appropriate measurements taken. If the score obtained from the equation is less than the section point then the individual is estimated to be female; if the score is greater then the individual is more likely to be male (Giles 1970; van Vark and Schaafsma 1992) .

There are a number of requirements of the data before this calculation can be carried out. Firstly, the difference between the metric attributes of the male and female samples must be statistically significant. This is not always the case in every sample. However, a simple student t-test comparing the means can establish this. Secondly, there should be no significant difference in the means of the metric attributes between the sample of individuals of known sex and those individuals of unknow sex. This is especially important if the known sex sample is from a different population. When only one variable is being measured a section point is obtained simply as the average from the male and female means of a variable from a related population or subsample from within the sample studied. In a test of this method against using pelvic and cranial criteria, consistency was estimated at 90.6% (Black 1978; DiBennardo and Taylor 1979) .

One of the most definitive methods of determining sex from skeletal remains is through DNA analysis. This technique eliminates the reliance on the plastic skeletal phenotype and places it with the genotype through analysis of the Y chromosomal sequences (Gotherstrom *et al.* 1997) . The extraction of ancient NA though, is not straightforward (Stone *et al.* 1996) . Contamination is always a possibility (Hummel and Herrmann 1991; Gotherstrom *et al.* 1997) , and the result may in fact represent the sex of the excavator or researcher rather than the skeleton in question. These errors can be eliminated through careful handling of bone and use of bone powder from an inner core of bone (Gotherstrom *et al.* 1997) . However, ancient bone cannot always provide the amount of bone required for DNA amplification (Stone *et al.* 1996) .

When skeletal material is in good condition, morphological observations can be easily (and cheaply) carried out, and with a high level of reliability (up to about 95%). The estimation of sex using mitochondrial DNA is relatively expensive. This method would be useful in estimating sex in fragmentary and/or juvenile remains, and in those individuals that fall into the overlapping range of male and female morphological variation (Stone *et al.* 1996) .

Morphognostic sex differences in the ilium and mandible have been proposed as a method for estimating the sex of those aged from birth to five years of age (Schutkowski 1993). Deciduous tooth measurements have also been postulated as sexually dimorphic (Devito and Saunders 1990 cited in Douglas 1996). However, these have not been widely tested. No attempt either using DNA analysis, or other analyses, has been made to estimate the sex of infants and children in these Thai skeletal series, given the expense of the former and the problems associated with both methods.

Two further areas of the pelvis may be useful in the estimation of sex of adults and also in establishing the level of fertility in a skeletal sample. Putschar (1931 cited in Stewart 1970) first identified pits adjacent to the joints of the pelvis of those women who had given birth. The areas involved are the preauricular groove on the ilium adjacent to the sacro-iliac joints, and the dorsal pubis adjacent to the pubic symphysis. These areas are both sites of ligament attachment that hold the respective joints in place. During pregnancy, the ligaments of the pelvis start to relax in order to allow an expansion of the birth canal and an active resorption of bone occurs at their sites of attachment leaving pits and depressions in the juxta-articular regions (Houghton 1975) . The presence of these pits will, therefore, generally be indicative of a female pelvis. However, there have been some inconsistencies associated with these features. Stewart (1970: 133) found that "...some women can bear children with a minimum of scarring, or even without any scarring". On the other hand pelvic changes may occasionally be seen in allegedly nulliparous females (Kelley 1979; Suchey *et al.* 1979; Cox and Scott 1992) . Pregnancy may not be the only cause of these changes. Trauma and pubic osteitis are examples of other factors, although relatively rare, that can cause these pelvic changes (Kelley 1979).

The accuracy of results will also depend on how much bony remodelling has occurred in this area. Remodelling occurs throughout life and the remodelling of the pelvic areas resorbed during pregnancy are no exception. The level of remodelling may, in part, be determined by the length of time evolved since the woman gave birth before dying. The older the woman at death, the more likely the pregnancy scars have been obliterated (Kelley 1979) .

**Table 3.5:** Pelvic features used to estimate sex

| Feature | Female appearance | Male appearance |
|---|---|---|
| 1. Pubic arch angle | wide (open) | narrow (steep) |
| 2. Ventral arc | present | absent |
| 3. Subpubic concavity | present | absent |
| 4. Sciatic notch angle | wide | narrow |
| 5. Ischio-pubic ramus | ridge | broad |
| 6. Sacrum morphology | wider, shorter | narrow, longer |
| 7. Pregnancy pitting - preauricular groove and dorsal pubis | often present | never present |

1, 4, 6. Buikstra and Mielke 1985; 2, 3, 5. Phenice 1969; 7. Houghton 1974.

**Table 3.6:** Cranial features used to estimate sex

| Feature | Female appearance | Male appearance |
|---|---|---|
| 1. Supraorbital ridge and glabella region | not prominent | prominent |
| 2. Orbital rim | sharp | blunt or rounded |
| 3. Occipital markings - nuchal lines and external occipital protuberance | smaller | larger and more marked |
| 4. Mastoid size | smaller | larger |
| 5. Anterior mandibular shape (chin) | rounded | square |
| 6. Mandibular gonial angle | little or no eversion | more everted |
| 7. Orbital shape | round | square |
| 8. Parietal bossing | present | absent |

1, 3, 4, 5, 7. Acsádi and Nemeskéri 1970; 2, 8. Bass 1987; 6. Buikstra and Mielke 1985.

A number of studies have discovered that the development of a preauricular groove is the most sensitive indicator of parity (Houghton 1975; Kelley 1979), more so than the dorsal pubis. The sacro-iliac joint is more in the line of weight transfer from the spine to the lower limb, compared with the pubic symphysis, and thus would be particularly under stress with the weight gain during pregnancy (Houghton 1975).

The overall conflicting results from studies of female pelves of known parity status may be the effect of interpopulation variability or denial and concealment of having given birth (Cox and Scott 1992). However, other factors also differ. The level of obstetrical care, the diameter of the maternal birth canal and the circumference of the foetal head, as well as the age at time of death, may all influence the degree of pelvic changes (Kelley 1979). Overall the best results may be obtained by using a combination of pelvic features, placing particular weight on pitting in the preauricular groove and medium to large dorsal pubic pits in the estimation of parity status and, therefore, sex.

In terms of estimating a general level of fertility for a sample, these observations may also be useful taking into consideration the problems outlined above. A number of studies have attempted to take these observations further in order to estimate the actual number of full term pregnancies experienced by each woman (Kelley 1979; Suchey *et al.* 1979). Suchey *et al.* (1979) did not find a strong correlation between the number of full term pregnancies and the degree of dorsal pubic pitting. Due to the extreme variability in this region they suggest predictions of parity status cannot be made from observations of the dorsal pubis. Kelley's (1979: 545) study concluded that "...it is doubtful that more precise statements concerning parity status (i.e., number of pregnancies and children born) can be made" other than the labelling of no parturition or parturition. The no parturition category may, however, include not only those women who have not given birth but also those who have given birth without scarring. Therefore, this may overestimate the number of women who had not given birth.

# Methods

The estimation of sex of the adults from Khok Phanom Di was carried out by Dr N Tayles and the results were obtained from Tayles (1992) and the associated database. There was no interobserver error in the estimation of sex between Tayles and myself in a subsample tested (Appendix A). The analysis of the skeletal sample of Nong Nor was begun by Dr Tayles and then continued by myself. A report on the people of Nong Nor has been published (Tayles *et al.* 1998a).

Independent assessment of the pelvis and the cranium were used as the major determinants in the estimation of sex of those aged 15 years and over in all samples. Sex estimates were not attempted for children (<15 years). Priority was given to the pelvic features outlined in Table 3.5. These features are presented in order of priority, but emphasis was placed on the multiple assessment of these features before making an estimate. Table 3.6 provides a list of the cranial and mandibular features used to estimate sex. These were relied on in the absence of the pelvis or to assist in the estimation when an equivocal pelvis was present. A subjective indication of robusticity was also determined.

In the absence of conclusive pelvic and/or cranial features, section point analyses of some metric variables were carried out. This, and the exact procedure for estimating sex in each sample, is explained below.

### Khok Phanom Di

The following describes the method followed from the previous study on the Khok Phanom Di skeletal material (Tayles 1992). The pelves from the Khok Phanom Di adults were observed for all of the features listed in Table 3.5, except that of the sacrum. The entire sample was well preserved and consequently all adults had pelves, most with little damage. With the exception of two adults, all other adults had crania present including mandibles. The features listed in Table 3.6 were analysed, excluding features 2, 6, and 8. In those cases where there was some discrepancy of sex, either within the pelvis or cranium, or between the pelvis and cranium, measurements of the diameter of the femoral head were analysed. The femoral head measurement from individuals in which sex was unequivocal, was compared with this measurement in individuals of known sex. Two individuals were only provisionally estimated to be males as data available for them was not entirely conclusive.

### Nong Nor

The following is derived from the report on the people of Nong Nor (Tayles *et al.* 1998a). The methods followed in order to obtain sex estimates for the Nong Nor people were hampered by the poor preservation of material. Standard morphological criteria of the pelvis and the cranium were used in the first instance for sex estimation where possible. On the pelvis, all of the features listed in Table 3.5 were included, but not many were actually observable, given the condition of the material. The same applies to the list of cranial and mandibular characteristics in Table 3.6. However, these were used where possible. General gracility and robusticity (subjective assessment of size and development of muscle attachments) were also noted, relative to the rest of the sample.

In many cases, other sexing criteria were also used to support what little morphological assessment could be carried out. The measurements of the anterior-posterior and medio-lateral tibial shaft diameters, taken at the level of the nutrient foramen, were used in a discriminant function analysis based on data from the Khok Phanom Di population. While there is debate about the reliability of this landmark (Steele 1970; Andermann 1976) it was used in the absence of other possibilities. As detailed in the introduction of this chapter, a discriminant function analysis requires a group of individuals of known sex and the discriminant function is calculated from this group (Giles 1970; Rosing 1983; van Vark and Schaafsma 1992), Khok Phanom Di in this case. The discriminant function obtained (Figure 3.1) was applied to the individuals of unknown sex from Nong Nor who could have the appropriate measurements taken. An estimate of sex was thus calculated depending on whether the result was greater or less than the section point determined from the reference population, Khok Phanom Di. Obviously this is not the optimum situation. Results would be more reliable if the discriminant function was derived from a group of known sex individuals from within the Nong Nor population. However, this was not possible. It was decided that the use of the Khok Phanom Di population was justified because the difference between the means of the tibial measurements from the two populations was not statistically significant. An accuracy score of 81% was obtained for this discriminant analysis. The 19% that would be incorrectly estimated could be attributed to the overlap in size of males and females that could be found in any population.

Discriminant Score =
(anterior-posterior diameter x 0.4581572) + (medio-lateral diameter x 0.1629583) - 18.10747

Section Point = 0.10235
Male if > section point
Female if < section point

**Figure 3.1:** Discriminant function for sexing adults from Nong Nor using tibial nutrient foramen measurements. The equation was derived from data (n=58) from the Khok Phanom Di sample

Section point analyses were used on other metric variables. This method is employed for single variable analyses where, as detailed in the introduction, the mid-point between male and female averages for the trait are used as the section point for sex assignment (DiBennardo and Taylor 1979). The section point was calculated from the Khok Phanom Di sample. Before this analysis was carried out, each measurement was statistically analysed to justify the use of the comparison between the two populations and to determine that the measurement was actually sexually dimorphic in the reference population. t-tests were performed on the sample means and those that were significantly different were excluded. Three measurements which were not significantly different between the two samples were used: the diameter of the femur head taken in the anterior-posterior plane, the maximum breadth of the mandibular condyle, and the width of the elbow (i.e. the maximum distance between the lateral and medial epicondyles of the distal humerus).

The section points obtained from the Khok Phanom Di sample are listed in Table 3.7. If the measurement from an individual from the Nong Nor sample was greater than the section point it was deemed male, if less then the individual was estimated to be female.

There were a number of individuals who had very few features from which to estimate sex in the Nong Nor sample. These individuals were assigned '?female' or '?male' to indicate them as 'possible' females or males as opposed to the males and females who were more confidently assigned to each sex. There were also a number of individuals unable to be assigned a sex.

**Table 3.7:** Section points used in the sex estimates of adults from Nong Nor. These are based on data from the Khok Phanom Di sample

| Measurement | Section Point (mm) | N |
|---|---|---|
| Femur Head | 43.40 | 47 |
| Mandibular Condyle | 21.48 | 54 |
| Elbow Left | 59.56 | 43 |
| Right | 59.65 | 43 |

N = number of measurements

**Table 3.8:** Section points used in the sex estimates of adults from Ban Lum Khao

| Measurement | Section Point (mm) | N |
|---|---|---|
| *Joint dimensions* | | |
| Humerus head diameter | 42.80 | 29 |
| Distal humerus width | 59.89 | 35 |
| Femur head diameter | 44.60 | 34 |
| Distal femur width* | 74.83 | 22 |
| *Long Bone Lengths* | | |
| Clavicle | 138.5 | 24 |
| Radius | 237.0 | 28 |
| Ulna | 259.0 | 26 |
| Humerus | 299.5 | 36 |
| Femur | 425.5 | 32 |
| Tibia | 355.5 | 33 |
| Fibula | 351.5 | 19 |

*measurement taken as the maximum distance between the two femoral epicondyles N = number of measurements

**Table 3.9:** Section points used in the sex estimates of adults from Ban Na Di

| Measurement | Section Point (mm) | N |
|---|---|---|
| *Joint dimensions* | | |
| Humerus head diameter | 41.49 | 12 |
| Distal humerus width | 58.47 | 21 |
| Femur head diameter | 43.98 | 14 |
| Distal femur width* | 74.89 | 12 |
| *Long Bone Lengths* | | |
| Clavicle | 132.8 | 9 |
| Radius | 246.9 | 9 |
| Ulna | 272.8 | 7 |
| Humerus | 304.0 | 11 |
| Femur | 448.2 | 11 |
| Tibia | 371.2 | 7 |

*measurement taken as the maximum distance between the two femoral epicondyles

Ban Lum Khao

Pelvic and/or cranial morphology was used in the estimation of the sex of adults at Ban Lum Khao using the lists provided in Tables 3.5 and 3.6. Pelvic morphology was given priority over cranial evidence. In the absence of these features, or when equivocal results arose, section point analysis of various metric variables was used. These are provided in Table 3.8.

The section points used in this case were obtained from within the Ban Lum Khao sample, from those sexed using pelvic and/or cranial morphology. Four joint dimensions as well as all major adult long bone lengths were used.

Ban Na Di

Pelvic and/or cranial morphology, as listed in Tables 3.5 and 3.6, were used, as the first and second priority in the estimation of the sex of adults at Ban Na Di, respectively. In the absence of these observations, or for equivocal results, metric variables were used in section point analysis. As with the Ban Lum Khao sample method, the section points were able to be obtained from the group of 'known' sex individuals from within the Ban Na Di sample. Four joint dimensions and six long bone lengths were able to be included. The fibula was omitted as only one measurement from an individual of known sex was available. The section points are listed in Table 3.9.

# Results

### Age at death and sex estimation

Khok Phanom Di

The skeletal material from the site of Khok Phanom Di was very well preserved with most of the 154 individuals represented by the majority of bones. This enabled a reliable estimate of age and sex to be obtained through an assessment of multiple factors.

The distribution of individuals in each age category is provided in Table 3.10 and Figure 3.2. Over 50% of the sample were subadults, the majority of which (40.9% of the total sample) were aged less than one year old at death. By comparison, relatively few children (5-14 years) were present in the sample. Sixty-eight adults (≥ 15 years) were identified, with the majority aged, at death, between 20-39 years. Relatively few individuals were classified as older adults (≥ 40 years).

The adult age at death distribution showed some differences between males and females (Table 3.11). More males than females died between the ages of 15-19 years, although both sexes showed the majority of adults died between 30-39 years. However, many more females than males survived past young to mid adulthood to die as older adults over 40 years of age.

The sex ratio of females to males was 1.13:1 (females = 52.9%, males = 47.1% of the adult group). This was close to the expected 1:1 ratio and therefore indicates it is reasonable to assume the sample is representative of the living population from which it came.

Nong Nor

The preservation of the material from Nong Nor was very poor and has made the age at death and sex estimation difficult. This was reflected in the large number of adults unable to be aged (23.9%) or sexed (23.7%). As discussed in the methods, the assignment of adult ages to this population was only given in the relative terms of young, middle, and old aged. However, for comparative purposes with the other samples, these three categories were tentatively assigned age ranges. Young is equivalent to 15 - 29 years, middle aged is 30 - 39 years and old aged is 40+ years. The age at death distribution in each of these categories is provided in Table 3.12 and Figure 3.3.

The proportion of infants and children, at 21.3% of the total sample, represented a low level of child mortality for this prehistoric group. Excluding the nine infant burial jars in which no bone was found, 14.8% were less than five years old and 6.5% were between five and 14 years. If the extra burial jars were taken into consideration the level of child mortality was still low at 25.6% (19.5% infants and 6.1% children).

The distribution of the 122 adults from Nong Nor among the age categories indicated the majority were dying between the ages of approximately 20 and 39 years, similar to the Khok Phanom Di sample's pattern. However, the 23.9% of adults unable to be aged placed a caveat on the reliability of these results. The adult age distribution of each sex was reasonably similar (Table 3.13). However, apart from those unaged, there were also 23.7% of unknown sex not contributing to this distribution.

A further problem with the estimated sex distribution was that the data from the 'possible' males and females have been combined with those more confidently sexed. Of the 49 females, only 35% were classed with the same level of confidence as in the other samples, the other 65% were more tentatively assigned to the female sex and classed as 'possible' females. Slightly more were more confidently assigned as males, 45% of the 44 males; 55% were classed as only 'possible' males. With this caveat acknowledged, the two categories within each sex were combined for subsequent analyses.

There were a similar number of females and males, a ratio of 1.1:1 (females = 52.7%, males = 47.3% of the adult group of known sex, n=93). This near-equal ratio was expected of a cemetery sample and, despite the unsatisfactory nature of the basis of sex estimation in many cases, there was no systematic bias evident. This sample, therefore, may be representative of the entire cemetery. The near one quarter of adults unable to be sexed may have altered the ratio if their sex had been able to be determined.

**Table 3.10:** Khok Phanom Di age at death distribution (Tayles 1992)

| Age | Number | % |
|---|---|---|
| 0 - 0.9 years | 63 | 40.9 |
| 1 - 4 years | 11 | 7.1 |
| 5 - 9 years | 7 | 4.5 |
| 10 - 14 years | 5 | 3.3 |
| Subtotal (children) | 86 | 55.8 |
| 15 - 19 years | 8 | 5.2 |
| 20 - 29 years | 23 | 14.9 |
| 30 - 39 years | 27 | 17.5 |
| 40 - 49+ years | 10 | 6.5 |
| Subtotal (adults) | 68 | 44.2 |
| Total | 154 | 100 |

**Table 3.11:** Khok Phanom Di adult age and sex distribution (Tayles 1992)

| Age Range | Females | | Males | | Total | |
|---|---|---|---|---|---|---|
| | N | % | N | % | N | % |
| 15-19 years | 3 | 8.3 | 5 | 15.6 | 8 | 11.8 |
| 20-29 years | 12 | 33.3 | 11 | 34.4 | 23 | 33.8 |
| 30-39 years | 13 | 36.1 | 14 | 43.8 | 27 | 39.7 |
| 40-49+ years | 8 | 22.2 | 2 | 6.3 | 10 | 14.7 |
| Total | 36 | ~100.0 | 32 | ~100.0 | 68 | 100.0 |

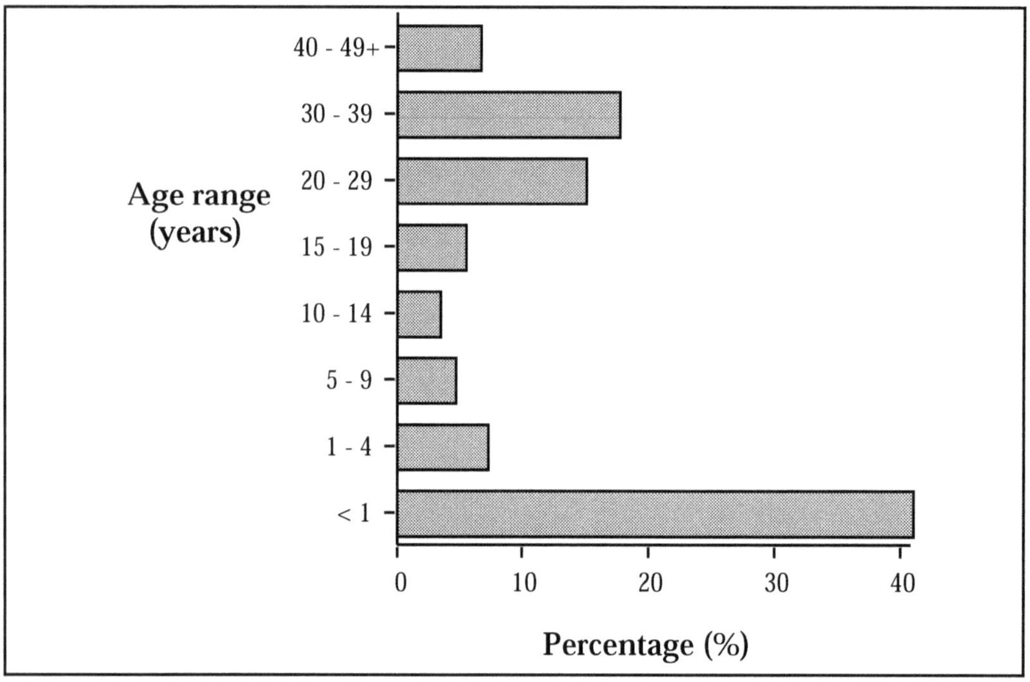


**Figure 3.2:** Khok Phanom Di age at death distribution

**Table 3.12:** Nong Nor age at death distribution

| Age | Number | % |
|---|---|---|
| 0 - 0.9 years | 17 | 10.9 |
| 1 - 4 years | 6 | 3.9 |
| 5 - 9 years | 4 | 2.6 |
| 10 - 14 years | 6 | 3.9 |
| Subtotal (children) | 33 | 21.3 |
| 15 - 19 years | 2 | 1.3 |
| 20 - 29 years | 26 | 16.8 |
| 30 - 39 years | 43 | 27.7 |
| 40 - 49+ years | 14 | 9.0 |
| unknown | 37 | 23.9 |
| Subtotal (adults) | 122 | 78.7 |
| Total | 155 | 100 |

**Table 3.13:** Nong Nor adult age and sex distribution

| Age Range | Females | | Males | | Unknown | Total | |
|---|---|---|---|---|---|---|---|
| | N | % | N | % | N | N | % |
| 15-19 years | 1 | 2.6 | 0 | 0 | 1 | 2 | 1.6 |
| 20-29 years | 12 | 31.6 | 12 | 31.6 | 2 | 26 | 21.3 |
| 30-39 years | 20 | 52.6 | 18 | 47.4 | 5 | 43 | 35.2 |
| 40-49+ years | 5 | 13.2 | 8 | 21.0 | 1 | 14 | 11.5 |
| Subtotal | 38 | 100.0 | 38 | 100.0 | 9 | 85 | 69.7 |
| Unknown | 11 | | 6 | | 20 | 37 | 30.3 |
| Total | 49 | | 44 | | 29 | 122 | 100 |

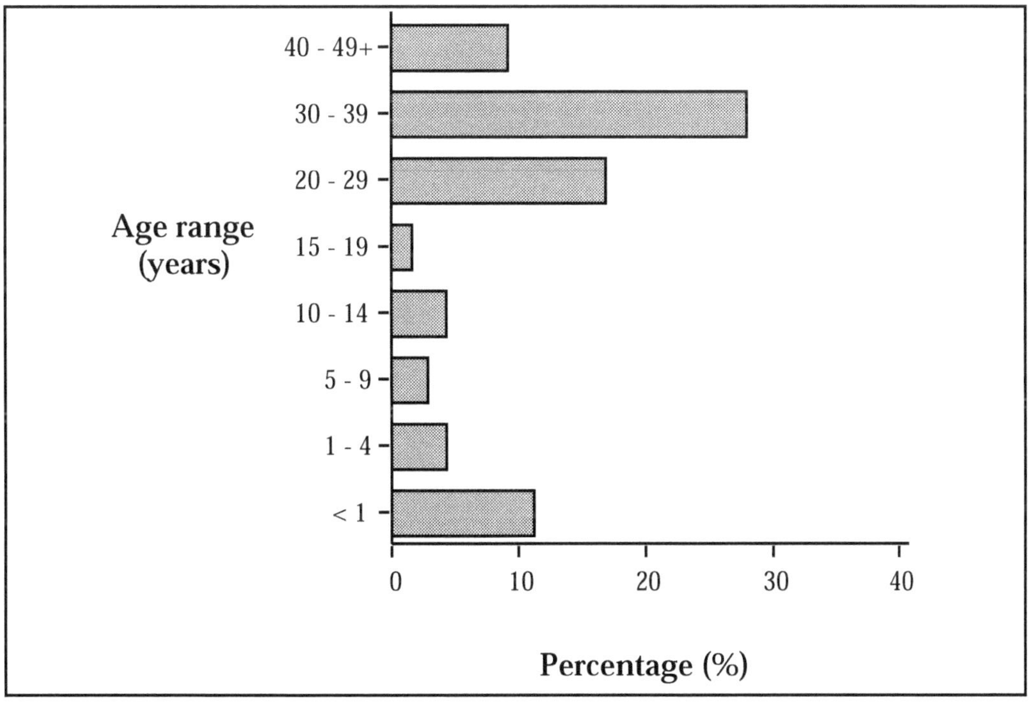


**Figure 3.3:** Nong Nor age at death distribution

Ban Lum Khao

The skeletal material from the cemetery of Ban Lum Khao was relatively well preserved. This enabled relatively reliable estimates of age at death and sex to be obtained. All 110 individuals were able to have their age at death estimated. The results of this, as divided by age groups, are provided in Table 3.14 and Figure 3.4.

Nearly 50% of the sample are subadults, with just under half of these (21/51) having died at less than one year of age (19.1% of the total sample). A further 12.8% of the total sample died between the ages of 1-4 years, totalling the infant mortality level to 31.9%. This level of mortality continued to be quite high up until nine years of age. Individuals between the ages of 10 and 19 years were not well represented reflecting a low mortality for this age range.

Fifty nine adults (≥ 15 years) were identified and just over 40% of these were aged between 20-29 years. The majority died between this age and into the next decade, 30-39 years. The number of adults surviving past this age and dying as older adults (40+ years) was 10.9% of the total sample.

Analysis of the age at death distribution by sex indicated some striking differences between males and females (Table 3.15). Over twice as many females died between 20-29 years (56.3% of adult females) than did males (23.1% of adult males). In comparison the 40-49+ year category contained over twice as many males than females, which indicated males had a greater chance than females of surviving this long.

Because these differences were somewhat striking, further analysis was carried out. There was the possibility that the sex estimation methods were too often based purely on the size differences between males and females and not the more reliable indicator, pelvic morphology. However, of the 20 young females (15-29 years), only three were sexed using solely cranial and metric variables. Three of the nine males were also sexed in this way. It would appear, therefore, that given the small proportion aged in this manner, these results are sound.

The sex ratio of the Ban Lum Khao adults was 1.23:1 females to males (females = 55.2%, males = 44.8% of the adult group of known sex, n=58). Only one individual (Burial 22) could not be given an estimate of sex as it was represented by a pair of feet only.

Ban Na Di

The preservation of the skeletal material from Ban Na Di was quite varied in both bone quality and completeness of individuals. The latter was a particular problem for obtaining multifactorial estimates of age and sex. Of the total sample, 17.9% were unable to be aged and 10.0% of the adults were unable to be estimated as male or female.

The results of the age at death estimates are presented in Table 3.16 and Figure 3.5. The previous analysis of the people of Ban Na Di is also provided in Table 3.16 for comparison and will be discussed later. Of the 78 individuals identified, 28 (35.9%) were classed as subadults (≤ 15 years old). As with the other samples, the infants, those aged between 0 and 4 years, contributed the most to the subadult sample, with 29.5% dying at this young age. Relatively lower levels of mortality were evident in childhood (5-14 years of age).

There was a lack of 15-19 year old individuals in the Ban Na Di sample. While this age group does have a low mortality level in the other samples, the complete lack of representation was unexpected. It may indicate that this age was not associated with a high risk of death in this sample, but it is also possible there was a systematic sampling error.

The remainder of the adult age categories were well represented with little difference between them. Slightly more individuals died at 20-29 years and 30-39 years compared with the older age range, 40-49+ years. The presence of nearly 20% unaged adults limited the reliability and interpretation of these results.

The results of this study and that of Houghton and Wiriyaromp (1984) were generally quite comparable. However, Burial T.22, a seven year old child, could not be assessed as it remains in a museum in Thailand. The present study found a number of extra individuals in with the remains of single individuals. However, differences are particularly obvious in the identification of five foetal aged skeletons by Houghton and Wiriyaromp (1984) while this study could not do so with confidence and aged them as newborns. Houghton and Wiriyaromp (1984) also identified three 15-19 year olds (aged of 17, 18 and 19 years). The 17 year old was aged much younger in this present study due to many unfused epiphyses, there was also a piece of adult pelvis that clearly did not belong with this 14 year old. The 18 year old individual could not be confidently assigned a single year age because of the limited completeness of this individual. The present study aged this individual between 20-24 years. The 19 year old was reasonably complete but aged slightly older, 25-29 years, from the partial fusion of the first and second sacral bodies.

Apart from this, the general composition of the sample was very similar, with similar proportions of subadults and adults and similar proportions of individuals within the adult age ranges. Of the adults Houghton and Wiriyaromp (1984) identified, 22 were estimated to be male, 19 female and nine unknown. This is not greatly different from the results presented in Table 3.17, although this study was able to estimate the sex of a few extra individuals.

The analysis of the age distribution by sex (Table 3.17) found some differences present between males and females. Of particular note was the considerably higher numbers of males within the 20-29 year and 30-39 years age groups, compared with females. Then in the 40-49+ year age group it was the females who predominate. The distribution of 10% of adults (5/50) of unknown sex may rebalance these results if their sex could be determined. Another factor affecting these results was the small sample sizes within each age and sex category. These may not enable an accurate representation to be obtained.

**Table 3.14:** Ban Lum Khao - Age Structure

| Age Range | Number | % |
|---|---|---|
| 0 - 0.9 years | 21 | 19.1 |
| 1-4 years | 14 | 12.8 |
| 5-9 years | 11 | 10.0 |
| 10-14 years | 5 | 4.5 |
| Subtotal (children) | 51 | 46.4 |
| 15-19 years | 5 | 4.5 |
| 20-29 years | 25 | 22.7 |
| 30-39 years | 17 | 15.5 |
| 40-49+ years | 12 | 10.9 |
| Subtotal (adults) | 59 | 53.6 |
| Total | 110 | 100.0 |

**Table 3.15:** Ban Lum Khao adult age and sex distribution

| Age Range | Females | | Males | | Unknown | | Total | |
|---|---|---|---|---|---|---|---|---|
| | N | % | N | % | N | % | N | % |
| 15-19 years | 2 | 6.3 | 3 | 11.5 | | | 5 | 8.5 |
| 20-29 years | 18 | 56.3 | 6 | 23.1 | 1 | 1.7 | 25 | 42.4 |
| 30-39 years | 8 | 25.0 | 9 | 34.6 | | | 17 | 28.8 |
| 40-49+ years | 4 | 12.5 | 8 | 30.8 | | | 12 | 20.3 |
| Total | 32 | ~100.0 | 26 | 100.0 | 1 | 1.7 | 59 | 100 |

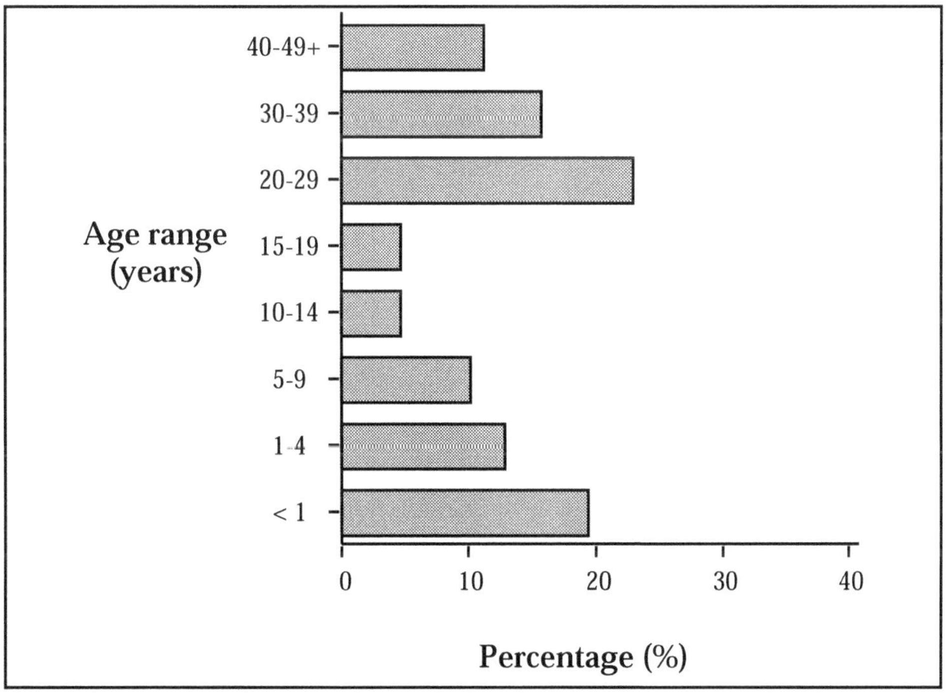


**Figure 3.4:** Ban Lum Khao age at death distribution

**Table 3.16:** Age at death at Ban Na Di

| | This study | | (Houghton and Wiriyaromp 1984) | |
|---|---|---|---|---|
| Age | Number | % | Number | % |
| 0 - 0.9 years | 15 | 19.2 | 14* | 18.9 |
| 1 - 4 years | 8 | 10.3 | 5 | 6.8 |
| 5 - 9 years | 2 | 2.6 | 4 | 5.4 |
| 10 - 14 years | 3 | 3.8 | 1 | 1.4 |
| Subtotal (children) | 28 | 35.9 | 24 | 32.4 |
| 15 - 19 years | 0 | 0 | 3 | 4.1 |
| 20 - 29 years | 13 | 16.7 | 11 | 14.9 |
| 30 - 39 years | 13 | 16.7 | 13 | 17.6 |
| 40 - 49+ years | 10 | 12.8 | 9 | 12.2 |
| Unknown | 14 | 17.9 | 14 | 18.9 |
| Subtotal (adults) | 50 | 64.1 | 50 | 67.6 |
| Total | 78 | 100 | 74 | 100.0 |

*includes five foetal skeletons

**Table 3.17:** Ban Na Di adult age and sex distribution

| Age Range | Females | | Males | | Unknown | Total | |
|---|---|---|---|---|---|---|---|
| | N | % | N | % | N | N | % |
| 15-19 years | 0 | 0 | 0 | 0 | | 0 | 0 |
| 20-29 years | 4 | 30.8 | 9 | 40.9 | | 13 | 26.0 |
| 30-39 years | 3 | 23.1 | 9 | 40.9 | 1 | 13 | 26.0 |
| 40-49+ years | 6 | 46.2 | 4 | 18.2 | | 10 | 20.0 |
| Subtotal | 13 | ~100.0 | 22 | 100.0 | 1 | 36 | 72.0 |
| Unknown | 7 | | 3 | | 4 | 14 | 28.0 |
| Total | 20 | | 25 | | 5 | 50 | 100 |

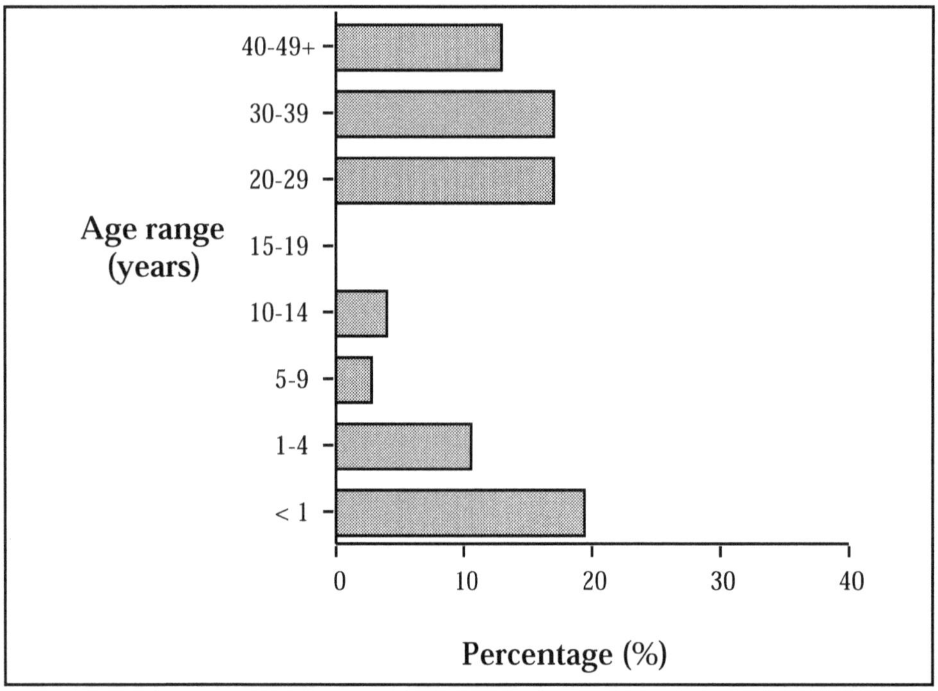


**Figure 3.5:** Ban Na Di age at death distribution

**Table 3.18:** Number of adult females with evidence of parity

| Site | No. females | No. females observed # | No. females parous | % | Youngest age | Oldest age |
|---|---|---|---|---|---|---|
| Khok Phanom Di* | 36 | 36 | 34 | 94.4 | 19 y | 52 y |
| Nong Nor | 49 | 2 | 0 | 0 | - | - |
| Ban Lum Khao | 32 | 23 | 11 | 47.8 | 20-24 y | 40+ y |
| Ban Na Di | 20 | 8 | 7 | 87.5 | 25-29 y | 45-49+ y |

* Data from (Tayles 1992); # number of females observed with the preauricular and dorsal pubis areas preserved

The sex ratio of adults at Ban Na Di was 0.8:1, females (44%) to males (56%). Overall, therefore, there were slightly less females than males. This is in contrast to the other samples, but again the presence of adults of unknown sex puts this result in question.

# Fertility

Estimates of fertility were made for each sample through observations of two areas on the pelvic bones. The preauricular groove on the ilium, adjacent to the sacro-iliac joint, and the region of the dorsal pubis, adjacent to the pubic symphysis. The pits and grooves are smooth-walled depressions of varying size and number. Pitting in one or both of these regions in female pelves was taken to indicate the women had given birth at least once and was, therefore, identified as 'parous'. Nulliparous women were considered to be those with no evidence of pitting. This has already been outlined in more detail in the introductory section on estimating the sex of an adult skeleton.

The results for the women in the prehistoric Thai samples are presented in Table 3.18. Both Khok Phanom Di and Ban Na Di females showed high levels of parity with over 94.4% and 87.5% of females respectively, with evidence of experiencing at least one pregnancy. This indicated almost universal fecundity which is expected in a prehistoric population. However, the results of the Ban Lum Khao females were not as clear. Of the 23 females with pelves intact, less than half showed evidence of parity. It is unlikely this means that less than half of the women had given birth as this would have had significant implications on the maintenance of this population. It is possible that bony remodelling had occurred in many women, obscuring the evidence for parity. A further possibility is that the women of Ban Lum Khao were not as prone to scarring as a consequence of pregnancy in comparison with the other samples. This will be further discussed later in this chapter.

The data from the sample of Nong Nor is excluded from further analysis as the number of females with intact pelves was too small to give an accurate indication of the sample's parity status.

Tayles (1992) found that amongst the parous Khok Phanom Di women, the majority were more commonly affected in the preauricular area. This was also found to be the case in both the Ban Lum Khao and Ban Na Di samples, both showing parous women with more pits in the preauricular area than the dorsal pubis. As mentioned earlier, this was also found in other studies (Houghton 1975; Kelley 1979) and was explained by the fact that the sacro-iliac joint takes more of the weight during pregnancy being in a more direct line of weight transfer from the spine to the lower limbs (Houghton 1975).

# Census Comparison

### Subadults

Overall the Khok Phanom Di sample had the highest proportion of subadults at 55.8% (Table 3.10). This was statistically higher than the subadults present within both the Nong Nor sample, 21.3% (Table 3.12) and the Ban Na Di sample, 35.9% (Table 3.16) (Fisher's Exact p-values both <0.01). The proportion of subadults present in the Ban Lum Khao sample was 46.4% (Table 3.14). This was not statistically different from the Khok Phanom Di or Ban Na Di sample, but was significantly higher than the Nong Nor sample (Fisher's Exact p-value <0.01).

The proportion of subadults, as distributed among age subgroups in the four prehistoric samples, are compared in Figure 3.6. All samples showed a relatively similar pattern of distribution, with the less than one year age group being the most vulnerable in all samples. However, the Khok Phanom Di sample had the highest proportion, perhaps as a result of the better preservation of these more fragile skeletons at Khok Phanom Di. In all samples a steady decline in mortality was evident as children grew older. The Ban Lum Khao and Ban Na Di subadult samples had a similar distribution of individuals among the age groups, with the exception of the 5-9 year range when more Ban Lum Khao children died.

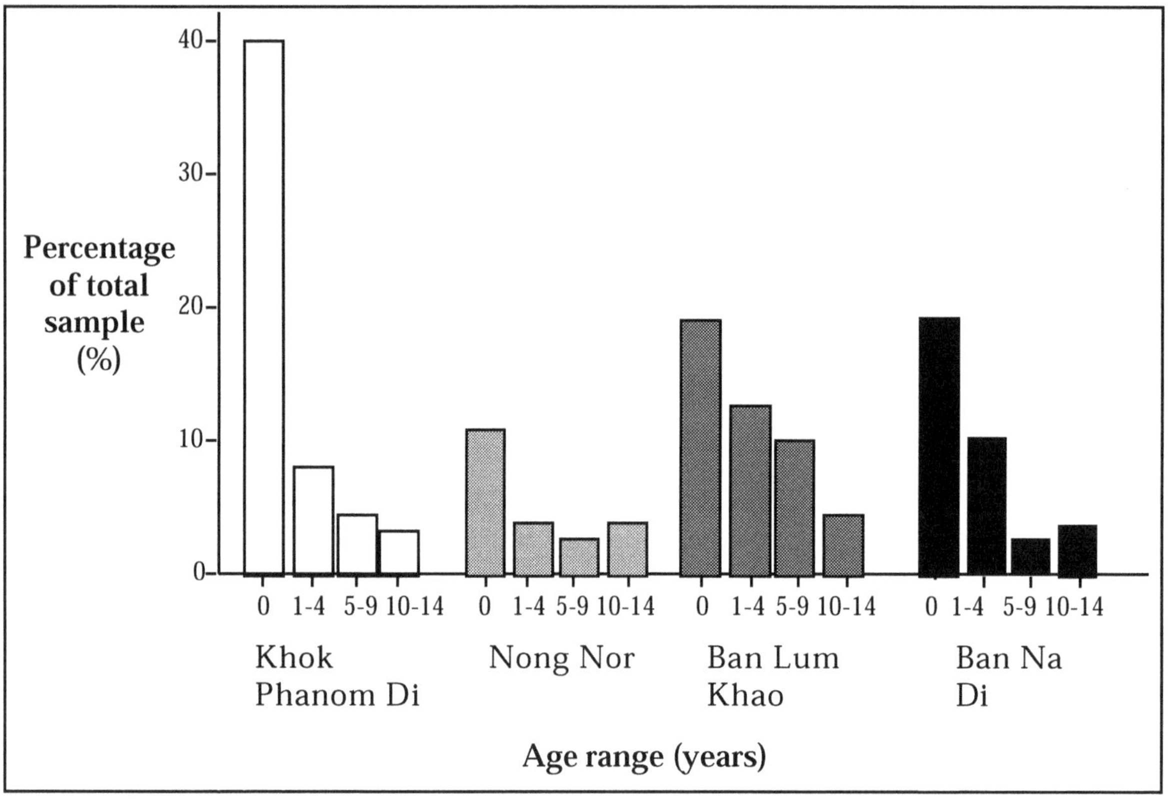


**Figure 3.6:** Age at death distribution of subadults

The Nong Nor subadult sample did not fit the general pattern quite as well as the other samples, but this can most probably be explained by the preservation problems of this sample.

### Adults

The adult age at death distributions for each skeletal sample are presented in Figure 3.7. This highlights differences and similarities in these profiles amongst the samples. There were no statistically significant differences in the age distribution amongst each sample, including separate analyses by each sex, and for the total sample of adults. There were some non-statistical differences apparent that are discussed as they may have implications for further analyses, particularly when age-related pathologies are considered.

There was a scarcity of 15-19 year olds in Nong Nor and Ban Na Di which may partly be explained by sampling factors, although this is not a high risk period of life anyway. These two sites were the least well preserved and Ban Na Di had an overall small sample size. Young adults (20-29 year olds) ranged in prevalence among the adult samples from 21.3% for Nong Nor to 42.4% for Ban Lum Khao. Ban Lum Khao young females were present in a higher proportion than in any other group.

The sex ratios within some age groups were markedly uneven for all samples except in the Nong Nor sample. Both the Nong Nor and Khok Phanom Di samples showed reasonably equal proportions of males and females for the 20-29 and 30-39 year age group. The Ban Na Di sample, however, showed much higher numbers of males than females in these age groups. In the older age range (40-49+ years) all samples showed unequal distributions of males and females with the Khok Phanom Di and Ban Na Di samples showing higher numbers of females, while the Ban Lum Khao and Nong Nor samples, showed higher number of males.

The overall sex ratios for each sample are provided in Table 3.19. All samples showed slightly more females than males present except for the Ban Na Di sample which had more males than females. However, there were no significant differences of proportions of males and females amongst the samples ($Chi^2$= 1.303, p-value = 0.73).

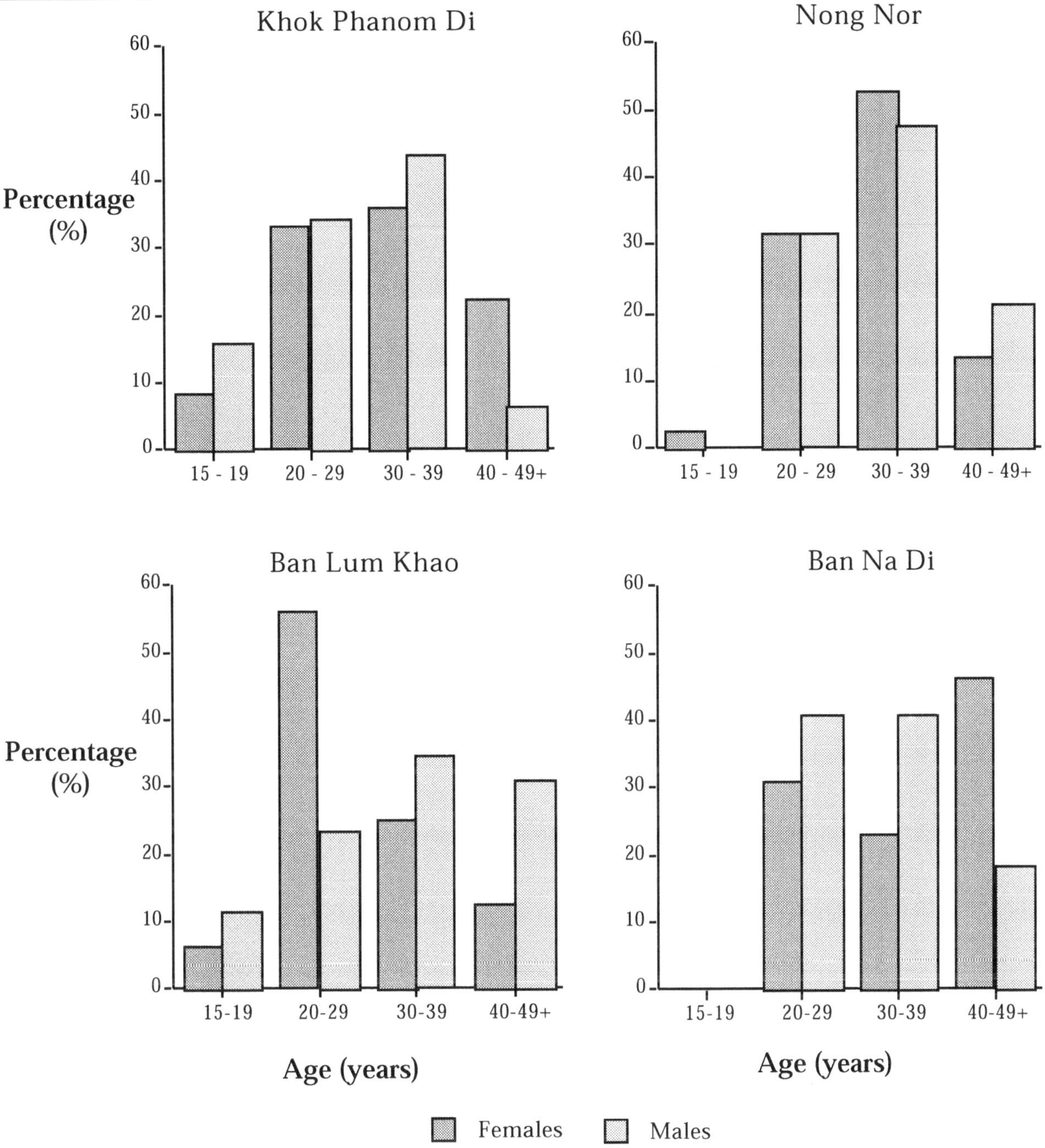


**Figure 3.7:** Adult age distribution for each population, divided by sex. (Data in Tables 3.11, 3.13, 3.15, and 3.17)

**Table 3.19:** Sex ratios among the samples

| Sample | Sex Ratio (F:M) | Females N | Females % | Males N | Males % | Total N | Total % |
|---|---|---|---|---|---|---|---|
| Khok Phanom Di* | 1.1:1 | 36 | 53 | 32 | 47 | 68 | 100 |
| Nong Nor | 1.1:1 | 49 | 53 | 44 | 47 | 93 | 100 |
| Ban Lum Khao | 1.2:1 | 32 | 55 | 26 | 45 | 58 | 100 |
| Ban Na Di | 0.8:1 | 20 | 44 | 25 | 56 | 45 | 100 |

* Tayles (1992)

# Discussion

As discussed in Chapter 1, there are certain limitations with all skeletal samples. In a comparative study such as this, it cannot be assumed the limitations, biases, and assumptions will be the same for each sample. Each sample comes from an archaeological site that differs in environmental conditions, such as soil composition and the level of the water table. These directly influence the preservation of skeletal material. Each sample also represents an incomplete sample of the cemeteries excavated. These factors influence the quality and quantity of the material excavated, but the biological anthropologist can only work with what was excavated and the extent of an excavation can be under financial, political, spatial, and temporal constraints.

However, the analysis of the distribution of the age and sex of individuals within each sample can indicate if the sample structure fits expectations and in what aspects the samples may differ. This analysis relies on the accuracy of age at death and sex estimation methods. There has been much debate in recent years concerning the effect of aging methods on palaeodemographic analyses, such as the calculation of life tables (Bocquet-Appel and Masset 1982, 1985, 1996; Buikstra and Konigsberg 1985). Subsequently, much renewed attention has been given to aging methods in order to improve their reliability (for example Lovejoy *et al.* 1985). The aging methods used on the four prehistoric Thai samples are the same and in this way the age distributions can be considered to be directly comparative. However, are they representative of the entire cemetery population?

There are two particularly important parameters that may assist in answering this as discussed by Waldron (1994). The first is the number of subadults present in the sample. For a pre-industrial society, a substantial proportion of the sample will be of subadult age. Waldron (1994) suggests a figure of around 30% and Weiss (1973), from a survey of the relevant literature, estimates a range of 19-60% for a variety of 'primitive' groups and early 18th and 19th century populations. The four Thai samples studied here, therefore, conform to this as they range from 21.3% (Nong Nor) to 55.8% (Khok Phanom Di) in the proportion of subadults in the samples. The Nong Nor sample is only just within the lower range as a result of sampling bias from the preservation of the subadult material. The age distributions are not identical among the Thai samples and differences have been summarised, however no statistical differences exist. Therefore, it is unlikely the differences that are evident are of a magnitude that will significantly influence interpretations.

A second parameter in determining how representative a sample is, is the consideration of the adult sex ratios (Waldron 1994). It is expected that the proportion of males and females in a cemetery sample should be near equal (Waldron 1994). The sex ratios of each of the four Thai samples are all very close to unity (Table 3.19) and no significant differences were evident in the sex ratios among the samples. It can therefore be assumed that a representative sample in terms of male and female proportions may have been obtained with no evidence of differential mortality or burial practices.

No life table calculation has been attempted for this present study. They may be valid in some samples, such as those with large sample sizes, excellent preservation, and no potential sampling errors. It was considered to be 'forcing' this Thai data too far to gain information that can only be gained with an array of assumptions. The assumptions necessary include the sample being derived from a stationary or stable population, as well as the sample being representative of the population from which it is derived (Bocquet-Appel and Masset 1982; Sattenspiel and Harpending 1983). The amount of migration is unknown for these Thai populations, therefore its effect on the former problem is unknown. Skeletal data can only provide data on the proportion of the population that had died, therefore it is never random and its representativeness to the living population is unknown. The further factor of assuming the age at death and sex estimation methods are reliable (Bocquet-Appel and Masset 1982; Jackes 1992) has been discussed already and with the introduction of multifactorial and seriation techniques this assumption may be more readily acceptable than the others.

### Age at death as a measure of health

The assessment of the age at death distribution of a skeletal sample is an important indicator of a sample's level of health.

> ...diet and health status are inextricably linked to demographic patterns, as are elements of social behaviour, ecological setting, and settlement organization. (Buikstra and Mielke 1985: 408)

These aspects are central to the aim of this thesis and some general comments will be made here with respect to the implications of these results. The diagnosis of particular diseases or dietary deficiencies may further elucidate the reasons for such age at death distributions. Therefore, the reasons why, for example, infant mortality is higher in the Khok Phanom Di sample, or why the Ban Lum Khao sample has a two to three-fold higher proportion of young children (5-9 year olds) will be discussed in further detail in conjunction with other evidence of health and disease in Chapter 8.

Infant and child mortality has a significant effect on population size as well as reflecting infant health (Molnar 1992). Children up until the age of approximately seven years are dependent on older individuals for both food and protection (Bogin 1999). Infant nutritional requirements are initially met by breastfeeding which also provides immunity to some diseases. After about six months of age the diet must be supplemented. In industrialised and post-colonial societies, mild to moderate energy undernutrition is a common problem (Bogin 1999). Children require a specialised diet, rich in energy, lipids, and protein for growth, but low in total volume as they have small stomachs (Bogin 1999). When these nutritional demands are not met a child becomes more vulnerable to infectious disease. Prehistoric populations may also have imposed a high workload on children, compromising their energy balance (Bogin 1999). Prehistoric populations experienced high disease loads, with respiratory infections and gastroenteritis being common causes of infant death (Roberts and Manchester 1995). The significance of these as a cause of the infant and child mortality levels in the prehistoric Thai samples is discussed in Chapter 8. Children are also prone to accidents and this, along with malnutrition, account for most childhood deaths in hunter-gatherer societies (Bogin 1999). Childhood mortality may also be the result of abuse and neglect (Roberts and Manchester 1995) and this is often overlooked as a contributing cause of subadult mortality levels, probably because of the nonspecific nature of injury and pathology observeable in skeletal remains. Infant health also reflects maternal health with undernourished women frequently giving birth to low birth weight, and therefore high risk, babies (Bogin 1999).

In general, subadult mortality can provide an excellent indication of the risks of infancy and childhood in a population. However, in order to make inferences such as those discussed above, the census information provided in this chapter needs to be considered in conjunction with information in the following chapters, especially those concerning childhood growth and disturbances of growth and evidence for nutritional adequacy.

The large proportion of young adult females present in the cemetery sample from Ban Lum Khao suggests they were exposed to a higher level of risk than in the other samples. This may well be explained by the risks of childbearing in prehistoric times. This may have had implications on the population's ability to maintain itself. Pregnancy and childbirth is a significant cause of death in all developing countries. In these countries a woman's probability of dying from pregnancy ranges from 1:15 to 1:70 (WHO 1987 cited in Ortner 1998). If the mortality level is considered in conjunction with the evidence for pregnancy, an association is evident between the low fertility for Ban Lum Khao women and a high risk of death during the reproductive years. One possible explanation could be that the women suffered from a high rate of miscarriage, early on in pregnancy, before bony changes had occurred, and this resulted in death from haemorrhage. Another alternative, is that some deformations of the maternal pelvis can cause great difficulties during childbirth. Rickets and osteomalacia are known to distort the pelvic inlet. No such deformations were observed in the people of Ban Lum Khao. However, even a *slight* deformation can significantly alter the birth canal and seriously increase the risk of death of the mother or infant (Molnar 1992), and may go unnoticed in a skeleton unless specifically investigated.

However, why would the risks of childbearing be any greater for the Ban Lum Khao people than any of the other prehistoric groups? There are, no doubt, many other explanations for this level of young females in the skeletal sample of Ban Lum Khao not necessarily related to childbearing. The suggestion that low numbers of pregnancies occurred in the Ban Lum Khao women may also be unrelated to pathological problems, but a reflection, as mentioned earlier, of their lack of bony scarring and/or higher bony remodelling rates.

# Summary

### Subadults

The Khok Phanom Di sample had the highest proportion of individuals dying as subadults, 55.8% of the total sample. By far the majority, 73.3% of these were less than one year old at death. The Ban Lum Khao sample had the second highest subadult mortality rate, 46.4%. However, unlike the Khok Phanom Di subadult sample, only 41.2% were less than one year old at death and 68.6% were less than five years at death. The 5-9 year olds at Ban Lum Khao were also present in higher numbers than in any of the other samples. The proportion of subadults in the Ban Na Di sample was third highest, 35.9% of the total sample, and just over half of these had died before reaching the age of one year. Over 80% had died before the age of five years. In comparison, the subadult mortality in the Nong Nor sample was very low, 21.3%, although, again, over half of them had died before reaching their first birthday and two-thirds before the age of five years. Over all the samples there was no trend associated with time, although the earliest sample, Khok Phanom Di, had the highest subadult mortality. However, if the evidence from the Nong Nor sample is ignored, which could be considered valid given the poor preservation and therefore lack of representivity, there is a decrease in subadult mortality with advancing time.

### Adults

All samples showed a generally similar age distribution of adults. The age structures all showed few or no adults dying between 15-19 years. All samples showed the majority of the adults in each sample were aged between 20-39 years at death. However, there was a particular predominance of females aged between 20-29 years in the Ban Lum Khao sample. Although this was not statistically significantly different from the others, it may be related to the women of Ban Lum Khao being unable to cope with risks associated with childbirth. All samples showed few adults aged between 40-49+ years at death. However, there was a consistent increase with time in the proportion of adults surviving to this age. The proportion of adults aged 40-49+ years in the Ban Na Di sample was nearly twice as high as in the earliest sample, Khok Phanom Di.

The sex ratios were all relatively equal indicating good representation of both males and females in each sample.

Fertility, based on pelvic evidence, appeared high in both the Khok Phanom Di and Ban Na Di samples. This was not so in the Ban Lum Khao sample, and was unknown in the Nong Nor sample.

# 4 Growth and Disturbances of Growth

## Introduction

This chapter is concerned with the use of growth as an indicator of health. This includes assessing whether the environment provided adequate nutrition for optimum growth and also addresses the influence of morbidity on growth.

> Growth, by definition, is a graded pattern of various changes related to age, occurring mainly by size increase and shape differentiation of each part of the living body. Its most important aspect is variation, which is due to the combined interaction of hereditary and environmental factors. (Ribot and Roberts 1996: 67)

There is a strong polygenetic component for growth (Ulijaszek 1998) . However, the genetic influence on the rate and pattern of growth in these prehistoric Thai skeletal samples is unknown. Studies of growth on modern Southeast Asia peoples are of some use, although the biological relationship between these modern and prehistoric populations is also unknown. This study assumes that these prehistoric samples are relatively genetically homogeneous in terms of growth, both among the samples and within each sample. That is, they all had a similar genetic potential for, and genetic control over, growth. However, if large differences are found among the samples, genetic differences cannot be ruled out as a contributing factor.

The external environment, including the adequacy of nutrition and morbidity (Bielicki 1998) , also influences growth and the end result, stature. The age at which malnutrition or disease occurs is important as there is the possibility of a period of 'catch-up' growth in infancy and adolescence. Catch-up growth may also occur post-illness (Stini 1985) . This is when the body attempts to compensate for a disruption in growth once favourable conditions have returned (for example, a period of seasonal malnutrition ends, or an infection is successfully overcome). If the level of nutrition is inadequate during or after this period, catch-up growth is unlikely. The extent to which it can compensate for the period of reduced growth depends on the severity and duration of the condition. Environmental factors very early on in life, such as the intra-uterine environment, can also influence growth and development (Cameron 1998) . This is a reflection of maternal health and nutrition, not of child health and nutrition.

An individual can respond differently to external factors depending on their level of frailty. The latter is established partly by an individual's genetic composition and partly by their response to the available diet and the presence of underlying disease. Therefore, there are individual differences as well as population differences in susceptibility to disease (Wood et al. 1992) . Males and females may also differ in their response to disease and nutritional deficiencies.

Analysis of the degree of sexual dimorphism can reflect the relative effects of the environment on the growth of males and females during childhood (Stinson 1985) . There is always some degree of sexual dimorphism in stature in all populations, but this level is not always the same suggesting some level of genetic input (Stinson 1985) . It has been suggested that male growth is more affected by environmental stress than females. A population experiencing nutritional stress, for example, may have a lower level of sexual dimorphism than otherwise, reflecting the more retarded growth of the males compared with female growth (Stinson 1985) .

Stinson's (1985) review of the sex differences in environmental sensitivity concluded that the available evidence only provides weak support for males being less buffered than females against the environment during growth. Given the importance of females for maintaining the species, undertaking a much higher burden than males, this difference is expected to be greater. Stinson (1985) reviews many studies and acknowledges that some have problems in their experimental method.

Buffering against the environment is partly due to the efficiency of an individual's immune system. Female adults are known to have a greater capacity to resist infectious diseases (Stini 1985) . This provides indirect evidence of sex differences in immunity (Ortner 1998) . If this differential begins in childhood it may confer better immunity to females during the growing years.

Diagnosing the cause of reduced growth within a population or individual is not straightforward. It cannot easily be stated whether reduced growth was due to an inadequate diet, either in quantity or quality, or disease (assuming genetic homogeneity). This is because it has been demonstrated that a synergistic relationship exists between malnutrition and infection (Scrimshaw *et al.* 1959) . Both have an impact on growth, and both are promoted in the synergistic relationship (Saunders and Hoppa 1993) .

A malnourished child is more vulnerable to infection "...due to a generally weakened state of health and a reduced host resistance" (Lunn 1991 cited in Saunders and Hoppa 1993: 134). Many infections will be of little clinical significance in a well-nourished child, but the same infection in a malnourished child can lead to further nutritional disease and increase the risk of growth disturbances and/or death (Scrimshaw 1981) . When adequate food is available and the infection is chronic or repeated, the disease can still be serious. For example, diarrhoeal infections can interfere with the absorption of nutrients (Scrimshaw 1981; Saunders and Hoppa 1993) . This, combined with a diminished appetite that results from nearly all infections (Scrimshaw 1981) , can lead to more serious nutritional inadequacies developing.

Acute infections in well nourished children will probably be overcome and catch-up growth will compensate for the reduced growth rate during the illness. Acute infections in malnourished children may not be as easily compensated for as nutrition is inadequate to sustain the necessary accelerated growth rates after recovery (Scrimshaw 1981) . It is this latter case, and the case of chronic infections and/or chronic malnutrition, that can lead to visible changes in the skeleton.

There are a number of methods available for assessing growth during childhood in skeletal remains. Dental development is relatively unmodified by the environment, in contrast with skeletal development (Saunders 1992) . Therefore, a comparison of the dental developmental age with the skeletal size age can indicate if the genetic potential for growth has been retarded due to environmental effects. Childhood growth may also be assessed through measurements of adult stature. Although the genetic potential for maximum growth is unknown in these Thai samples, some information can be obtained from modern Thai (for example Sangvichien *et al.* 1985) . Large differences in mean stature estimates among the prehistoric samples may indicate that the populations suffered more morbidity and nutritional inadequacies during childhood. While low stature may reflect chronic undernutrition, it may also be seen as an adaptive feature that indicates an optimisation of growth allowing more individuals to survive when under reduced dietary resources (Rodriguez Martin and Beranger Meteos 1992) . This 'small but healthy' concept is further discussed in Chapter 8.

Disturbances in growth can also be observed through skeletal and dental defects, called Harris lines and enamel hypoplasia, respectively. These can be induced by a variety of problems including starvation, severe malnutrition, and severe infection (Cohen 1989) . While Harris lines and enamel hypoplasia do not allow the identification of a specific disease, they can provide an indication of the frequency of problems in individuals (Cohen 1989) and, therefore, the populations from which they came.

Growth studies are also combined with the mortality patterns of the samples under study. These two factors are recognised as having a strong correlation with one another (Saunders and Hoppa 1993) .

Linear growth ceases at adulthood, but bone is continually remodelled. Bone mass is also addressed in this chapter. There is a strong relationship between nutritional quality and bone mass, as adequate nutrition is required for skeletal maintenance (Larsen 1995) . Poor bone mass may be indicative of nutritional deficiencies, and may have pathological consequences such as an increased risk of fracture.

# Childhood Growth

## Skeletal size versus dental age

### Introduction

Subadult skeletal material is used in this section to determine the influence of the environment on growth patterns. The development, particularly the formation of the dentition is thought to be less influenced by external factors such as the environment, including the quality and quantity of the diet, compared with the growth of the long bones (Ubelaker 1987; Saunders 1992) . Dental age is thought to provide a closer approximation to chronological age, while skeletal age can be a marker for alteration and defects of growth (Saunders 1992) . Therefore, a comparison of the stage of development of the dentition with the length of the long bones can provide evidence that skeletal growth has been disrupted which may indicate the influence of external factors.

"Growth of the physical dimensions of the body is a highly regulated process" (Karlberg 1998: 108). Growth is characterised by particular periods of rapid and slow intervals of growth rate. Immediately following birth there is a very high growth rate which rapidly decelerates until about three years of age (Karlberg 1998) . From this age until puberty, growth slowly decelerates. The onset of puberty is marked by an increased growth rate which reaches a peak velocity and then continues to decelerate until linear growth ceases in early adulthood (Karlberg 1998) .

Some studies (for example Wall 1991) have found a lag in growth occurring between the ages of two to three years of age and have correlated this with the age at weaning. Weaning may not only reduce the nutrient intake but is also associated with systemic disturbances such as weanling diarrhoea (Wall 1991) . Both of these factors may be severe enough to retard growth.

> Environmental influences on postnatal growth include cultural practices such as subsistence strategies (e.g. hunting and gathering versus agriculture) and age at weaning as well as ecological parameters (e.g. subsistence base and subsequent diet composition). (Wall 1991: 9)

A longitudinal study on Guatemalan children compared the growth of low socioeconomic status Mayan children with high socioeconomic status *ladino* children (Bogin *et al.* 1992) . The Mayan children had a higher morbidity associated with infectious and chronic disease and were more likely to be undernourished in comparison with the *ladino* children who were much healthier and well nourished. The Mayan children were both shorter and had a reduced growth rate at many ages compared to the *ladino* children. Bogin *et al.* (1992) concluded that this difference may be partly explained by genetic differences. However, the Mayan growth pattern is very similar to other disadvantaged populations from other regions of the world, implying that environmental differences have also contributed to the growth pattern.

Comparing growth rates and patterns with those of modern studies on living children cannot be undertaken without acknowledging that the prehistoric data are from children who died. Therefore, they probably represent children who were sick (unless they were accidentally killed) and more likely to have a reduced growth rate than those who survived to adulthood, especially if the illness was chronic.

Ribot and Roberts (1996) compared growth in individuals with and without 'stress indicators' such as enamel hypoplasia, Harris lines, and cribra orbitalia, in two Medieval skeletal populations. They found no association between shorter long bone lengths and the number of 'stress' indicators. They concluded that the stress episodes were probably not of long enough duration to affect growth.

As growth has a genetic component, growth rates can be population specific (Saunders 1992) . There is some information on the eruption of teeth in modern Thai children (Kamalanathan *et al.* 1960b) in comparison with other parts of the world and, as discussed in Chapter 3, has been incorporated into the aging methods used here. Bailey *et al.* (1984) studied growth patterns in over 1000 children in 29 rural villages in northern Thailand. They found that these Thai children were consistently behind international standards for growth in all the anthropological measurements taken. For example, recumbent length measurements at three years of age were almost one year behind US standards in females and 16 months behind in males. Additionally, skeletal age, assessed from hand/wrist radiographs was graphed against chronological age and by nine years of age males were 36 months behind the Gruelich-Pyle means and females were 48 months behind. Growth retardation became particularly noticeable after the age of 18 months. These children were also smaller than many other Asian and local Thai children. It is possible the northern Thai children were genetically smaller, especially considering they were not experiencing general food shortages. However, Bailey *et al.* (1984) do not believe genetics fully explains the growth retardation. They suggest that the undernutrition may be a problem caused by children's small gastrointestinal tracts unable "...to digest sufficient quantities of a calorically sparse food such as rice" (Chandrapanond *et al.* 1972; Gershoff *et al.* 1977 both cited in Bailey *et al.* 1984: 555). Thus, once weaning has begun, usually around the age of 18 months, caloric insufficiency may begin.

## Methods

The methods detailing the estimation of age at death from the dentition and measurements of skeletal size (diaphyseal bone lengths) have previously been presented in Chapter 3. The humeri provided the largest sample of bones able to be measured. As a general picture of growth rates was required and other bones provided much smaller samples, only this bone was used in the following assessment. Both left and right humeri measurements were used separately in order to increase sample sizes. An individual had to have both a dental age estimate and a humerus measurement to be included.

For each age level in each sample the humerus length is averaged when more than one individual is represented to obtain the mean humeri length for that age. The 'growth rate' is therefore obtained by using dental age ($x$ axis) as representing the closest approximation to chronological age and graphing it against a measure of linear growth, in this case humerus diaphyseal length ($y$ axis). Any deviations among the samples or in comparison to other growth studies can be identified and addressed.

## Results

The results of the mean humerus measurements are graphed against dental age in Figure 4.1. Appendix C provides the number of measurements available from each sample for each age group. No results were possible for the subadults of Nong Nor due to poor preservation of skeletal material. There were no individuals from Khok Phanom Di under five years of age who had been aged solely from dental development; therefore they were excluded and only those from Khok Phanom Di older than five years included.

The growth rate is quite similar over all samples. The line gradients showed the highest growth rate between birth and three years in both the samples that have data for this age, Ban Lum Khao and Ban Na Di. This corresponds with the rapid period of growth typical of this age group. No growth lag was evident at any stage of infancy during which weaning may have occurred. However, small sample sizes may have obscured such detail.

After three years of age, the slope levels off to a lower rate, again typical of this age period. Among the samples, after three years of age, the growth rates diverge very slightly until age 11 where the three samples are all quite similar again. However, the differences between 3 and 11 years among the samples is not large. Ban Lum Khao subadults have only slightly shorter humeral lengths for their dental age than either Khok Phanom Di or Ban Na Di subadults. This may have also been influenced by the small sample sizes available.

After the age of approximately 11 years, another small acceleration is evident. This is particularly apparent in the Ban Lum Khao sample, as this sample has more data for this age period. This increase in growth may be equivalent to the adolescent growth spurt that typically occurs during puberty.

The absence of data from the Khok Phanom Di subadults aged less than five years and small sample sizes from both Ban Lum Khao and Ban Na Di, makes interpretation difficult. Considering the differences among the samples are quite minimal, they will not be overly emphasised.

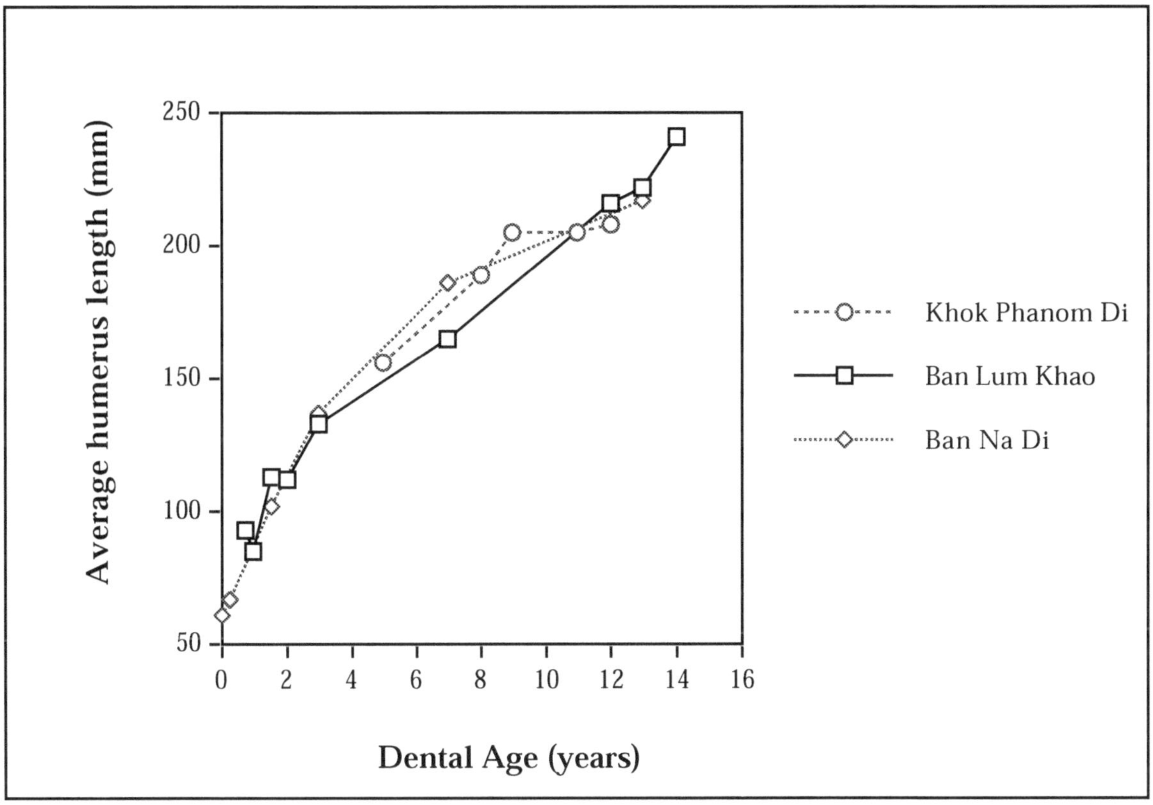


**Figure 4.1:** Mean humerus diaphyseal length plotted by dental developmental age for the subadult samples from Khok Phanom Di (n= 13 m/m, 10 individuals) (Tayles 1992), Ban Lum Khao (n = 21 m/m, 13 individuals), and Ban Na Di (n = 10 m/m, 8 individuals)

## Stature

### Introduction

"...Every human being has a genetically determined upper limit to adult stature..." (Bogin 1999: 333); this is referred to as their genetic potential. As discussed in the introduction to this chapter, adult stature is highly variable and can be influenced not only by genetic make-up, but also nutritional status, the environment, and disease during the period of growth (Roberts and Manchester 1995).

Stature estimates from skeletal remains can be compared among genetically similar populations to determine whether the environment, including diet and disease, affected childhood growth. Adult stature measurements are from those who actually survived childhood and this period's potential insults, while the children who suffered the most would have died as children.

Another consideration in interpreting achieved population stature is the timing of any childhood growth disturbances. Growth may well be reduced during short periods of nutritional inadequacy, however a return to adequate nutrition may result in catch-up growth (Huss-Ashmore *et al.* 1982) with no detectable affect on ultimate adult stature.

Social inequalities, including access to food and vulnerability to disease, within and between communities can create stature differences that reflect these differences. As Tanner (1978 cited in Lindgren 1998: 381) stated growth can act "...as a mirror of conditions in Society...".

A linear relationship exists between stature and the length of the long bones. This relationship is the basis for estimating stature in skeletal remains. Regression equations were first reliably developed from the cadavers of predominantly white and black Americans, but also including a smaller group of American Indians and other nationalities, by Trotter and Gleser (1952, 1958). These equations are still widely used today. However, there has been some suggestion that regression equations tend to overestimate the shortest individuals and underestimate the tallest, and therefore narrow the range of variation (Sjøvold 1990).

Body proportions differ among populations, therefore equations established using one population should not be applied to another (Trotter 1970). If inappropriate equations are used different bones from one individual will give a wide range of stature estimates. If equations are not available for the population in question, or the population relationship is unknown, those equations that provide the least variation within individuals for different bones should be used. This means that the individual has the most similar limb proportions to the individuals used to establish the equations, therefore a more accurate stature estimate will be obtained than if other equations are used.

Regression equations are available from a variety of populations including White Americans, Negroes,

'Mongoloid' peoples (Trotter 1970), Javanese (Bergman and The 1955) and Thai/Chinese cadavers (Sangvichien *et al.* n.d., 1985). The equations established from the cadavers of modern Thai and Chinese people by Sangvichien *et al.* (1985, n.d.) are of most interest to this study. However, because the exact genetic relationship between the prehistoric people under study and the modern Thai is unknown, it needs to be established if these equations are the most appropriate to use.

Before the Thai/Chinese equations were available the stature of prehistoric Thai people were estimated using Trotter and Gleser's American equations. In comparison, the Thai/Chinese equations produce, in general, shorter stature estimates than the American equations. However, the Thai/Chinese equations produced the least variation in stature estimates for different long bones in individuals from Khok Phanom Di (Tayles 1992). Therefore, this indicates the prehistoric Thai people have the most similar limb proportions to the sample on which the modern Thai/Chinese equations were based. These equations are, therefore, used to estimate stature in all of the four samples analysed here. It is acknowledged that these modern population based formulae may bias the stature estimates when used on earlier populations (Brothwell 1981), but this is unavoidable.

There is always some degree of sexual dimorphism in stature in all populations, as males and females have different body builds as discussed in Chapter 3. Average male size can exceed female body size by 5 to 10% (Molnar 1992). However the effects of sex are eliminated in stature estimates by using sex-specific equations. Sangvichien *et al.* (n.d., 1985) provide regression equations for both males and females for all the major long bones.

The Thai/Chinese equations for the lower limb bones are presented with their standard errors (Sangvichien *et al.* 1985). This enables the identification of the most reliable long bone to estimate stature with. Unfortunately, no standard errors are available for the upper limb equations. Trotter's (1970) regression equation found lower standard errors in all lower limb bones compared with upper limb bones. This means that lower limb bones are more highly correlated with stature than upper limb bones.

Some studies have taken various approaches in estimating stature from regression equations. For example, Houghton *et al.* (1975: 330) presents regression equations for prehistoric Polynesians for left and right bones separately in order to deal with "...the problem of physical asymmetry". Others, such as Trotter and Gleser (1951, 1952), have estimated the stature from paired left and right bones and taken the average as the stature estimate. However, the average of stature estimates obtained from different bones should not be averaged, rather it is better to choose the equation with the least standard error (Trotter and Gleser 1958; Waldron 1998). Trotter and Gleser (1951, 1952) state there is very little difference in the standard deviation between any left and right pair. This apparent lack of difference between left and right bones seems to also support the use of one side only; if there is little variation, then using either left or right is acceptable and is supported by Genovés (1967) study of stature among Mesoamericans. This is a useful finding because of the problem of missing bones in skeletal material.

Waldron (1998) emphasises the need to consistently use the same bone, for example the femur, in all estimations. This will produce the least amount of error as shown in a study of two historic English skeletal populations using Trotter's (1970) equations (Waldron 1998). While this may be the optimum method it is yet to be confirmed in other studies and is not always practical for the analysis of skeletal remains. Problems with preservation mean that not all skeletons will consistently have the same bone with which to estimate stature. Analysis following this method would reduce the sample size markedly in the populations in question in this study. The statement of this problem here is recognition of the level of error it may produce.

> ...the most accurate estimates of stature are obtained when the equation applied to the unknown has been derived from a representative sample of the population of the same sex, race, age, geographical area, and time period to which the unknown is believed to belong. (Trotter 1970: 82)

However, rarely are all of these aspects known for a prehistoric skeletal sample.

## Methods

For this study, stature has been estimated using equations derived from modern Thai and Chinese cadavers (Sangvichien *et al.* n.d., 1985) as these are the most appropriate. The equations are provided in Appendix D. The standard errors of each estimating equation (sex and bone specific) provided a ranking system for determining the bone or combination of bones best used to estimate stature. The order of preference, based on known standard errors, was slightly different between males and females; for consistency the male ranking was used. The order of preference was: 1. femur and tibia, 2. tibia, 3. femur, 4. humerus and radius, 5. humerus, 6. radius, 7. ulna (Tayles 1992).

The long bones of all adults (≥ 15 years with long bone epiphyses fused) from each sample were measured using either an osteometric board or, when that was unavailable, sliding calipers. In a report by Sangvichien *et al.* (1985), a combination of the osteometric board and sliding calipers were used without identifying which tools were used for which measurements. It is acknowledged that the use of different measuring instruments may have introduced a small, but insignificant error between measurements. In the present study, many bones have had to be reconstructed in order to obtain estimates. This may influence results by one or two millimetres, which is assumed to make little significant difference to the final results.

An osteometric board was used for the skeletons of Khok Phanom Di and Ban Na Di. Ban Lum Khao estimates were made using sliding calipers as analysis was carried out in Thailand and no board was available. Nong Nor estimates were made using a combination of sliding calipers on bones in the ground and an osteometric board for those complete enough or reconstructed once out of the ground.

**Table 4.1:** Summary statistics of female adult stature for the prehistoric Thai samples (cm)

| Site | n | Mean (cm) | S.D.(cm) | Range (cm) | Sexual Dimorphism (% of male stature) |
|---|---|---|---|---|---|
| Khok Phanom Di* | 36 | 154.3 | 4.5 | 141.1 - 163.2 (22.1) | 4.9 |
| Nong Nor | 14 | 156.1 | 3.6 | 150.7 - 162.1 (11.4) | 6.6 |
| Ban Lum Khao | 25 | 154.7 | 3.7 | 147.9 - 162.2 (14.3) | 6.1 |
| Ban Na Di | 13 | 155.9 | 4.0 | 150.0 - 164.4 (14.4) | 7.2 |

n= number of individuals; *(Tayles 1992). Comparison of female mean stature among samples: ANOVA F ratio = 0.85, F probability = 0.47, n.s.

**Table 4.2:** Summary statistics of male adult stature for the prehistoric Thai samples (cm)

| Site | n | Mean (cm) | S.D. (cm) | Range (cm) |
|---|---|---|---|---|
| Khok Phanom Di* | 30 | 162.2 | 5.1 | 153.8 - 170.7 (16.9) |
| Nong Nor | 19 | 167.2 | 6.5 | 158.8 - 181.6 (22.8) |
| Ban Lum Khao | 19 | 164.7 | 6.3 | 152.4 - 174.9 (22.5) |
| Ban Na Di | 17 | 168.0 | 4.9 | 159.5 - 176.0 (16.5) |

n= number of individuals; *(Tayles 1992). Comparison of male mean stature among samples: ANOVA F ratio = 4.25, F probability <0.01 (significant), Student-Newman-Keuls test show significant differences are between Khok Phanom Di with Ban Na Di and with Nong Nor.

**Results**

Summary results for stature are presented in Table 4.1 (females) and Table 4.2 (males) and Figure 4.2.

Stature estimates were possible on all adults with epiphyses fused from Khok Phanom Di. Only 27% (n=33) of adults from Nong Nor were able to be measured for stature estimates and 24 of those individuals had measurements taken in situ, the other nine in the laboratory. From Ban Lum Khao and Ban Na Di, 74% and 60% of adults respectively, had their stature estimated. Both of these samples had all measurements taken in the laboratory.

Some individuals showed extreme values relative to others in their respective samples. These are mentioned here as a cautionary note as they were included in the following analyses. One female from Khok Phanom Di was very short (Burial 56, 141.1 cm), 6.5 cm shorter than the next shortest female. Two males from Nong Nor were exceptionally tall (Burials 47 and 112, 180.7 cm and 181.6 cm respectively), a minimum of 5.8 cm taller than the third tallest individual. Neither population of Ban Lum Khao or Ban Na Di showed any obvious outliers. These outliers are identified in Figure 4.3.

Within any population a certain level of sexual dimorphism is to be expected. Sexual dimorphism was measured by calculating the percentage that the difference in the average stature between each sex was of the male stature. Within each sample the subsample of males was statistically significantly (p-value < 0.0001) taller than each sample of females. These values are given in Table 4.1. Sexual dimorphism appeared to increase through time, with Khok Phanom Di people having the lower level and the later population of Ban Na Di the higher degree of dimorphism. There were no significant differences among the female samples in all populations as indicated by the ANOVA F-ratio in Table 4.1. Male stature amongst these populations showed differences with Khok Phanom Di males being significantly shorter than both Nong Nor and Ban Na Di males. The tallest individual from Nong Nor was 10.9 cm taller than the tallest person at Khok Phanom Di.

The range of variation in stature within the samples is presented graphically in Figure 4.3. Within the Ban Lum Khao sample, males had a range nearly twice as wide as the female sample. This was also evident in the Nong Nor population. Even with two male stature estimates as outliers the range is wider than the females in the Nong Nor sample. Khok Phanom Di females showed a range of variation, one third greater than their male counterparts but this was probably due to the presence of one very short women whose measurement was an outlier.

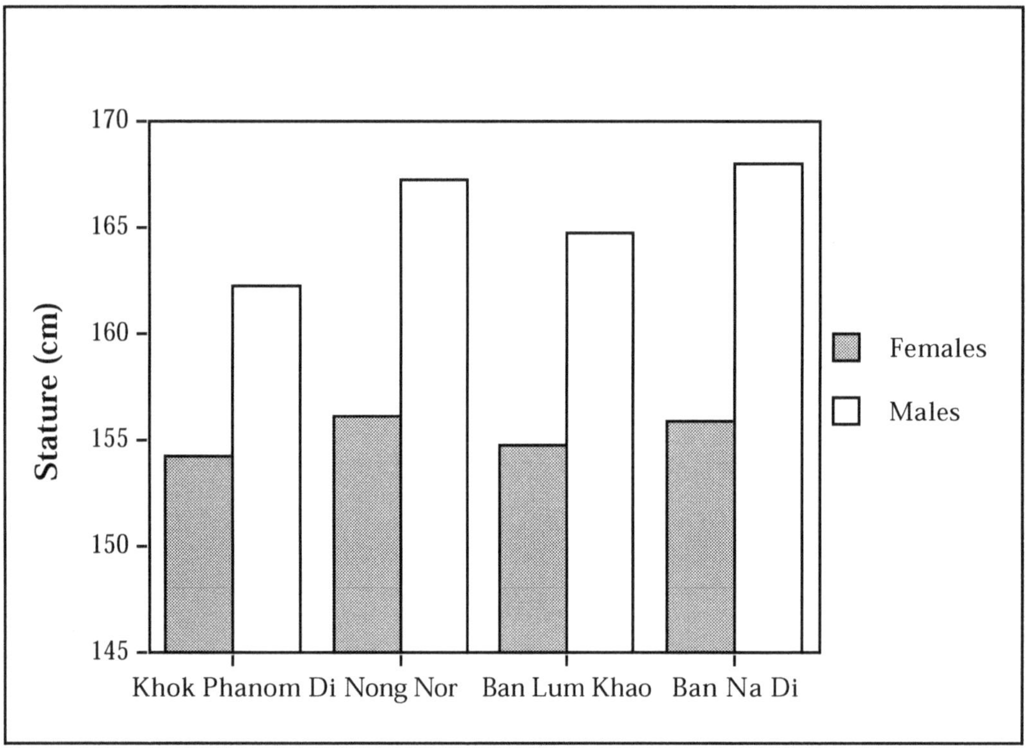


**Figure 4.2:** Mean adult stature of males and females for the prehistoric Thai samples. Sample size and standard deviations are in Tables 4.1 and 4.2

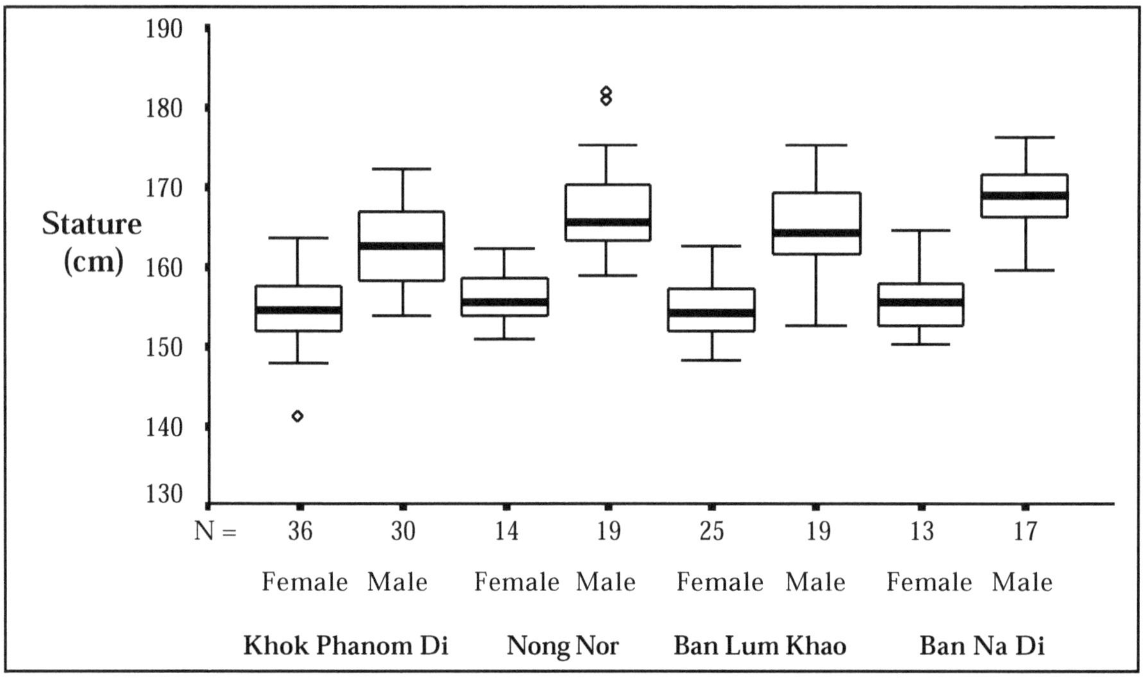


**Figure 4.3:** Comparison of stature among the prehistoric Thai samples. Boxplot indicates median, range, quartiles and outliers of the stature estimates

# Childhood Growth Disruptions

## Harris Lines

### Introduction

Harris (1926) was one of the first researchers to identify 'bony transverse striations' in human long bones as indicators of growth disturbances. These lines are seen as transverse lines of greater radio-opacity in long bones. They have been termed a variety of names including 'lines of arrested growth', 'transverse lines' and 'Harris lines'. The latter term is used here as the other terms imply an aetiology and structure which is not strictly accurate.

The presence of a Harris line indicates the resumption of growth after a period of reduced growth. For example, during periods of nutritional deficiency, bone continues to grow, but at a less than normal rate (Saunders and Hoppa 1993). There is some degree of recovery after a period of malnutrition or illness in childhood when favourable conditions return. In the palaeopathological literature, Harris Lines are considered to be indicators of nonspecific stress in childhood (Martin *et al.* 1985; Grolleau-Raoux *et al.* 1997). Such problems as diarrhoea, infection, seasonal malnutrition, weaning, and other illnesses can induce a short period of disturbance. The disturbance may be either a reduced growth rate or the total arrest of the growth of a long bone (Martin *et al.* 1985; Larsen 1997; Lewis and Roberts 1997). Through the growth hormone-somatomedin axis such problems as mentioned above can result in failure of the growth cartilage and the normal balance of activity of chondroblasts and osteoblasts is disturbed (Hummert and Van Gerven 1985; Grolleau-Raoux *et al.* 1997). Structurally, Harris lines are trabeculae both denser and thicker than normal (Hummert and Van Gerven 1985), hence the greater opacity seen radiographically.

This activity takes place at the epiphyseal disc of a growing long bone, but as the bone grows, the lines remain the same distance from each other (Hummert and Van Gerven 1985) and the midshaft. Attempts have been made to determine chronologies for the ages of line formation (Maat 1984; Hummert and Van Gerven 1985; Byers 1991; Grolleau-Raoux *et al.* 1997). The most vulnerable time during growth and development is during infancy, but this also creates a long time frame in which lines are more likely to remodel.

There are a number of problems associated with the interpretation of Harris lines. Firstly, the severity of the insult is not directly related to the formation of a line, as this is dependent on an individual's response to the insult. One study found no difference in the frequency of Harris lines between two groups, one of normal subjects, the other showing signs of stress (Driezen 1964 cited in Grolleau-Raoux *et al.* 1997). A study by Gindhart (1969), indicated that not all severe diseases result in Harris line formation; only 25% of severe diseases resulted in Harris lines in immature individuals. There also appears to be no consistent link between the number of lines and the number of illnesses occurring in any one individual (Grolleau-Raoux *et al.* 1997). Lines can even occur when no disease is observed, perhaps as the result of psychological stress, making the establishment of an aetiology and interpretations difficult (Ribot and Roberts 1996).

The rate of bone remodelling is another problem. Remodelling has been shown to occur at different rates in males and females (Lewis and Roberts 1997) and is likely to be individually variable. As the remodelling and disappearance of lines takes time, this problem could be reduced by limiting the study to juvenile populations (Lewis and Roberts 1997).

A Harris line represents the resumption of growth and thus the return of favourable conditions. This means that an individual with no Harris lines has not necessarily never been malnourished or ill. It could also be interpreted as an individual who was chronically subjected to minor problems and saw no return to favourable conditions to allow a normal growth rate. In this case reduced stature may also be evident and would therefore help in the interpretation of the Harris lines. This emphasises the importance of considering all available evidence concerning growth to obtain an accurate picture as possible.

Other problems include a high inter- and intra-observer error in recording. This is due to considerable architectural variation and changes occurring during resorption (Macchiarelli *et al.* 1994) although Clark and Mack (1988) did not come to the same conclusion in a similar study.

This study of prehistoric Thai skeletal samples considers only one bone and only the lines at the distal end of that bone. Lewis and Roberts (1997) suggest to score the presence or absence of these lines and not compare counts of lines with other populations until better methods have been established, hence in the following analysis only tentative comparisons will be made among the populations.

### Methods

Antero-posterior radiographs were taken of as many adult tibiae as possible in each skeletal series. The tibia was chosen as this is one of the most commonly affected bones (Roberts and Manchester 1995). Because of the variation in preservation and remineralisation of each sample, there was a wide range of sample sizes. No radiographs were able to be taken of the Nong Nor bones, compared with 65% of Khok Phanom Di individuals being able to be radiographed. Of the Ban Lum Khao and Ban Na Di samples, 22% and 14%, respectively, of tibiae were radiographed.

This analysis was restricted to the adult samples only. Although in hindsight analysis of the juvenile population may have also been useful, but the subadult sample sizes were small.

The number of lines in the distal end of each tibia were counted. Lines extending across at least one quarter of the medullary cavity were included. This follows the method established in the analysis of the Khok Phanom Di tibiae by Tayles (1992).

**Table 4.3:** Prevalence of Harris lines on the distal tibiae of adult males and females for the prehistoric Thai samples

| Site | Female | | | | Males | | | |
|---|---|---|---|---|---|---|---|---|
| | N | % with lines | lines mean | s.d. | N | % with lines | lines mean | s.d. |
| Khok Phanom Di* | 20 | 100 | 2.9 | 2.7 | 24 | 92 | 1.8 | 1.1 |
| Nong Nor | no data possible | | | | no data possible | | | |
| Ban Lum Khao | 4 | 25 | 4.0 | - | 6 | 17 | 1.0 | 1.4 |
| Ban Na Di* | 2 | 100 | 9.0 | 11.3 | 5 | 80 | 1.0 | 0.0 |

N = number of individuals with radiographed tibiae; *(Tayles 1992).

> The degree of remineralization of the skeletons from Khok Phanom Di has compromised the definition of structures in the bones most highly remineralized, and highlighted the presence of lines in those least remineralized. (Tayles 1999: 119)

Counting the lines that extend across at least one quarter of the medullary cavity ensures the majority of lines will be included, despite the of level of mineralisation of the bone.

### Results

The number of males and females with Harris lines and the mean number of lines for each sex for each sample are presented in Table 4.3. No statistical tests were carried out to compare these results because of the insufficient sizes of some of the samples.

The data from the Ban Lum Khao sample represents only two individuals with Harris lines, one male and one female. The data for Ban Na Di females was represented by only two individuals in total, and one has a minimum of 17 lines, while the other has only one line. This has clearly distorted the mean numbers of lines. Four of the five male tibiae (80%) from Ban Na Di, that were able to be radiographed, had one line each. The Khok Phanom Di data was probably the most reliable, as reasonable sample sizes were able to be collected. Only two individuals, both males, did not have lines.

In general, the percentage of males with lines and their mean number of lines is slightly less than in females within each sample. The differences, however, are probably not significant. Ban Lum Khao adults show less Harris lines while both Khok Phanom Di and Ban Na Di samples have a very high number of lines. In consideration of the problems with the methodology in all assessments of Harris lines and the small sample sizes in some of the samples, these results will be only be used in conjunction with other evidence.

## Enamel Hypoplasia

### Introduction

Enamel hypoplasia is a nonspecific indicator of adverse events that occurred during childhood. Such events include general malnutrition, specific nutritional deficiencies including rickets and scurvy, childhood illnesses such as measles, smallpox, diarrhoea, and other bacterial and viral infections (Hillson 1996). So many different diseases and nutritional deficiencies can lead to enamel hypoplasia that it is now considered a generally non-specific phenomenon (Hillson 1996).

Enamel hypoplasia is a deficiency of enamel thickness which occurs due to a systemic disturbance during the formation of enamel. The first stage of enamel calcification is the formation of the matrix by ameloblasts (enamel forming cells) and if this process is disturbed, a reduction in enamel thickness results (Huss-Ashmore *et al.* 1982). Most hypoplastic defects are seen macroscopically as a transverse line or pits across the external surface of the tooth crown (Hillson 1996). Most studies record only linear enamel hypoplasia (LEH), with pit-type defects (non-linear enamel hypoplasia) less often recorded.

Amelogenesis follows a fixed plan in the timing of the laying down of enamel. The location of a hypoplasia reflects the "...relative completeness of crown development at the time of the stress" (Goodman and Armelagos 1988: 936), therefore the age at which a hypoplasia developed, can be determined (Goodman *et al.* 1984b). There are studies that present these details, for example Goodman *et al.* (1980). The age at which the hypoplasia occurred is calculated by measuring the distance from the cemento-enamel junction to the defect and converting this to the biological age at which the crown had developed to this stage (Lewis and Roberts 1997). Observations in the deciduous dentition coincide with events during the last five months *in utero* through to one year of age (Goodman *et al.* 1984b), while observations in the permanent teeth of adults correlate to the period when the permanent dentition was forming during about the first seven years of life (Lewis and Roberts 1997). This allows a study of childhood to be carried out in a skeletal sample of adults (Goodman and Armelagos 1988), as subjects are those individuals who survived childhood.

"Early childhood is a period of increased risk for individuals" (Wood 1983 cited in Goodman and Armelagos 1988: 936). Some studies have observed peak periods of stress through the analysis of enamel hypoplasia. According to Lewis and Roberts (1997), most indicate a peak period of stress between 2-4 years of age. This correlates to the time of weaning in developing societies, but other environmental factors should not be ruled out.

Other studies (Goodman *et al.* 1980; Larsen 1995; Ribot and Roberts 1996) have concluded that the presence of enamel hypoplasias spaced by one year is an indication of an annual cycle of environmental stress. This could be related to annual food shortages and/or a high number of contagious

**Table 4.4:** Prevalence of linear enamel hypoplasia in the permanent dentition of the subadults from the prehistoric Thai samples

| Site | Tooth | |
|---|---|---|
| | A/N | % |
| Khok Phanom Di* | 9/186 | 4.8 |
| Nong Nor | 9/63 | 14.3 |
| Ban Lum Khao | 30/106 | 28.3 |
| Ban Na Di | 22/58 | 37.9 |
| Chi$^2$ | **47.542** | |
| p-value | **<0.01** | |

A= affected, N= total number of teeth observed, **Bold** indicates statistical significance among samples at 1% level. *(Tayles 1992).

**Table 4.5:** Fisher's Exact p-values for significance of the enamel hypoplasia prevalence among the subadult permanent teeth samples

| | Nong Nor | Ban Lum Khao | Ban Na Di |
|---|---|---|---|
| Khok Phanom Di | 0.02 | **<0.01** | **<0.01** |
| Nong Nor | - | 0.04 | **<0.01** |
| Ban Lum Khao | - | - | 0.22 |

**Bold** indicates statistical significance at 1% level

infections such as viral infections like influenza and colds that can be seasonal. In comparing the mean age at death with the presence or absence of enamel hypoplasia in individuals from the Dickson Mound prehistoric population, Goodman and Armelagos (1988) found that those with two or more enamel hypoplasia had a lower mean longevity than those with one or no hypoplasia. "Low status during childhood may promote disease and undernutrition, which leaves individuals less able to rally from future insults" (Goodman and Armelagos 1988: 942).

Identifying the age at which a lesion occurs is not straightforward and a recent study on subadults has outlined methodological problems (Lewis and Roberts 1997). Lewis and Roberts (1997) caution against using the chart provided by Goodman *et al.* (1980) as it is partly based on modern children. This assumes that tooth development rates in diseased prehistoric children are similar to today's modern healthy populations. The Goodman *et al.* (1980) chart also assumes that the development rate of the dentition is consistent between individuals and populations, which may not be a reality (Lewis and Roberts 1997). Kamalanathan *et al.* (1960b) exemplified this fact through their study of the eruption of teeth in modern Thai children who differ in their timing compared with other populations.

A recent study by Boldsen (1998) sought to determine how episodes of illness during childhood (evident from enamel hypoplasia) affected the growth and stature in a medieval village skeletal population. No association was found, but this may not be the case in all populations and is an interesting approach.

It would be logical to think that there would be an association between Harris lines and enamel hypoplasia as both can occur as a result of similar adverse events during childhood. A study by McHenry and Schultz (1976) investigated this but found no association in prehistoric California Indians. Enamel hypoplasia is deemed to be the more accurate indicator of growth disturbance during childhood than Harris lines for a number of reasons. For example, the analysis of teeth has advantages over bone because it does not remodel like bone and therefore provides a 'memory' of earlier events (Goodman *et al.* 1984b). Teeth are also more likely to survive burial due to the hardness of enamel.

## Methods

Enamel hypoplasia was identified macroscopically as a transverse line or groove across the tooth enamel. Only linear defects were recorded for the Ban Lum Khao, Nong Nor, and Ban Na Di samples. This type of defect is considered to be the easiest to identify and, since two observers were involved in recording this data, linear defects are the least likely to show a significant interobserver error. The Khok Phanom Di teeth were observed for linear and non-linear defects by Tayles (1992). However, only the linear data are used. The measure of interobserver error between the two researchers for Khok Phanom Di is provided in Appendix A.

Every deciduous and permanent tooth was assessed for linear enamel hypoplasia and the data are presented in the following three categories: deciduous, subadult (<15 years) permanent, and adult (≥15 years) permanent teeth. The methodological approach used in this comparative study considers only the presence and absence of one or more hypoplastic defects per tooth. The age of occurrence has not been included for reasons mentioned above.

## Results

Subadults

No deciduous teeth in any of the four populations had linear enamel hypoplasia. However, 7% of deciduous teeth had non-linear hypoplasia in the subadults of Khok Phanom Di (Tayles 1999). Results of the permanent teeth belonging to subadults (<15 years) are given in Table 4.4. Statistical analyses, reasons for their use and the set level of statistical significance are described in Appendix B.

**Table 4.6:** Prevalence of linear enamel hypoplasia, by sex, in the adult teeth for the prehistoric Thai samples

| | Khok Phanom Di* | | Nong Nor | | Ban Lum Khao | | Ban Na Di | | Comparison among samples | |
|---|---|---|---|---|---|---|---|---|---|---|
| | A/N | % | A/N | % | A/N | % | A/N | % | Chi $^2$ | p-value |
| Total | 265/1086 | 24.4 | 101/843 | 12.0 | 94/799 | 11.8 | 70/473 | 14.8 | **75.5** | **<0.01** |
| Female | 145/568 | 25.5 | 47/429 | 11.0 | 56/452 | 12.4 | 17/135 | 12.6 | **49.6** | **<0.01** |
| Male | 120/518 | 23.2 | 54/414 | 13.0 | 38/347 | 11.0 | 53/338 | 15.7 | **28.4** | **<0.01** |
| *Male vs Female Fisher's Exact p-values* | | | | | | | | | | |
| | 0.40 | | 0.40 | | 0.58 | | 0.47 | | | |

A= affected, N= total number of teeth observed; **Bold** indicates statistical significance among the samples at 1% level. *(Tayles 1992).

**Table 4.7:** Fisher's Exact p-values for the significance of the linear enamel hypoplasia prevalences in the adult teeth among the prehistoric Thai samples

| *Total* | Nong Nor | Ban Lum Khao | Ban Na Di |
|---|---|---|---|
| Khok Phanom Di | **<0.01** | **<0.01** | **<0.01** |
| Nong Nor | - | 0.94 | 0.15 |
| Ban Lum Khao | - | - | 0.12 |
| *Females* | | | |
| Khok Phanom Di | **<0.01** | **<0.01** | **<0.01** |
| Nong Nor | - | 0.43 | 0.64 |
| Ban Lum Khao | - | - | >0.99 |
| *Males* | | | |
| Khok Phanom Di | **<0.01** | **<0.01** | **<0.01** |
| Nong Nor | - | 0.43 | 0.35 |
| Ban Lum Khao | - | - | 0.07 |

**Bold** indicates statistical significance at 1% level

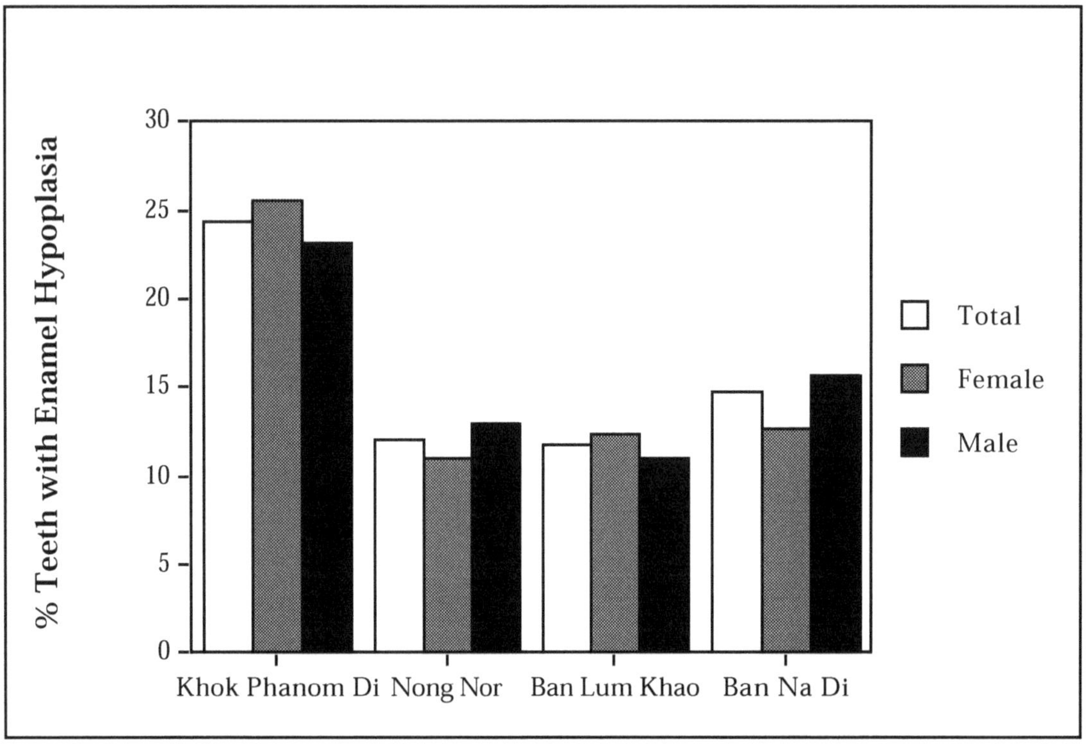


**Figure 4.4:** Percentage of teeth with linear enamel hypoplasia for each prehistoric Thai sample

The Ban Na Di subadults have the highest percentage of enamel hypoplasia in the permanent teeth (37.9%) compared with the other three samples. This difference is statistically significant compared with both the Khok Phanom Di and Nong Nor samples of subadults, as the Fisher's Exact p-values indicate in Table 4.5. Ban Lum Khao subadults also have a high percentage of enamel hypoplasia (28.3%) and this level is statistically significant compared with the Khok Phanom Di subadults only. Overall these results show a trend of increasing enamel hypoplasia in subadults with time.

Adults

The prevalence of linear enamel hypoplasia in the permanent teeth of the adult samples are presented in Table 4.6 and Figure 4.4.

The Khok Phanom Di people have the highest levels of enamel hypoplasia overall and in both the female and male samples. This is statistically significant compared with all other samples as the Fisher's Exact p-values indicate in Table 4.7.

The other three samples, Nong Nor, Ban Lum Khao, and Ban Na Di, all have very similar percentages of teeth affected with enamel hypoplasia. These results differ by only 3% in the total samples, 1.6% among the female samples and 4.7% in the male samples. No statistically significant differences are evident among these three samples.

All samples show very similar percentages of enamel hypoplasia between males and females within each sample, with no statistically significant sex differences as the Fisher's Exact p-values indicate in Table 4.7 (1% level of significance).

# Adult Health

This chapter has so far focused on presenting the evidence of health during childhood from subadult and adult skeletal remains. This next section presents the evidence for health during adulthood. Once adulthood is reached bone length will not significantly change as linear growth has ceased. However endosteal and periosteal bone growth can continue and can consequently change bone mass throughout life (Plato and Norris 1980), as new bone is laid down and old bone resorbed.

## Cortical Bone Mass

### Introduction

When linear bone growth and development ceases, the cellular activity of bone does not. The mature skeleton undergoes constant remodelling through a balanced process of bone resorption and formation. These processes are essential for maintaining the biomechanical competence of bone tissue and maintaining calcium homeostasis in the body. An imbalance of these processes may lead to a clinical problems, such as osteoporosis or osteopenia, where bone resorption has outweighed the rate of bone formation in the remodelling process. Osteoporosis is a reduction in bone mass that is usually irreversible and may lead to an increased risk of fracture (Weaver 1998). Osteopenia is also defined as less than normal bone mass, but is differentiated from osteoporosis because it can potentially be reversible and thus does not necessarily have as important clinical consequences for an individual as osteoporosis (Weaver 1998).

The structure of normal mature bone consists of two morphological types. Dense, cortical or compact bone forms the outer surface enclosing a meshwork of trabecular bone termed spongy or cancellous bone. The cortical bone has a subperiosteal (outer) surface and an endosteal (inner) surface, the latter being continuous with trabecular bone. Trabecular bone appears as honeycomb with large spaces (marrow space) and there is a paucity of bone in comparison to the cortex (Resnick *et al.* 1995a). The cells that are involved in the maintenance of both types of bone are osteoblasts (bone-forming cells) and osteoclasts (bone resorbing cells). The other major cell type is the osteocyte, the mature form of the osteoblast which is concerned with the proper maintenance of the bone matrix (Resnick *et al.* 1995a).

During linear growth, bone mass increases, but after childhood and adolescence, when linear growth ceases, up to 90% of peak bone mass is reached (Parfitt 1988; Francis 1990). The remaining 10% is achieved during the process of bone consolidation and is completed at about 35-40 years of age for cortical bone (Parfitt 1988). A larger person has a higher bone mass, therefore males and females have different peak bone mass levels, about 25-30% higher in men than women (Parfitt 1988; Resnick and Niwayama 1995c). Once peak bone mass is reached, a gradual loss of skeletal mass occurs. This begins earlier in women than in men, usually during the fourth decade, and at a rate within the range of 0.75-1.0% loss of peak adult bone mass per year. This is further increased to 2.0-3.0% loss for about five years after menopause due to the reduced level of estrogen. Estrogen plays a critical role in bone maintenance. In men, bone is lost at a rate of about 0.3-0.4% per year, and doesn't usually begin until the fifth to sixth decade (Parfitt 1988; Resnick and Niwayama 1995c).

Changes in the density of cortical bone are often surface-specific and can be assessed differentially; the measurement of the total width of the bone accounts for changes to the subperiosteal surface, while changes in the width of the medullary cavity relate to changes at the endosteal surface (Garn 1970). In general, the subperiosteal surface is primarily involved in bone formation while the endosteal surface is mostly involved in bone resorption. These processes are under the humoral control of the parathyroid hormone, calcitonin, and 1,25-dihydroxyvitamin D (Resnick *et al.* 1995a).

Bone mass is a valuable indicator of how well bone density has been maintained throughout life, both as a result of nutritional status and as an index of the strength of bone. Between 80 and 90% of the strength of bone can be explained by its bone mass or density (Martin *et al.* 1985; Aloia 1994; Resnick and Niwayama 1995c; Stini 1995).

Therefore a loss of bone equates to a loss of bone strength. This can leave an individual with an increased risk of fracture and osteoporosis (Mazess 1982; Stini 1995). Osteoporosis is a common clinical problem in the modern world, particularly in women. The high percentage of older aged people present today is thought to be one reason for the high amount of osteoporosis. Prehistoric populations did not have the same longevity as is experienced today. Therefore, prehistoric people are assumed to be less likely to have suffered from severe osteoporosis solely due to aging.

Bone mass is influenced by a number of other factors, in addition to age, and at any given time during life the level of bone mass represents the interaction between "...(a) the genetic component of maximal possible bone mass for an individual and (b) all the environmental factors that influence bone gain and loss" (Aloia 1994: 2). The genetic factors include sex, as discussed previously, and environmental factors include diet, physical activity, hormonal factors, and socioeconomic factors, but it can be difficult to differentiate between these influences (Pfeiffer and King 1983; Parfitt 1988; Francis 1990). Both nutrition and biomechanical forces are important in skeletal maintenance and therefore should also be considered when analysing bone thickness and bone distribution (Larsen 1995).

It is assumed that prehistoric populations were involved in physical activity from an early age, as they relied on their own physical labour for subsistence. This level of physical activity can help lead to an enhanced development of bone (Pfeiffer and Lazenby 1994). Therefore, evidence of reduced bone mass in prehistoric skeletal samples would more likely be related to factors such as diet rather than age or lack of biomechanical stress.

Calcium is essential in order to maintain physiological homeostasis. The intake of calcium during the growing years has been shown to be positively correlated with peak bone density (Stini 1995). More general nutritional stress results in an increase in bone resorption and a decrease in formation, resulting in a net loss of bone (Martin *et al.* 1985). Measures of bone mass are deemed useful indicators of dietary and nutritional stress in prehistoric populations and age- and sex-specific loss may provide insight into biases within or among populations (Buikstra and Cook 1980; Huss-Ashmore *et al.* 1982; Martin *et al.* 1985).

Cortical bone, being more dense, preserves better than trabecular bone after burial and therefore provides the most reliable measure of mass. Most commonly, radiographs of the femur and the second metacarpal are used in determining bone mass in skeletal populations. Some studies state that the measurement of the metacarpal may be less accurate than that of the femoral cortex because the actual dimensions of the metacarpal are so much smaller (Barnett and Nordin 1960). However, a complete and intact metacarpal is more likely to survive than an intact femur. A complete bone is needed to ensure that there is consistency in the location of the measurement between individuals. Sample size is also important for statistical analyses. The cylindrical model which bone mass estimation is based on, is thought to be best approximated in the metacarpal compared to the femur. Intraobserver error has also be found to be very low in the measurement of this bone (Garn 1970). It has been shown that it is not necessary to measure more than one metacarpal on each hand. The second metacarpal is nearly as highly representative of bone area as a larger number of metacarpals (Garn *et al.* 1991).

Barnett and Nordin's (1960) method is most commonly used to estimate bone mass in skeletal remains. This method involves the measurement of the total width (T) and medullary cavity (M) width on the radiograph of each metacarpal. The equation $C = T - M$ gives a measure of the cortical thickness (C). The presence of size differences among individuals will affect the cortical thickness, therefore bone mass must be calculated as an index to eliminate the size factor and the equation $CI = (C/T) \times 100$ gives the cortical index (CI) as a percentage (also known as 'Nordin's score'). This is a two-dimensional measurement and others have suggested a three-dimensional analysis of the actual cortical area (percent cortical area PCA) which may be more useful. However, both PCA and Nordin's score are acceptable (Garn 1970).

In modern clinical situations more sophisticated equipment than radiography is now used to calculate bone density. This includes dual X-ray analysis (DEXA), computed tomography, and single- and dual-photon absorptiometry. Some of these techniques are not appropriate to use on prehistoric bone due to problems such as soil embedded within the medullary cavity (Pfeiffer and Lazenby 1994) that interferes with the x-rays. Financial constraints and the lack of availability of machines were also limiting factors for this study.

### Methods

Cortical bone mass was measured in each population from antero-posterior radiographs of the second metacarpal following the method of Barnett and Nordin (1960) and Garn (1970). Measurements to determine the cortical index (CI) of each metacarpal were carried out using digital calipers. The following measurements were taken:

1. **Maximum length of the second metacarpal**. This measurement was then halved so the midpoint along the shaft could be located. The two following measurements were subsequently taken at this point.
2. **Total bone diameter** (T) (included the medullary cavity and cortical bone).
3. **Medullary cavity width** (M).

The equation $C = T-M$ was used to determine the thickness of the cortical bone (C) which is then converted into a percentage using $CI = (C/T)*100$ to determine the cortical index. Both left and right metacarpals were used. This method follows that used by Tayles (1992) on the population of Khok Phanom Di. No measurements were possible from the skeletons of Nong Nor due to poor preservation.

### Results

The results for each site able to be analysed are presented in Tables 4.8 (females) and Table 4.9 (males) and illustrated in Figure 4.5. There were no significant differences between the means of each population (by sex) as indicated by the ANOVA p-values presented below each table. There were also no significant differences between the sexes within each

**Table 4.8:** Summary statistics of female adult bone mass (cortical index) for the prehistoric Thai samples

| Sample | n# | Mean (CI) | S.D. | Range |
|---|---|---|---|---|
| Khok Phanom Di* | 45 | 59.2 | 10.2 | 35.7 - 75.0 (39.3) |
| Ban Lum Khao | 14 | 60.2 | 6.5 | 51.0 - 71.7 (20.7) |
| Ban Na Di | 5 | 61.4 | 8.3 | 47.8 - 68.5 (20.7) |

# Number of bones radiographed; *(Tayles 1992).
Comparison of populations: ANOVA p-value = 0.85, n.s.

**Table 4.9:** Summary statistics of male adult bone mass (cortical index) for the prehistoric Thai samples

| Sample | n# | Mean (CI) | S.D. | Range |
|---|---|---|---|---|
| Khok Phanom Di* | 41 | 63.0 | 7.9 | 43.4 - 76.4 (33.0) |
| Ban Lum Khao | 16 | 65.3 | 8.6 | 47.0 - 80.9 (33.9) |
| Ban Na Di | 11 | 62.6 | 7.8 | 51.9 - 77.3 (25.4) |

# Number of bones radiographed; *(Tayles 1992).
Comparison of populations: ANOVA p-value = 0.57, n.s.

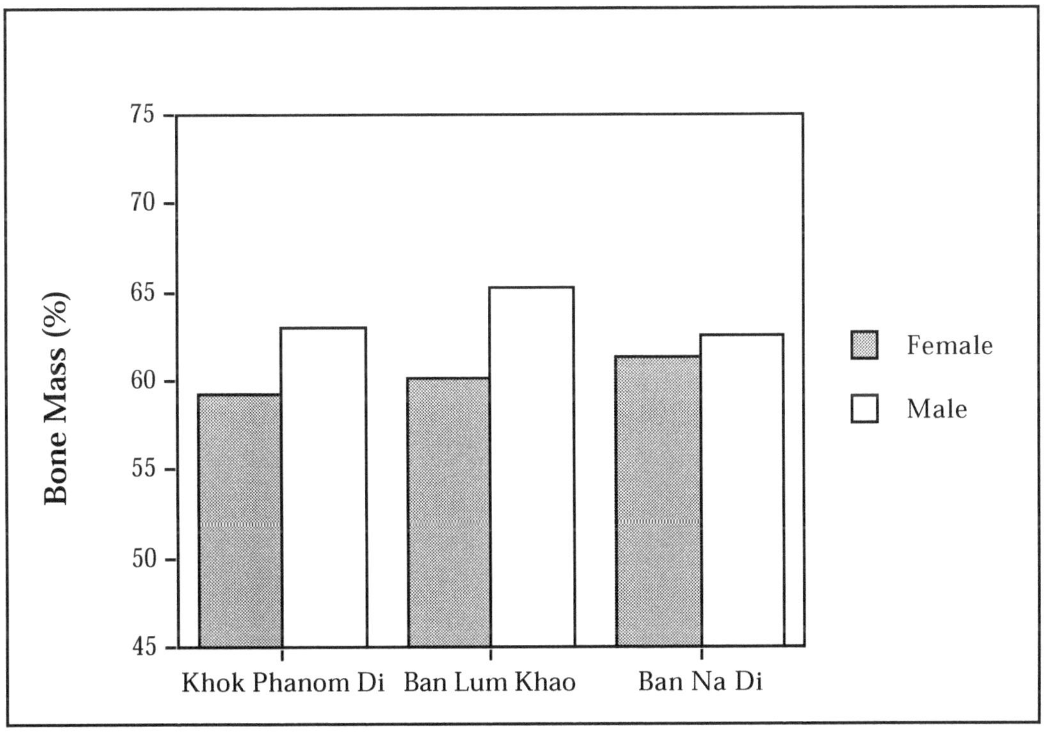


**Figure 4.5:** Mean adult bone mass percentage (cortical index) for the prehistoric Thai samples

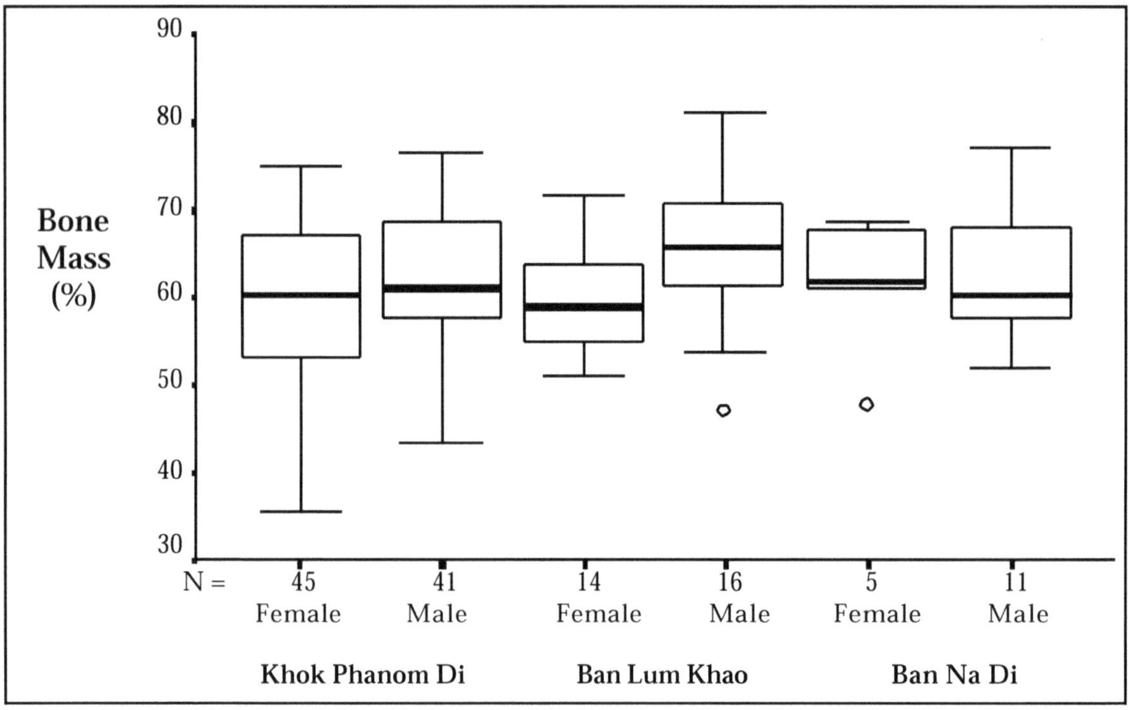


**Figure 4.6:** Comparison of bone mass among the prehistoric Thai samples. Boxplot indicates median, range, quartiles and outliers of the cortical index measurements

**Table 4.10:** Adult bone mass (cortical index) by age, combined sexes

| Sample | < 30 years | | | | ≥ 30 years | | | |
|---|---|---|---|---|---|---|---|---|
| | n | Mean | s.d. | Range | n | Mean | s.d. | Range |
| Khok Phanom Di* | 39 | 63.6 | 8.0 | 49.4 - 76.3 | 47 | 58.9 | 9.9 | 35.7 - 76.4 |
| Ban Lum Khao | 15 | 60.6 | 7.4 | 51.0 - 75.8 | 15 | 65.2 | 8.3 | 47.0 - 80.9 |
| Ban Na Di# | 4 | 64.2 | 7.1 | 55.8 - 71.9 | 10 | 61.6 | 8.9 | 51.9 - 77.3 |

n= number of individuals with second metacarpal radiographed, s.d.=standard deviation; *(Tayles 1992), significant at the 2% level; # Two adults were unaged and could not be included in this analysis.

population (Khok Phanom Di: t=-1.92, p=0.06; Ban Lum Khao: t=1.82, p=0.10; Ban Na Di: t=0.26, p=0.80).The difference that was apparent was the range of bone mass measurements within and between each population. Within Khok Phanom Di and Ban Na Di, males and females had similar ranges, but the males of Ban Lum Khao showed a much wider variation in the cortical index than females, as they did in measurements of stature. This is illustrated in Figure 4.6.

Bone mass decreases with age in modern populations and this age related loss can also be investigated in prehistoric populations. If there is no apparent change with age and cortical index values are low, then there may be other environmental reasons for this. Results divided by age for each sample are presented in Table 4.10. Each sample was divided into two age groups, those aged less than 30 years and those aged equal to or greater than 30 years. This division was used to provide as near equal sample sizes as possible between the two age groups. Restricting the analysis to two adult age groups was necessary as further divisions would have produced very small sample sizes in each division.

It is evident that only the samples from Khok Phanom Di and Ban Na Di showed a decrease in bone mass maintenance with age. This decrease was statistically significant at the 2% level in the Khok Phanom Di sample only (t-value = -4.70, p-value = 0.019). The Ban Lum Khao sample showed an increase in bone mass with age, although the difference between the two age groups was not statistically significant.

# Results Summary

No substantial differences were identified among the samples in their growth rates. The other main points regarding childhood growth and growth disturbances of the four samples are summarised in Table 4.11. The terms low, moderate, and high are used as relative terms only in order to compare these samples. Bone mass is considered later as it is an indicator of adult health.

Subadult enamel hypoplasia was low in the southeastern samples, Khok Phanom Di and Nong Nor. Significantly

**Table 4.11:** Relative measures of childhood growth and growth disturbances in prehistoric Thailand

| Sample | Subadult EH | Adult EH | Harris lines | Adult Stature | Stature variation |
|---|---|---|---|---|---|
| Khok Phanom Di | low - moderate | high | high | low (short) | wide in females* |
| Nong Nor | low | moderate | - | high (tall) | wide in males |
| Ban Lum Khao | moderate - high | moderate | low | low - moderate | wide in males |
| Ban Na Di | moderate - high | moderate | high | high (tall) | none |

* explained by the presence of one woman with a very short stature; EH = enamel hypoplasia

higher levels were seen in both the northeastern samples, Ban Lum Khao and Ban Na Di. Enamel hypoplasia in the adult skeletons followed a different pattern than in the subadults with Khok Phanom Di showing significantly higher prevalences compared to the other three samples. Harris lines were very frequently observed in the Khok Phanom Di and Ban Na Di adults, but were minimal in the Ban Lum Khao adults. However, small sizes of some samples may have been a factor in these results.

Stature was short (low) in Khok Phanom Di, particularly in the males whose mean was significantly lower than both Nong Nor and Ban Na Di. These two latter samples stood out as being taller (moderate - high). A wide range of variation was evident in the males from both Ban Lum Khao and Nong Nor.

Bone mass results, as an indication of adult health, did not show any marked differences among the three samples that could be assessed (excludes Nong Nor). A range of variation within males and females from Khok Phanom Di and for males from Ban Lum Khao was evident.

# Discussion

### Childhood Health

The discussion of each of the parameters analysed has been left until the end of this chapter in order for the information to be integrated to provide a more complete picture of growth as a measure of childhood health in each sample. Mortality data presented in Chapter 3 are also of relevance here. Isolated comments on individual parameters are of much less significance when the aim is to present a comparative picture of health in prehistoric Thailand.

There are a number of factors to consider in the interpretation of health indicators that are of relevance here. Within most populations there is a *hidden heterogeneity* (Wood *et al.* 1992) in the response to any one disease. One individual may have a different response to an infectious disease, for example, than the next individual. Some unknown proportion of individuals in a population will have an underlying frailty that may have arisen from a variety of factors. This could include their genetic composition, age, microenvironment (for example, diet), and the socioeconomic group they belong to within the community (Wood *et al.* 1992) which can influence their access to food, healthcare, and the work load required of them.

Within any one group there will probably be a proportion of frail individuals, strong individuals (Wood *et al.* 1992) and those in between of medium frailty (after Lukacs comment on Wood *et al.* 1992). It cannot be assumed that hidden heterogeneity is normally distributed within populations (Eisenberg comment on Wood *et al.* 1992). However, the relative proportions of frail, medium frail, and strong children present in a prehistoric skeletal population may be inferred from the measures of growth and disturbances of growth presented here.

The four prehistoric skeletal samples in question show a range of prevalences of Harris lines, enamel hypoplasia in adults and subadults, as well as a range of mean adult statures and different mortality structures. These results can help indicate the general state of health of the subadults from each population. The results may possibly identify the proportions of frail and strong children, providing an indication of their general quality of life relative to the other populations. Of course, some subadult deaths will be accidental, but it possible that malnourished and/or frail individuals are more likely to not only have accidents, but also to suffer more significantly from them.

Other limitations to the interpretation of skeletal remains have been presented in Chapter 1. It also needs to remembered that the majority of lesions visible in a skeleton represent those of a chronic nature, that the individual has not succumbed to, but has survived. However, matters may be further complicated by the circular and synergistic relationship between nutrition and infection. Individuals who are malnourished have a depressed immune response to pathogens which increases their likelihood of dying before the disease becomes chronic, leaving no signs on the skeleton (Ortner 1998).

For any results there is likely to be an array of possible explanations and due consideration to each possibility is recommended by Wood *et al.* (1992). In order to help overcome the osteological paradox all possible evidence should be considered. Here the influence of hidden heterogeneity and mortality are taken into account in the interpretation of the skeletal evidence in order to present the most likely picture of childhood health in prehistoric Thailand.

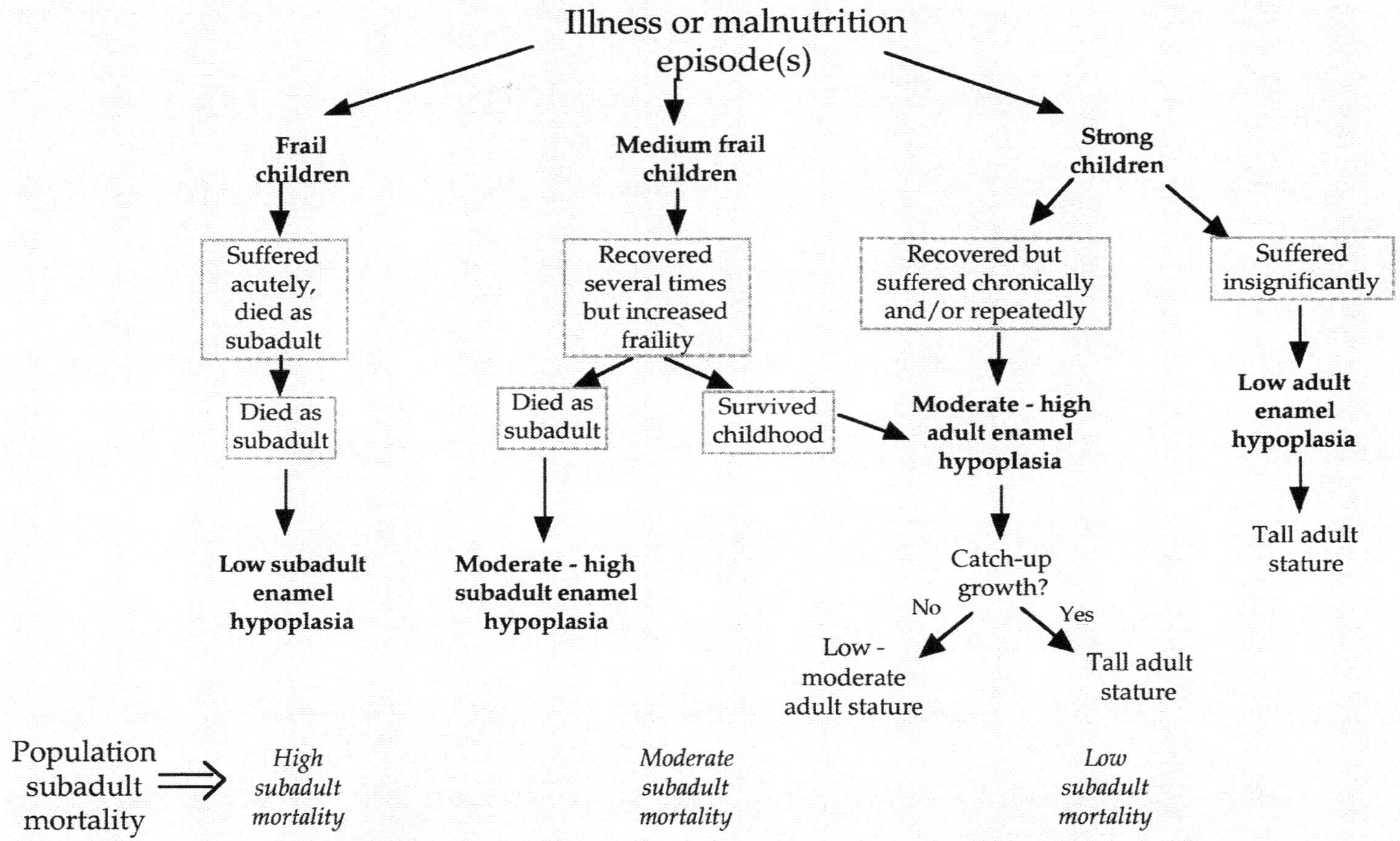


**Figure 4.7:** Model for the significance of the effects of episodes of illness and/or malnutrition on children - the effects on enamel hypoplasia levels in subadults and adults, subadult mortality, and adult stature

The model (Figure 4.7) has been developed based on issues discussed by Wood *et al.* (1992). It is used here in order to combine and integrate all available evidence concerning subadult growth. The likely outcome of an illness or episode of malnutrition on each type of child, based on their ability to cope with such 'stressors', is presented. Then, the Thai prehistoric samples will be analysed to see how each fits with this theory and to estimate the proportions of frail and strong children within each sample. There are additional influences on child health, such as the frequency of exposure to disease, especially infectious pathogens that are rampant in tropical regions. The diet is also a major influence on child health. These issues and how they may have differed among the prehistoric sites are addressed in Chapter 8.

**1. Frail children**. Some children may be more susceptible to disease and suffer more due to a variety of factors. It may be from chronic or seasonal undernutrition, genetic makeup, or social factors. When a frail child is ill, they are most likely to suffer acutely and severely and are more likely to die as a consequence. In this case the length of time exposed to the problem will not be long enough for a disturbance of growth to occur and it would be expected that these subadult skeletons would have a low prevalence of enamel hypoplasia.

**2. Strong children.** Strong children exposed to the same illness or period of malnutrition as a frail child may not suffer at all. Their immune systems were strong enough to fight off the illness with no disturbance to their health and growth. Few of these children will die as subadults (although accidental death may occur). These children would show low enamel hypoplasia prevalences as adults and stature that is tall relative to other adults, as they were able to reach their genetic potential, uninhibited by significant growth disturbances during childhood. Other strong children may suffer during childhood but will recover sufficiently after each insult to reach adulthood. These children would show moderate to high prevalences of enamel hypoplasia as adults. Their stature as adults would depend on their ability to compensate for a growth arrest with a period of accelerated growth post-illness. If the accelerated growth fully compensated for the reduced growing time (nutrition being adequate) then they may reach their genetically determined stature. Those individuals who only partly compensated for their missed growth, perhaps due to inadequate nutrition or the timing of the problem (no catch-up growth possible), will present with shorter statures as adults.

**3. Medium frail children.** The division between a child's susceptibility to illness and malnutrition is not clear cut, therefore there would also be children of an intermediate susceptibility. These children, when exposed to the same illness or episode of malnutrition, are likely to rally several times against the illness. Over time, the effects of the nutrition-infection synergism may increase their frailty and ultimately see some of them die as subadults from other causes. Their skeletons (subadults) would therefore show moderate to high levels of enamel hypoplasia indicating the recovery from some disturbances. Those subadults from this group that managed to recover and reach adulthood (but with a subsequent increased frailty) would exhibit a reduced short adult stature. This is because they had suffered severe or chronic growth disturbances during their childhood illnesses and the period of catch-up growth did not allow them to fully compensate for the growth arrest because of their increased frailty even under adequate nutrition. When 'stress' is chronic the end result is often reduced size of the adult (Stini 1985).

As mentioned previously, the range of statures within a sex, and a population, is important. A wide range of statures indicates some individuals reached their growth potential while others did not (assuming genetic homeogeneity in this respect; otherwise it may indicate a wide gene pool). Those with short statures may represent those of medium frailty as children who survived through to adulthood after some periods of growth disturbance. Those within the taller range of statures may represent the stronger children. Steckel (1995 cited in Larsen 1997) indicates stature is a product of nutritional adequacy and, to a lesser extent, disease history. Genetics also have an influence. The growth of an individual is linked strongly with the quality of their diet, particularly during childhood and adolescence. A division within a group can be based on social factors such as the establishment of a hierarchy of wealth or rank, with the more wealthy individuals receiving a better diet for example.

Environmental stressors do appear do have at least a slightly greater effect on male growth, development and survival than on females (Stini 1985).

> The range of variation in continuous traits is more tightly clustered around the mean in females than in males. Virtually every measurable value shows a higher frequency of extreme values occurring in males. (Stini 1985: 221-2)

Other social factors may also come into play in determining who is exposed to certain pathogens. Those that are required to work with the soil are more likely to contract infectious diseases such as some fungi that reside in the soil (Ortner 1998). While the effects of fungal disease are not usually evident skeletally, the period of illness may disturb growth and present as a non-specific enamel hypoplasia, Harris line, or shorter stature. In this way a division of labour may influence the exposure to disease.

Using the above model (Figure 4.7) regarding the differential susceptibility of children, the following inferences may be made regarding each prehistoric Thai sample. The aim is to estimate which cultural and natural environment provided for a healthier childhood. This will also be further discussed in Chapter 8.

Khok Phanom Di subadults suffered relatively low prevalences of enamel hypoplasia (Table 4.4), compared to the other samples, indicating many of the subadult skeletons died from acute conditions. Subadult mortality was also the highest in this sample at 55.8% (Table 3.10). Therefore we can assume there was a higher proportion of frail children in the Khok Phanom Di sample than perhaps in the others. There was some evidence in the Khok Phanom Di sample for non-linear enamel hypoplasia in the deciduous teeth representing disruptions to growth between five months in utero and one year (Tayles 1992). Adult dentitions exhibited high prevalences of enamel hypoplasia (Table 4.6) and tibiae showed almost universal Harris lines (Table 4.3). This indicates there must also have been some medium frail children within this sample who survived numerous insults but reached adulthood. The fact that, as adults, they did not attain a very tall mean stature in relation to the other samples analysed, suggests that the illness(es) or malnutrition they suffered as children did have an effect on their growth. Perhaps conditions were not so favourable, for example inadequate nutrition, to allow a period of accelerated catch-up growth post-illness.

> The relatively short mean stature of the males at Khok Phanom Di compared with their prehistoric counterparts suggests that they suffered sufficient ill health in childhood to have disrupted growth to the extent that they failed to achieve genetic potential...The females at Khok Phanom Di either did not suffer to the same extent, or were able to recover more quickly. (Tayles 1999: 114)

Some studies (for example Stini 1969), although not all (for example Stinson 1985), have suggested that male growth is more vulnerable to environmental stress than female growth. Therefore this low level of dimorphism at Khok Phanom Di could indicate some type of nutritional stress. The male sample showed retarded growth and a subsequent reduction in the level of sexual dimorphism.

The subadult skeletons from Nong Nor also exhibited a low enamel hypoplasia prevalence, in comparison with the northeastern samples and higher than Khok Phanom Di (Table 4.4). The adult skeletons showed moderate levels of enamel hypoplasia, similar to the northeastern samples and lower than Khok Phanom Di (Table 4.6). Subadult mortality was very low in the Nong Nor samples, at only 21.3% (Table 3.12), but this is possibly more likely to be a reflection of the poor preservation than an indication of actual low mortality. It appears therefore, there was a mix of medium frail and strong children at Nong Nor. The medium frail children died after several insults although a small proportion died acutely. The strong children in this sample went on to grow tall after some experienced growth disruptions. This may indicate a difference in nutrition during growth compared to Khok Phanom Di people; that is, better nutrition for children helped them recover better from growth disturbances at Nong Nor. The Nong Nor sample showed a wide variation in stature in adult males. This may indicate a potential inequality in the males as children.

Ban Lum Khao and Ban Na Di subadults present a different picture as their skeletons showed moderate to high levels of enamel hypoplasia (Table 4.4). This may indicate they suffered from more growth disturbances than the other two samples. Eventually, a certain proportion, probably those of medium frailty, died as subadults. The subadult mortality is not significantly different between these two samples with 46.4% of the Ban Lum Khao total sample dying as subadults and 35.9% of Ban Na Di (Fisher's Exact p-value = 0.18). The moderate level of enamel hypoplasia in adults in both of these northeastern samples (Table 4.6), indicates that some children, although they suffered, they did recover and survive through to adulthood. The Harris line results contradict the enamel hypoplasia results with Ban Na Di individuals showing almost universal Harris lines in the few that could be assessed (Table 4.3). However, the Ban Lum Khao people showed very few Harris lines. This contradiction may have arisen given the problems, such as remodelling, already discussed. Considering the similarity in childhood growth disturbances, the differences between Ban Lum Khao and Ban Na Di individuals is unlikely to be the result of nutritional or morbidity differences.

Another difference between these two samples arises in their attained adult stature. Ban Lum Khao males only reached a relatively low to moderate mean stature while Ban Na Di males appeared to have reached a taller mean adult stature (perhaps their growth potential) indicating good nutrition during growth enabled them to catch up at Ban Na Di but not at Ban Lum Khao. The Ban Lum Khao sample perhaps comprised more children of medium frailty while the Ban Na Di sample had more stronger children and better nutritional quality.

The variation in male statures at Ban Lum Khao compared to females (Table 4.1 and 4.2) may indicate an inequality within this sex that resulted in some males attaining only a short stature and others relatively taller. The inequality must have been established in childhood (or earlier if it is the result of genetic differences). Some males were of medium frailty, and while able to rally from illness during childhood, were made frailer by the experience and were not able to obtain their growth potential. The relationship between stature and grave wealth has been investigated at Ban Lum Khao but no pattern was obvious (O'Reilly 1999). More socially complex groups are more likely to have inequalities and it is perhaps surprising that Ban Na Di does not also show this bias given that it was the later group of the four samples and supposedly more socially complex.

### Adult Health

The assessment of bone mass provides another dimension to interpreting the health of these people, this time of the adults. While no significant differences were found among the samples in their mean male and female bone mass indices or between the sexes within each sample, the range of variation in some samples is interesting. As Figure 4.6 indicates, the range of bone mass indices in both the males and females from Khok Phanom Di and the males only from Ban Lum Khao is relatively wide in comparison to the others. There is the possibility that because the Khok Phanom Di skeletal material had the larger sample size available, it has produced a wider range of measurements.

As with stature, bone mass results also support the existence of an inequality within the males from Ban Lum Khao. While the stature range indicated a bias may have existed during childhood, the bone mass results indicate this inequality continued through into adulthood. The quality of the diet in childhood can also affect an individual in later life.

## Summary

Low prevalences of subadult enamel hypoplasia in the permanent teeth were evident in both the Khok Phanom Di and Nong Nor subadults relative to the northeastern samples. However, Khok Phanom Di subadults had a lower prevalence than Nong Nor subadults. This indicates the Khok Phanom Di and Nong Nor frail children were dying from more acute conditions, but there were also more children of medium frailty at Nong Nor who died after several episodes of growth disruption. The prevalence of subadult permanent teeth with enamel hypoplasia in the Ban Lum Khao and Ban Na Di samples were both higher than the southeastern samples. This indicates a significant number of chronic, growth disruptive problems in children of medium frailty that eventually resulted in death occurring during the 0-7 year age range in both these samples. In the Ban Lum Khao sample mortality was high until 9 years of age which may indicate the presence of more chronic subadult sufferers whose growth was affected by illness or undernutrition and who finally succumbed to death at a slightly older age compared with other samples.

Enamel hypoplasia in Khok Phanom Di adults was present at quite high levels, indicating some children had suffered chronically. Given that this sample also had the shortest mean statures, many children, although surviving the insults, did not recover sufficiently well to reach their genetic potential (assuming genetic homogeneity). They were therefore probably of medium frailty. The prevalence of adult enamel hypoplasia in all the other samples was at a relatively moderate level, less than that in the Khok Phanom Di sample. This indicates there were a proportion of medium frail and strong children who suffered chronically in the other samples. However, these samples differed in their ability to compensate for the periods of reduced growth because they differed in the proportions of medium frail and strong children.

In the Nong Nor sample, children who overcame their chronic conditions were likely to compensate for periods of reduced growth, as the tall mean adult stature estimates suggest. These children were strong. Male statures, however, had a wider range than females and it is possible that a subsample of males, with the shorter statures, were not able to fully compensate for reduced growth periods, perhaps as the result of a non-egalitarian society.

In contrast, the Ban Lum Khao adults had a shorter mean stature indicating the likelihood of them suffering chronically as children and not having adequate resources to compensate for the periods of reduced growth. Therefore, many of these children were of medium frailty. There was, as in the Nong Nor males, a wide range of male statures in the Ban Lum Khao sample compared with females. This may indicate groups of males from these Bronze Age samples did not catch up in growth sufficiently post illness, perhaps as a result of differential treatment in terms of the nutrition required for such accelerated growth because of the presence of a hierarchical social structure.

Ban Na Di adults showed similarities to the Nong Nor people in that they generally must have had access to adequate nutrition for good catch-up growth to occur after chronic problems during childhood were overcome, given the relatively tall stature they attained as adults. Many children were, therefore, strong. No inequality was evident in either males or females in their range of statures.

Despite these wide variations in childhood morbidity and mortality, all adult samples maintained comparatively similar, and adequate, bone mass indices. Ban Lum Khao males did have a wide range compared to the females that appears not to be related to any sampling bias.

However, there are a number of confounding factors which may have had an unknown affect on all of these results. For example, the poor preservation of the skeletal material from Nong Nor may have influenced their low subadult mortality in the Nong Nor sample. Additionally, it has been assumed that these samples are genetically similar. In reality, this is unknown and may have had some affect on the relative proportions of frail and strong children.

# 5 Joint Disease

## Introduction

### Osteoarthritis

Osteoarthritis is the most common form of joint degeneration in modern populations. It is also the most commonly observed skeletal pathology in prehistoric remains. Osteoarthritis is the final culmination of multifactorial processes acting on a joint; it is not a single entity (Hutton 1987; Hough and Sokoloff 1989) . Many of the factors necessary for the development and progression of osteoarthritis are still unknown (Hutton 1987) . These unknown factors make it difficult to precisely define the aetiology of osteoarthritis. The end result is identifiable, as defined by Rogers and Waldron (1995: 32), osteoarthritis "...is characterised by a focal loss of articular cartilage and subsequent bony reaction of the subchondral and marginal bone". This definition has a structural emphasis, only part of which can be observed in dry bone. Other joint components, such as the synovium and the joint capsule may also be affected (Doherty *et al.* 1998) . Functional consequences are also an important aspect of osteoarthritis and these will be considered in relation to prehistoric life.

Alternative terminology has been suggested for osteoarthritis. One such alternative is 'degenerative joint disease' which emphasises its degenerative and wear-and-tear nature. However, osteoarthritis is not solely a degenerative process, or an inevitable consequence of aging, but a dynamic and essentially reparative process (Doherty *et al.* 1998) . The term 'osteoarthritis' implies an inflammatory condition ('-itis') that is not always evident with osteoarthritis and this has led to the adoption of the term 'osteoarthrosis' by some. It is now recognised that inflammation, although not as accentuated as in rheumatoid arthritis, does occur in many osteoarthritis cases (Doherty *et al.* 1998) . Thus the use of the term osteoarthritis is justified here.

Osteoarthritis is restricted to synovial (freely moveable) joints. In a normal synovial joint, hyaline cartilage, averaging 2-3 mm in thickness, covers the articulating ends of the bones (Resnick and Niwayama 1981) . In the process of developing osteoarthritis, the articular cartilage becomes fibrillated and begins degenerating, becoming rough and eroded and may eventually flake off (Hough and Sokoloff 1989; Resnick and Niwayama 1995a) . Cartilage has no vascular supply, therefore the healing of this damaged tissue can be a slow process.

Further changes with osteoarthritis occur in the subchondral bony plate beneath the cartilage. This becomes sclerotic and small subchondral cysts may form in areas surrounding the sclerosis. When both articulating ends have lost sufficient cartilage the process of bone rubbing on bone creates one of the only pathognomonic signs of osteoarthritis, eburnation (Rogers and Waldron 1995) . Rothschild (1997) has recently debated that eburnation is a nonspecific finding in any severe arthritis. However, the eburnation seen in other types of joint diseases, particularly in erosive arthropathies, is considered to be a sign of overlying osteoarthritic changes (Rogers and Waldron 1995) .

Osteophytes, or bony extensions, may form during the development of osteoarthritis. These are usually seen around the marginal areas of the joint as bony outgrowths, although they are not exclusive to this site. The formation of osteophytes are a bony response to mechanical stimulation (Doherty *et al.* 1998) . They may represent an attempt to enlarge the joint surface and thereby increase stability (Resnick and Niwayama 1995a; Rogers and Waldron 1995; Doherty *et al.* 1998) .

The development of each type of lesion in the progression of osteoarthritis does not necessarily occur in the order presented here. Cartilage degeneration may be the first lesion to develop, but there is an alternative theory that suggests the subchondral bone is the site of initial damage (Radin 1973 cited in Hough and Sokoloff 1989; Resnick and Niwayama 1995a).

Radiological imaging is most commonly used in modern clinical situations to diagnose osteoarthritis, with the narrowing of the joint space as one positive identifier. The loss of joint space represents the loss of articular cartilage (Rogers and Waldron 1995) . This is not useful in skeletal analyses but other radiological signs may be observed in dry bone. These include the sclerosis of the subchondral bone and the formation of cysts below it (Rogers and Waldron 1995) as well as osteophytic extensions marginally. Bony contour changes may also be observed (Rogers and Waldron 1995) .

It is apparent that the diagnosis of osteoarthritis in prehistoric populations differs in some respects from that in modern populations. Non-pathological degenerative changes are able to be observed in skeletal material, but as they do not cause clinical symptoms, these changes are often not discussed in modern clinical literature. Pitting or porosity of the articular surface is one such sign. The relationship between porosity and osteoarthritis has been challenged recently by Rothschild (1997) and, while it may not be of clinical significance, it does represent an abnormality that is commonly present in any analysis of joint disease in skeletal populations. The key is to consider it in conjunction with one or more additional signs of joint degeneration before diagnosing osteoarthritis (Rogers and Waldron 1995) .

The pain and disability associated with osteoarthritis in clinical cases is variable among individuals. Generally, the cardinal symptom of osteoarthritis is pain and stiffness in the affected joint. However, only about 30-50% of people with radiographic evidence of osteoarthritis complain of pain at the relevant sites (Moskowitz 1989; Altman 1997) . In those that do suffer, pain initially only occurs after the first joint movement of the day. As the disease progresses, pain is present even at rest and can be severely disabling. In attempting to interpret the level of disability in prehistoric people suffering from osteoarthritis, it is best to consider the severe cases only.

Numerous aetiological factors have been implicated in the initiation of osteoarthritis. In general, osteoarthritis develops from an interaction between the biological processes within the tissues and the mechanical uses being made of them (Sokoloff 1980) . There are two main ways in which joint degeneration may begin:

> ...articular degeneration may result from either an abnormal concentration of force across a joint with normal articular cartilage or a normal concentration of force across an abnormal joint. (Resnick and Niwayama 1995a: 1264)

An abnormal force can arise from such factors as chronic mechanical stress, activity, obesity, and altered function due to trauma such as a poorly healed long bone fracture. On the other hand, abnormal cartilage may be due to genetic, traumatic, or aging factors, or disease.

Some clinical work has attempted to divide osteoarthritis aetiology into two groups. Primary osteoarthritis arises as an initial disease and frequently for no immediately apparent reason other than age (Resnick and Niwayama 1995a) . This is a fairly generalised form not affecting any particular pattern of joints. Secondary osteoarthritis develops in joints initially affected by trauma, mechanical derangement, or another pre-existing articular lesion (Swanson and de Groot-Swanson 985) . It is usually restricted to the joint or joints that are altered by the injury. This categorisation serves to emphasise the limited knowledge of the aetiology of osteoarthritis. In reality, primary osteoarthritis probably does not exist, the exact aetiology is just unknown (Resnick and Niwayama 1995a) . Aetiological factors are often numerous and complex and primary and secondary divisions are too simplistic (Fergusson 1987) .

Age is a major aetiological factor. An autopsy study has shown that degenerative changes can begin as early as the second decade and almost all persons will have some abnormalities in their weight bearing joints by age 40 (Moskowitz 1984) . Therefore, it could be said that biological aging and all the mechanical stresses that the human body is exposed to will eventually, given enough time, produce degenerative changes in at least some joints. However, these changes may not always be symptomatic or debilitating. Osteoarthritis should not be considered an invariable consequence of aging (Resnick and Niwayama 1995a) . Not only is age not the sole initiating or accelerating factor, but osteoarthritic changes in the cartilage indicate a metabolically active tissue compared with 'normal' aging cartilage which has reduced cellular activity (Fergusson 1987) . Osteoarthritis is a dynamic process combining both degradation and repair (Dieppe 1990) .

Epidemiological studies indicate the prevalence of osteoarthritis increases particularly between the ages of 40 and 50 years (Altman 1997) , and increases exponentially after the age of 50 or 60 years (Resnick and Niwayama 1995a) . By the age of 65, at least 50% of the population will have at least one joint affected (Doherty *et al.* 1998) .

There are joints which are more vulnerable to developing osteoarthritis. There are also common patterns of joints that are differentially affected in men and women. There is also some variation among ethnic groups. Overall, the hand, knee, and hip joints are some of the most vulnerable to osteoarthritis (Rogers and Waldron 1995; Doherty *et al.* 1998), but the prevalence can differ in men and women. While mild osteoarthritis affects males and females almost equally, in general women more frequently develop severe osteoarthritis (Spector and Campion 1989; van Saase *et al.* 1989; Doherty *et al.* 1998) . In women, the joints of the hands, particularly the distal interphalangeal joints, and the knee joints are most vulnerable (van Saase *et al.* 1989; Doherty *et al.* 1998) . Hand osteoarthritis in women is most often seen as part of the polyarticular osteoarthritis described below. Obesity is one of the main risk factors for developing knee osteoarthritis, especially in women (Altman 1997) .

The hip joint shows relatively lower prevalences of osteoarthritis in men and women (Hochberg *et al.* 1991) . However, osteoarthritis in this large, weight bearing joint can be the most debilitating compared with osteoarthritis in other joints. The zygapophyseal joints of the lumbar and cervical spine are also vulnerable to osteoarthritis (Hutton 1987) .

There is a generalised form of osteoarthritis that is considered a subset of osteoarthritis. However, its characteristics are so well defined it has been said that "...there is nothing generalized about generalized osteoarthritis..." (Smythe 1983 cited in Fergusson 1987: 439). It typically occurs in women aged 40-50 years of age, its onset peaking around the time of menopause (Doherty *et al.* 1998) . The distribution tends to be symmetrical (Fergusson 1987) and polyarticular. The first joints to be affected are the interphalangeal joints of the hands and the appearance of Heberden's and Bouchard's nodes commonly follow (Doherty *et al.* 1998) . These nodes are bony lateral deviations (osteophytes) of the distal and proximal interphalangeal joints respectively (Doherty *et al.* 1998) . They frequently have associated mucoid cysts (Flatt 1993) . The involvement of these joints of the hands are considered to be characteristic of a predisposition to generalised osteoarthritis. Osteoarthritis in the larger weight bearing joints, the hip and knee, and the spine frequently follows (Doherty *et al.* 1998) .

Worldwide there are very few contrasting osteoarthritis patterns in different ethnic groups in terms of the prevalence and joints affected (Doherty *et al.* 1998) . However, the hip joint does appear to be substantially less affected in black and Oriental populations than among whites. Malaysian and African populations also appear to have less polyarticular hand osteoarthritis (Doherty *et al.* 1998) .

Only one study specifically focusing on the epidemiology of osteoarthritis in modern Thailand has been found. Chaiamnuay *et al.* (1998) studied a rural agricultural community northeast of Bangkok with the aim of identifying the epidemiology of musculoskeletal pain. The subjects were aged from 15 years through to 90 years. The most prevalent diagnosis was osteoarthritis and spondylosis, with that of the knee, hand, and lumbar and cervical joints most commonly affected (Chaiamnuay *et al.* 1998) . This study also concluded that rheumatic disorders, in general, causes considerable pain and distress in these individuals (Chaiamnuay *et al.* 1998) .

Van Saase *et al.* (1989) compared the prevalence of osteoarthritis in the Netherlands with a worldwide selection of populations previously reported. While no population was spared, differences did occur in the level of osteoarthritis. However, interobserver error in interpreting radiographs could not be ruled out as contributing to the differences as "...the freedom of interpretation of the standardising atlas is rather large" (van Saase *et al.* 1989: 279). Determining differences in risk factors or genetic differences throughout the world is also proving difficult.

The patterns presented represent the situation in modern times. This does not necessarily correlate with osteoarthritis patterns in historic or prehistoric periods. For example, a number of studies measuring osteoarthritis prevalences are considered in Rogers (1998) , mostly from Europe. The majority of the studies reviewed used the same operational definition of osteoarthritis. These studies show a predominance of spinal osteoarthritis and a lack of ankle osteoarthritis which is a common pattern found in studies of modern groups. However, during the premedieval period osteoarthritis was more common in the hip than the knee, and not until the postmedieval period did this reverse to approach the modern trend of higher osteoarthritis in the knee joint. Osteoarthritis in the shoulder joints was also quite common (Rogers 1998) . Each skeletal group mentioned showed a predilection for males to have the higher level of osteoarthritis compared with females (Rogers 1998) , in direct contrast to modern groups.

A further example is a review of the evidence for osteoarthritis among twenty-five prehistoric Amerindian skeletal groups. This demonstrates some significant differences compared with contemporary Americans (Bridges 1992) . The knee joint shows the highest prevalence in 17 of the groups, which is similar to modern populations, but the second most affected joint is the elbow in 15 of the groups. The joints of the hand show mixed results in those groups assessed. Also in contrast to modern groups is the sex differences of osteoarthritis. In the prehistoric Amerindians there was a wide variation in the sexual dimorphism of osteoarthritis, with males often showing higher rates, although rarely to significant levels (Bridges 1992) . Bridges (1992) attempted to link the differences between modern and prehistoric osteoarthritis prevalences with activities, but Bridges cautions against making a link between any specific economic activity and arthritis.

The relationship between osteoarthritis and physical activity is not straightforward. Assessments vary widely as to the extent to which mechanical overuse, particularly occupational overuse, contributes to the development of osteoarthritis (Maetzel *et al.* 1997) . Overuse through activity or occupation, a mechanical alteration of the joint, is only one aspect of the osteoarthritic process; an age and genetic predisposition are also going to affect the outcome in any one individual (Rogers and Waldron 1995) . Inactivity may also lead to articular degeneration (Resnick and Niwayama 1995a) .

The correlation of osteoarthritis and advancing age is not linear but increases exponentially after the age of 50 or 60 years (Resnick and Niwayama 1995a) and given the reduced life expectancy of prehistoric populations in comparison to modern populations, the cause of osteoarthritis in prehistoric populations is more likely to be due to functional stress, that is, the effects of aging are not going to be so relevant. Other factors unrelated to age, such as trauma, may also influence the development of osteoarthritis.

Rogers and Waldron (1995) make some recommendations when attempting to link activity with occupation, primarily that it should only be attempted on a population basis. If, for example, two populations with comparable age structures show different prevalences of osteoarthritis in particular joints, and it was *likely* the pattern of activity was different, then it is "...reasonable to infer that the differences in activity were related to the variation in the expression of OA [osteoarthritis]" (Rogers and Waldron 1995: 107).

Many studies of past populations have attempted to link osteoarthritis with specific activities. It may be possible to attempt to describe such a relationship in populations where some knowledge of behaviour is available, perhaps from the archaeological record or ethnographic work. However, this may not be possible in most studies because so many kinds of activities and injuries can lead to joint disease. Perhaps the best that may be determined is at the general level where, for example, increased osteoarthritis may be a sign of an activity change, rather than specifying the exact activity (Bridges 1992) .

Studies of workers involved in heavy manual labour or repetitive manual tasks have not consistently confirmed that the incidence of osteoarthritis is increased in prevalence and/or severity when a particular joint is strained over a long period of time. For example, in a review of the association of osteoarthritis of the hip and knee and occupational exposure, it was concluded that a particularly strong relationship was present only between work-related knee bending and knee osteoarthritis. The evidence between work related exposure, farming in particular, and hip osteoarthritis was consistently positive, but weak (Maetzel *et al.* 1997) .

Hadler *et al.* (1978) similarly found a positive association with joint use and osteoarthritis in the hands. Their study that considered the effects of patterns of usage on patterns of degeneration in the hand joints in a modern sample of women from a worsted mill. The women were separated into three groups based on the particular manual task they had been involved in over the last 20 years. All tasks were highly repetitive, stereotyped, and complex. They found that significant and consistent differences in such aspects as range of motion, radiographic degeneration changes, and osteophyte formation existed among the groups. Therefore they concluded that there is a role of usage in the pathogenesis of osteoarthritis of the hands (Hadler *et al.* 1978).

Alternatively, a study of the relationship between osteoarthritis and occupation in the Spitalfields collection showed no positive association (Waldron and Cox 1989) . Occupations for individual skeletons were obtained from coffin plates and church records, with the largest sample being male weavers. An analysis of the hand joints showed these individuals had no significant tendency to develop osteoarthritis in these joints compared with a control group (Waldron and Cox 1989) .

Studies with negative results are often more difficult to publish than those with positive results and this can lead to biased impressions. Many more studies showing the relationship of specific lesions, including osteoarthritis, have been published. Kennedy (1989) , for example, has listed over 140 skeletal markers of occupational stress including a number of osteoarthritic joint patterns, but presents few of the inconsistencies and negative associations of skeletal markers and occupation. Many researchers (Jurmain 1990, 1991a; Waldron 1994; Rogers 1998) do not accept the proposed relationship, and caution against linking specific activities with patterns of osteoarthritis as discussed above.

Attempts have been made to link specific skeletal lesions and patterns of joint degeneration with specific activities by relating present day activities to prehistoric activities. For example, lesions in the elbow are related to tasks in modern times such as javelin throwing, woodcutting, and archery; these are all possible activities within the archaeological context of the Neolithic Saharan people in Dutour's study (1986) . Material remains, such as tools, have also been included to link possible activities to skeletal pathologies.

Merbs (1983) extensive study on a Canadian Inuit skeletal population incorporated ethnohistorical accounts of activities with skeletal observations, including that of osteoarthritis, osteophytosis, and spondylolysis. A number of significant activity-induced pathologies were identified. For example, the patterning of osteoarthritis in males was suggested to correlate with the activities of harpoon throwing and kayak paddling, both known to have been part of these people's lives.

In an extensive review of arthritis in prehistoric America previously mentioned, the patterning of osteoarthritis amongst the major appendicular joints did not correlate with subsistence economies (Bridges 1992) . However, differences may be apparent between males and females reflecting variations in their daily tasks. In general, the agriculturalists of prehistoric Amerindians showed significant sexual dimorphism more often than hunter-gatherer groups (Bridges 1992) . The higher levels in the male agriculturalists are suggested to reflect a shift in the division of labour with the adoption of agriculture (Bridges 1992) . However, age and status differences between samples of males and females of the same population may also have been a factor in the patterns of osteoarthritis observed. The approach of the latter study is similar to that taken in this thesis. The aim is to present a general impression of how joint degeneration may have been related to workload.

**Vertebral Osteophytosis (Spondylosis)**

Vertebral osteophytosis is identified by the formation of new bone around the margins of the vertebral bodies. Osteophytes in this region are a characteristic feature of intervertebral disc degeneration, although they are also age progressive (Rogers and Waldron 1995; O'Neill *et al.* 1999) . The joints between the vertebral bodies are fibrocartilaginous and degeneration in these joints are considered separately from osteoarthritis as the latter is restricted to synovial joints. However, as with osteoarthritis, vertebral osteophytosis is extremely common in skeletal material, particularly in older adults.

The term vertebral osteophytosis is used here as it describes what is observed in skeletal material. Another common term is spondylosis, but this can be used more non-specifically to identify any degenerative lesion of the spine and may therefore cause confusion.

The vertebral bodies are both separated and united by fibrocartilaginous intervertebral discs. The disc is composed of an outer ring, the annulus fibrosus which surrounds the nucleus pulposus (Resnick and Niwayama 1995b) . A defect or abnormality in the outer fibres of the annulus fibrosus is generally accepted to be the initiating factor for degeneration of the disc and subsequent vertebral osteophytosis. This leads to a breakdown of the attachment of the disc to the vertebral margin. The corresponding part of the disc is no longer anchored sufficiently and can therefore become displaced. This most frequently occurs anteriorly or anteriolaterally. Displacement is particularly enhanced by weight-bearing and motion (severe osteophytes are rarely seen in immobile or inactive people). The displaced disc tissue can, in turn, displace the anterior longitudinal ligament and cause stress at the ligament's attachment site on the anterior vertebral body. This is the site at which osteophytes usually develop (Resnick and Niwayama 1995b) .

Osteophytes can vary considerably in shape and size (Rogers and Waldron 1995) but typically appear as shelf-like bony protrusions (Ortner and Putschar 1985) arising marginally around the vertebral body. Osteophytes tend to develop horizontally initially, but, if severe enough, may progress vertically (Rogers and Waldron 1995) . In advanced cases, osteophytes from adjacent vertebrae may even fuse, where an inferior osteophyte from one vertebrae fuses with a superior osteophyte from the vertebra below.

Relating the presence and severity of osteophytes in individuals to clinical symptoms is not straightforward. Osteophytes are often detected more frequently in pathologic studies rather than by clinical or radiologic examinations (Resnick and Niwayama 1995b) . This means that the presence of an osteophyte is not always associated with pain. Even large osteophytes may not always be associated with symptoms. Posteriorly placed osteophytes, extending into the spinal canal can be a more significant problem than those located anterior to the body. In those individuals who do suffer, symptoms may include dysphagia (difficulty in swallowing), stiffness, restricted movement, and pain (Resnick and Niwayama 1995b) . Evidence of osteophytes in dry bones cannot always be assumed to reflect pain and disability.

Table 5.1: Bones assessed in each joint for osteoarthritis

| Joint | Bones |
|---|---|
| Shoulder | glenoid, humerus head, acromion, and lateral end of clavicle |
| Elbow | distal humerus, proximal ulna, and proximal radius |
| Wrist | distal ulna, distal radius, and carpals |
| Hands | metacarpals and phalanges |
| Hip | acetabulum and femur head |
| Knee | distal femur, patella, and proximal tibia |
| Ankle | distal tibia, distal fibula, calcaneus, and talus |
| Feet | metatarsals, phalanges, and other tarsals not included in ankle joint |

Aetiological factors associated with the development of a vertebral osteophyte include both age and biomechanical stress. Generalised osteophyte formation is thought by some to be a normal accompaniment to aging similar to the aetiology of osteoarthritis (Buikstra and Ubelaker 1994; Rogers and Waldron 1995) . There is evidence to suggest that degeneration may be accelerated by physical activity, particularly heavy physical labour (Resnick and Niwayama 1995b) . In a recent study by O'Neill *et al.* (1999) , vertebral osteophytosis was found to be associated with heavy physical activity in men. Thus implying that "...mechanical factors appear to play a role in conferring susceptibility to osteophytes" (O'Neill *et al.* 1999: 847).

Trauma can also induce localised osteophytes to form within one to three months of an injury, but a period of one to three years is generally required before initial outgrowths are large enough to be seen radiographically (Resnick and Niwayama 1995b) . Other alternative aetiological factors include localised infection which may produce the initial defect in the annulus fibrosa fibres required for degeneration to begin. Genetics may vary the shape of the vertebral body (Resnick and Niwayama 1995b) , also predisposing vertebrae to degeneration.

One of the best known aetiological factors, however, is age. The prevalence of osteophytes in the spine has a strong association with age. Clinical studies indicate approximately 60% of women and 80% of men have osteophytes by the age of 50 years (Resnick and Niwayama 1995b) . Men are also more likely to have more severe osteophytes than women (O'Neill *et al.* 1999) . Vertebral osteophytosis is seen most commonly in the cervical and lumbar regions (Rogers and Waldron 1995) but any segment may be affected (Resnick and Niwayama 1995b) .

Schmorl's nodes are a common age-related condition (Buikstra and Ubelaker 1994) . They represent disc displacement inferiorly or superiorly resulting in the formation of cartilaginous nodes (Resnick and Niwayama 1995b) and appear as indentations in the inferior or superior surface of the vertebral body (Rogers and Waldron 1995) . These nodes are often found with osteophytosis because they are both common conditions; there is no evidence that they are causally related (Rogers and Waldron 1995) . Not solely related to degeneration, they may also be a consequence of osteoporosis and other systemic processes including infection and rheumatoid arthritis (Resnick and Niwayama 1995c) .
The association of vertebral osteophytes with heavy physical labour discussed above (O'Neill *et al.* 1999) has lead to the postulation of activities in prehistoric groups based on the presence of vertebral osteophytosis (for example Bridges 1992) . However, this relationship has been found to be inconsistent.

A recent study (Knüsel *et al.* 1997) cautions against the use of degeneration of the vertebral column for studying physical activity. Due to the functional constraints on the vertebral column it is suggested that it is not used in the study of activity-related lesions (Knüsel *et al.* 1997) . The vertebral column, being the central structure of the skeleton, is affected by all movements of the axial and appendicular skeleton and must support upright posture and bipedal locomotion. Therefore, "...the vertebral column must balance movement with support, flexibility with fixation" (Knüsel *et al.* 1997: 493).

However, Videman *et al.* (1990) did find a positive association between vertebral osteophytes and heavy physical work in a cadaveric study. This supports the conclusions of O'Neill *et al.* (1999) . Taking into consideration the normal functions of the vertebral column and the consistent pattern of the cervical and lumbar joints being more affected by joint disease than the thoracic region, any pattern contrary to this may reflect some unusual biomechanical influence.

# Methods

The skeletal material from Nong Nor could not be included in any part of this analysis because of the lack of almost any complete articular surfaces. As the data for the sample from Khok Phanom Di was obtained from another observer (Tayles 1992) , a test for potential interobserver error was carried out. The results are given in Appendix A. Some differences in the assessment of osteoarthritis and vertebral osteophytosis were found among the individuals reanalysed. Approximately 20-25% of all joints observed for osteoarthritis and vertebral osteophytosis were different. However, only four individuals were reanalysed for osteoarthritis and five for vertebral osteophytosis. The assessment of joint disease is subjective. This level of interobserver error was not deemed significant, but further investigation of more individuals would be necessary before these results are taken further than the general prevalences presented here. Further explanation is provided in Appendix A.

Osteoarthritis data were recorded only for the major appendicular joints. The components of each joint included in this analysis are listed in Table 5.1. Not all possible joints were assessed for osteoarthritis. The synovial joints not analysed include the temporomandibular joints, the sternoclavicular joints, and the zygapophyseal joints of the vertebral column.

Every possible complete joint surface was assessed individually for the presence or absence of osteoarthritis, indicated by the presence of two or more of the bony changes mentioned above. Next, an overall assessment of whether osteoarthritis was present or absent in each joint was determined. In a joint with all its components present, osteoarthritis was deemed present in the whole joint if one or more of the surfaces had osteoarthritis present. For example, if the distal humerus and proximal ulna did not have sufficient bony changes but the proximal radius did, then overall the elbow was described as osteoarthritic. In joints where not all joint surfaces were present, the same rule applied. However, for an incomplete joint without osteoarthritis to be included in the total denominator, it had to have at least two joint surfaces present. In this study, results from left and right joints are pooled to obtain a general indication of the most at risk joints. For a positive osteoarthritis result in the paired shoulder joints, for example, osteoarthritis had to be present in either one or both sides.

In the observation of osteoarthritis, emphasis was placed on the occurrence of osteophyte formation, porosity indicating subchondral cysts, and eburnation, although, sclerosis is difficult to identify without radiographic imaging of all joints. As discussed in the introduction, a diagnosis of osteoarthritis was made only when two or more of these bony changes were present on the one joint surface. Figure 5.1A-B shows two examples of osteoarthritis in the knee joint, this provides an indication of the variation of the severity of osteoarthritis.

In the assessment of vertebral osteophytosis every superior and inferior vertebral body surface present was assessed for the presence of osteophytes. These were assessed as either minimal, moderate, or severe. Examples of each of these are shown in Figure 5.1C. Minimal osteophyte development was not considered pathological and therefore only moderate to severe osteophytes were taken as representing the presence of vertebral osteophytosis.

The vertebrae were divided into the region from which they came, that is cervical, thoracic or lumbar. Vertebral osteophytosis in each region of the vertebral column from each individual was considered to be present if one or more vertebral body surfaces had moderate to severe osteophytes. An assessment was also made for the whole vertebral column on the same basis. However, it is acknowledged that not all individuals had complete vertebral columns. It has had to be assumed that the vertebrae present are representative of the region from which they come. It is uncommon for only a single vertebrae to have an osteophyte without an adjacent vertebrae also having osteophytes as the regions of the spine act as a functional unit.

Some detail has been lost as a consequence of this approach to both osteoarthritis and vertebral osteophytosis. However, it was used to obtain the maximum amount of information from the poorest material; a joint did not have to have all components present to be included, nor did a vertebral column have to have all vertebrae present to be included.

Preservation differences among the samples are a complicating factor in both the assessment of osteoarthritis and vertebral osteophytosis. The excellent preservation of the skeletal material from Khok Phanom Di has allowed a more reliable indication of the prevalence of joint diseases to be obtained. For example, the individual hand and foot bones were commonly well preserved at Khok Phanom Di, in contrast to both the Ban Lum Khao and Ban Na Di skeletal material. This may have increased the likelihood of an observation of osteoarthritis in the Khok Phanom Di hand and foot joints. This also applies to the observation of osteophytosis in the vertebrae. The Khok Phanom Di sample had excellent preservation of vertebrae, Ban Lum Khao vertebrae were relatively moderately preserved, and the Ban Na Di vertebrae the least well preserved of the these three samples.

Some individuals, through lack of alternative information, were aged in part, by the severity of their joint degeneration in appendicular and axial joints. As both these joint disorders can be strongly associated with age, no comparisons based on age, such as comparing prevalences among younger and older adults, were carried out.

In comparing joint degeneration among the different samples, the age structure of each group may affect the results. If one sample had a higher proportion of older adults they would be more likely to have higher levels of joint degeneration. This may subsequently lead to the erroneous interpretation of this group having, for example, higher mechanical stress levels. Age structures were therefore compared statistically among the three samples. Each sample was divided into two age groups, those aged less than 30 years and those aged equal to or greater than 30 years. This division was used to provide as near equal sample sizes as possible between the two age groups. Restricting the analysis to two adult age groups was necessary as further divisions would have produced very small sample sizes in each division that would not have been able to be statistically compared.

The prevalence of osteoarthritis and vertebral osteophytosis were determined following a method similar to that of Waldron (1994) . The calculation involves the assumption that the distribution of osteoarthritis or vertebral osteophytosis amongst the individuals with missing appendicular joints or missing vertebral regions to be the same as the distribution in the joints or regions present in other individuals. This attempts to avoid the over or under estimation of the presence of osteoarthritis or vertebral osteophytosis in a sample.

The prevalence of osteoarthritis is calculated for each of the major appendicular joints with right and left sides combined. The number of joints with osteoarthritis is divided by the number of adult individuals with both left and right joints present, plus the number of single joints, without a pair, which show evidence of osteoarthritis. For the vertebral osteophytosis analysis, the denominator is simply the total number of adult individuals able to be observed for vertebral

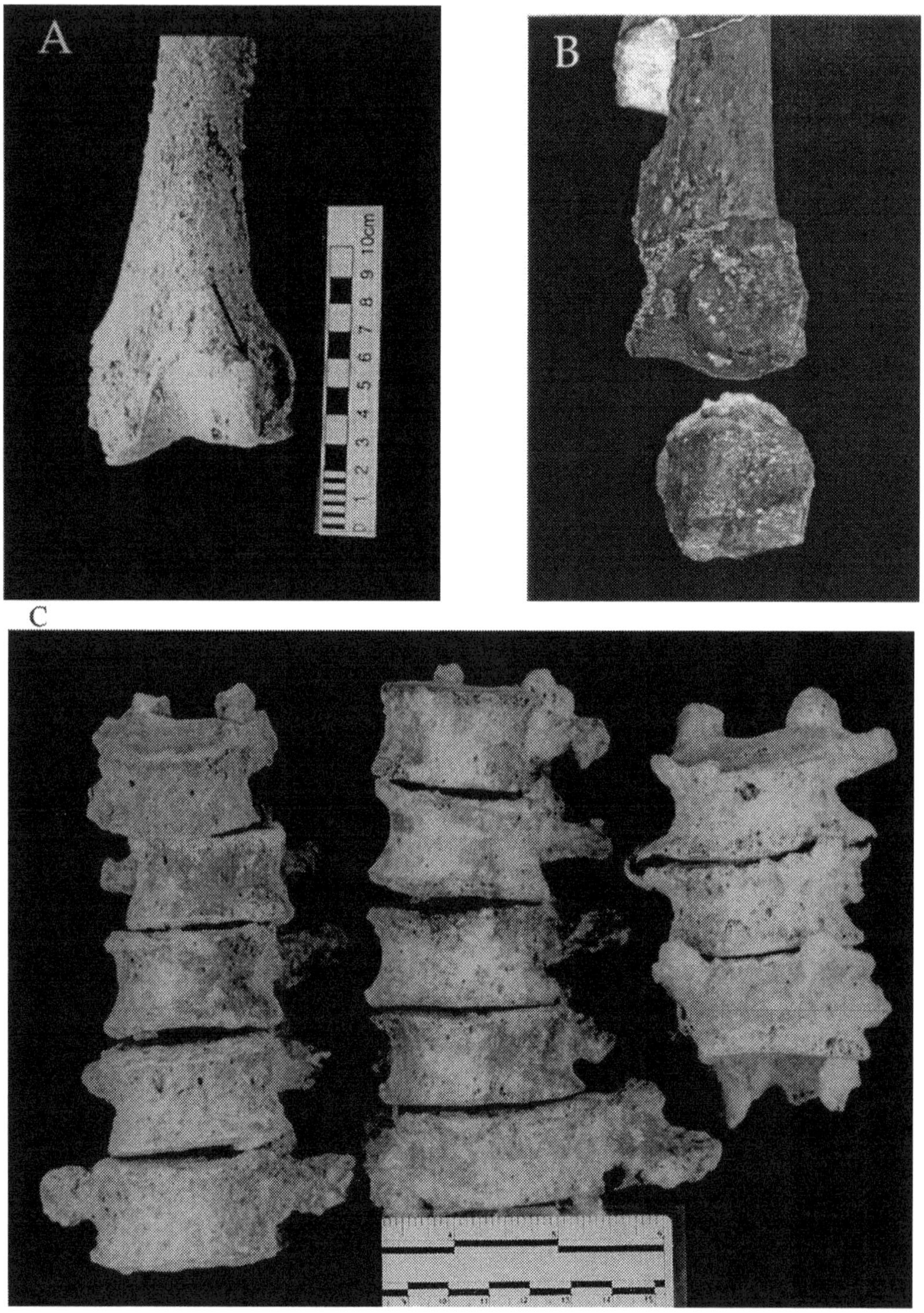


**Figure 5.1:** Osteoarthritis of the knee joint (A) with a lateral osteophyte (arrow). The condyles of the left femur from Ban Lum Khao, Burial 44, 30-39 yr old male (anterior view); (B) the distal condyle of the left femur (anterior view) and patella (posterior view) from Nong Nor, Burial 104, mid aged female. (length femur fragment = 75 mm). (C) Vertebral Osteophytosis. Three examples showing the range of severity in the lumbar vertebrae (anterior view). From left to right: Khok Phanom Di, Burial 83, 30 yr old female with minimal vertebral osteophytosis; Ban Na Di, Burial M.28, 40-44 yr old female with moderate vertebral osteophytosis; Ban Na Di, Burial T.17, 45-49+ yr old male with severe vertebral osteophytosis. Only the latter two are considered pathological

**Table 5.2:** Age structures of the female samples observed for osteoarthritis

| | Khok Phanom Di | | Ban Lum Khao | | Ban Na Di | |
|---|---|---|---|---|---|---|
| Age | N | % | N | % | N | % |
| < 30 years | 13 | 38.2 | 17 | 65.4 | 3 | 25.0 |
| ≥ 30 years | 21 | 61.8 | 9 | 34.6 | 9 | 75.0 |
| Total | 34 | | 26 | | 12 | |

**Table 5.3:** Age structures of the female samples observed for vertebral osteophytosis

| | Khok Phanom Di | | Ban Lum Khao | | Ban Na Di | |
|---|---|---|---|---|---|---|
| Age | N | % | N | % | N | % |
| < 30 years | 14 | 40.0 | 17 | 68.0 | 3 | 25.0 |
| ≥ 30 years | 21 | 60.0 | 8 | 32.0 | 9 | 75.0 |
| Total | 35 | | 25 | | 12 | |

osteophytosis. In this way the missing data are acknowledged by removing from the denominator those individuals with no observations. These would otherwise artificially inflate the prevalence. In both cases the results are presented as a percentage.

Samples sizes for the Ban Na Di people were relatively small which creates elevated prevalences that are probably not truly representative of these samples. Therefore the actual number of joints affected as well as the number observed are provided for each sample.

Statistical analyses were undertaken in order to compare the prevalence of osteoarthritis and vertebral osteophytosis among the samples in the total adult sample (excluding those adults with no estimate of age or sex), as well as divided by male and female groups. $Chi^2$ and Fisher's Exact Test calculations were performed on this data. A description of these tests with their relevant significance levels, is provided in Appendix B.

# Results

### Joint degeneration in females

The age structure of the female samples showed significant differences amongst the three groups for the subsample of individuals in both the osteoarthritis (Table 5.2) ($Chi^2$ = 6.89, p-value = 0.03) and vertebral osteophytosis (Table 5.3) ($Chi^2$ = 7.44, p-value = 0.02) calculations. Fisher's Exact p-values were then calculated for each pair of groups to determine which groups were significantly different. None of the Fisher's Exact p-values showed a significant level below 0.01 (1%) and are therefore not considered statistically different.

Ban Lum Khao females did show a non-statistical difference compared with the other two groups. There was a higher proportion of younger females compared with older females in the Ban Lum Khao sample. This was in contrast to the Ban Na Di and Khok Phanom Di samples, where the age distribution was the reverse to Ban Lum Khao. This difference is acknowledged but appears not to be great enough to have had a statistical effect on the results. One further potential problem with this data was the sample size of the Ban Na Di females. This was very small and caution will be taken with interpreting the results obtained from these individuals.

The osteoarthritis data for the upper limb joints of the female samples are presented in Table 5.4 and Figure 5.2. No statistically significant differences were calculated among the three samples in any of the upper limb joints of the females (Table 5.5). However, some non-statistical patterns were evident. The Khok Phanom Di female upper limb joints consistently showed higher prevalences of osteoarthritis than the other samples. The shoulder and hand joints had particularly higher prevalences of osteoarthritis compared with the Ban Lum Khao and Ban Na Di samples. However, the difference was not statistically significant (Table 5.5) although the very small sample sizes in these joints may have affected the comparisons with the Ban Na Di sample. The females from Ban Lum Khao showed very little osteoarthritis in any of the upper limb joints. The shoulders and elbows were the only joints affected, both at prevalences less than 15%. The wrist joints and joints of the hands were unaffected. The Ban Na Di females also showed very little osteoarthritis in the upper limb with only the wrists affected in two females.

**Table 5.4:** Osteoarthritis prevalences in the upper limb joints of the female samples

| | Khok Phanom Di* | | Ban Lum Khao | | Ban Na Di | |
|---|---|---|---|---|---|---|
| Joint | A/N | % | A/N | % | A/N | % |
| Shoulders | 9/26 | 34.6 | 1/11 | 9.1 | 0/1 | 0 |
| Elbows | 5/30 | 16.7 | 2/14 | 14.3 | 0/4 | 0 |
| Wrists | 6/22 | 27.3 | 0/14 | 0 | 2/6 | 33.3 |
| Hands | 4/15 | 26.7 | 0/21 | 0 | 0/8 | 0 |

A = number affected, N= number of individuals with both left and right joints + number of individuals with single joints with osteoarthritis. * Tayles (1992)

**Table 5.5:** Fisher's Exact p-values for comparing osteoarthritis prevalences amongst the female samples at each upper limb joint

| Joint | Khok Phanom Di vs Ban Lum Khao | Khok Phanom Di vs Ban Na Di | Ban Lum Khao vs Ban Na Di |
|---|---|---|---|
| Shoulders | 0.22 | >0.99 | >0.99 |
| Elbows | >0.99 | >0.99 | >0.99 |
| Wrists | 0.06 | >0.99 | 0.08 |
| Hands | 0.02 | 0.26 | - |

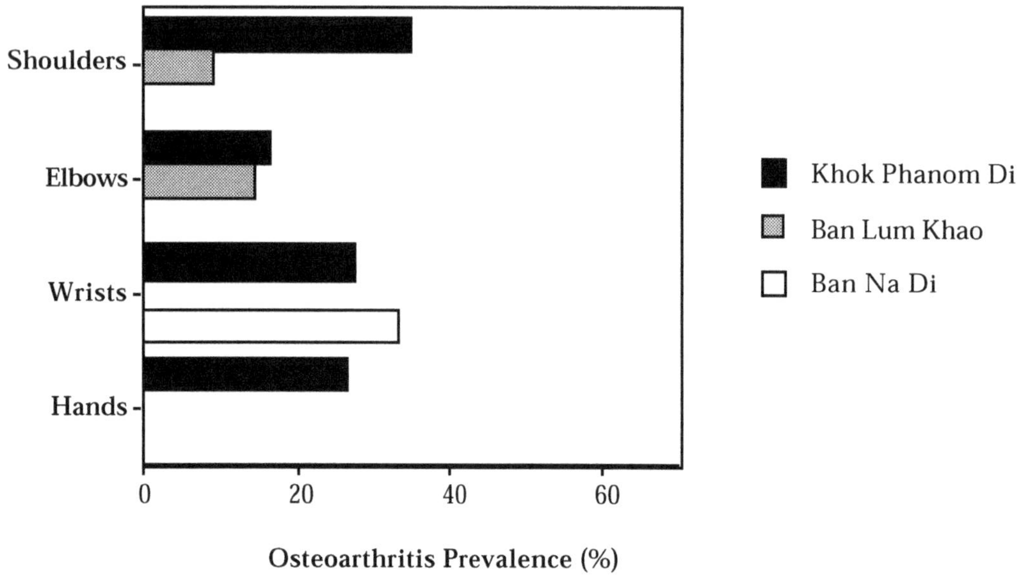


**Figure 5.2:** Osteoarthritis prevalences in the upper limb joints of the female samples

**Table 5.6:** Osteoarthritis prevalences in the lower limb joints of the female samples

| Joint | Khok Phanom Di* A/N | % | Ban Lum Khao A/N | % | Ban Na Di A/N | % |
|---|---|---|---|---|---|---|
| Hips | 6/17 | 35.3 | 1/13 | 7.7 | 0/2 | 0 |
| Knees | 17/28 | 60.7 | 0/11 | 0 | 1/4 | 25.0 |
| Ankles | 5/26 | 19.2 | 0/17 | 0 | 1/4 | 25.0 |
| Feet | 4/25 | 16.0 | 1/18 | 5.6 | 5/9 | 55.6 |

A = number affected, N= number of individuals with both left and right joints + number of individuals with single joints with osteoarthritis. * Tayles (1992)

**Table 5.7:** Fisher's Exact p-values for comparing osteoarthritis prevalences amongst the female samples at each lower limb joint

| Joint | Khok Phanom Di vs Ban Lum Khao | Khok Phanom Di vs Ban Na Di | Ban Lum Khao vs Ban Na Di |
|---|---|---|---|
| Hips | 0.10 | 0.54 | >0.99 |
| Knees | **<0.01** | 0.30 | 0.27 |
| Ankles | 0.14 | >0.99 | 0.19 |
| Feet | 0.38 | 0.03 | **0.01** |

Bold indicates statistical significance at 1% level

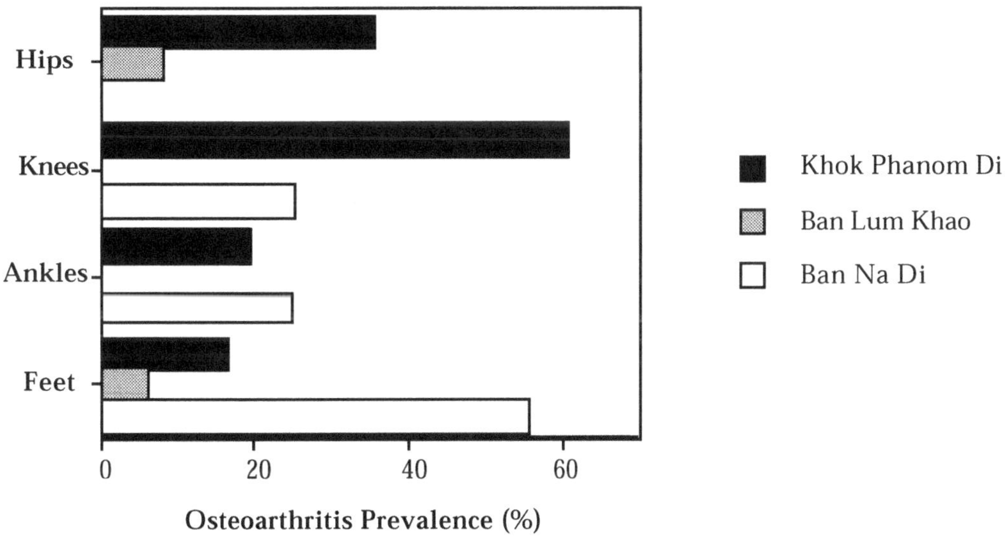


**Figure 5.3:** Osteoarthritis prevalences in the lower limb joints of the female samples

The osteoarthritis data for the lower limb joints of the female samples are presented in Table 5.6 and Figure 5.3. In the sample of females from Khok Phanom Di, each of the lower limb joints were affected by osteoarthritis; the knee joint exhibited a particular vulnerability. This was statistically higher than the prevalence of knee osteoarthritis seen in the Ban Lum Khao females (Table 5.7) who had no osteoarthritis in this joint. The ankle joints of the females from Ban Lum Khao also showed no sign of osteoarthritis and the hips and feet joints were affected in only one individual each. The Ban Na Di females exhibited a particular vulnerability to osteoarthritis in the joints of the feet and this result was statistically higher than the prevalence seen in the Ban Lum Khao sample, and close to significant compared with the Khok Phanom Di sample (Table 5.7). Only one ankle and one knee joint were affected with osteoarthritis in the Ban Na Di females.

**Table 5.8:** Vertebral osteophytosis prevalences in the female samples

| Spinal | Khok Phanom Di* | | Ban Lum Khao | | Ban Na Di | | Comparison among samples | |
|---|---|---|---|---|---|---|---|---|
| Level | A/N | % | A/N | % | A/N | % | $Chi^2$ Stat | p-value |
| Cervical | 5/35 | 14.3 | 0/18 | 0 | 2/9 | 22.2 | - | - |
| Thoracic | 5/34 | 14.7 | 0/24 | 0 | 3/11 | 27.3 | - | - |
| Lumbar | 18/34 | 52.9 | 6/21 | 28.6 | 5/11 | 45.5 | 3.14 | 0.21 |

A = number affected, N= number of individuals with vertebrae able to be observed. Stat - statistic; $Chi^2$ tests comparing the samples could not be performed on the cervical and thoracic data as samples sizes are too small, therefore FET are used. * Tayles (1992)

**Table 5.9:** Fisher's Exact p-values for comparing vertebral osteophytosis prevalences amongst the female samples at the cervical and thoracic level

| | Khok Phanom Di vs Ban Lum Khao | Khok Phanom Di vs Ban Na Di | Ban Lum Khao vs Ban Na Di |
|---|---|---|---|
| Cervical | 0.15 | 0.62 | 0.10 |
| Thoracic | 0.07 | 0.38 | 0.03 |

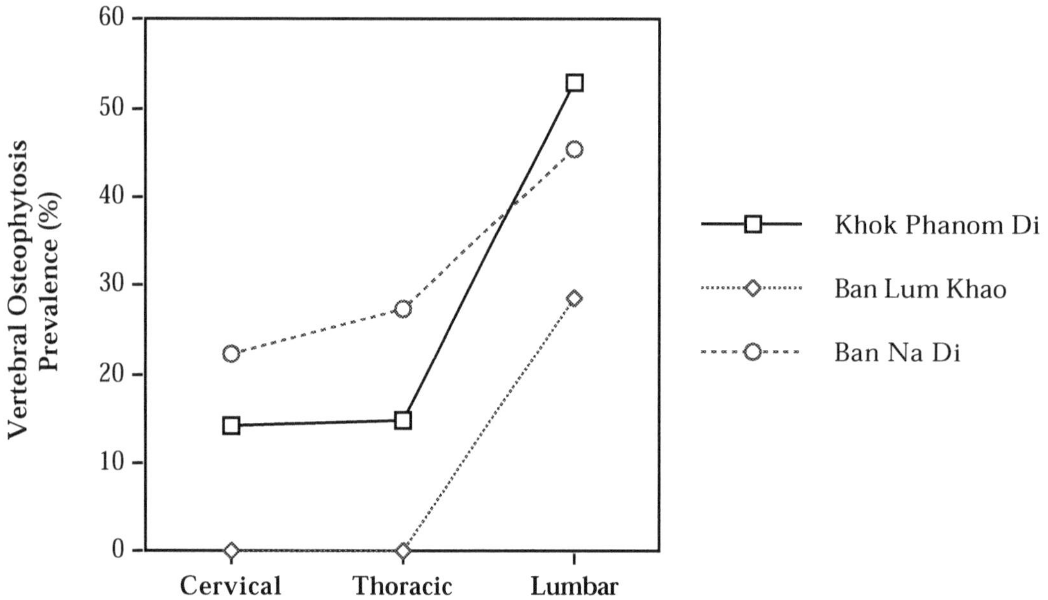


**Figure 5.4:** Vertebral osteophytosis prevalences in the female samples

The vertebral osteophytosis data for the females in each sample is presented in Table 5.8 and Figure 5.4. All samples showed the same pattern of increasing vertebral osteophytosis from the cervical region to the lumbar region. There were no significant differences in vertebral osteophytosis prevalences in any region of the vertebral column among the different samples (Table 5.8 and 5.9). However, non-statistical patterns among the samples are described. All samples had the highest levels in the lumbar region compared to their cervical and thoracic regions. Khok Phanom Di females had relatively moderate levels of vertebral osteophytosis in the cervical and thoracic regions, but the prevalence was highest, compared with the other samples, in the lumbar region. As with appendicular osteoarthritis, Ban Lum Khao females showed the lowest prevalences in all vertebral regions compared with the other two samples, with no vertebral osteophytosis in the cervical and thoracic regions of any adult female. Ban Na Di females showed higher prevalences in the cervical and thoracic regions compared with the other female samples and had the second highest prevalence of osteophytosis in the lumbar region.

**Table 5.10:** Age structures of the male sample observed for osteoarthritis

| | Khok Phanom Di | | Ban Lum Khao | | Ban Na Di | |
|---|---|---|---|---|---|---|
| Age | N | % | N | % | N | % |
| < 30 years | 14 | 48.3 | 8 | 36.4 | 9 | 42.9 |
| ≥ 30 years | 15 | 51.7 | 14 | 63.6 | 12 | 57.1 |
| Total | 29 | | 22 | | 21 | |

**Table 5.11:** Age structures of the male sample observed for vertebral osteophytosis

| | Khok Phanom Di | | Ban Lum Khao | | Ban Na Di | |
|---|---|---|---|---|---|---|
| Age | N | % | N | % | N | % |
| < 30 years | 13 | 44.8 | 6 | 37.5 | 9 | 42.9 |
| ≥ 30 years | 16 | 55.2 | 10 | 62.5 | 12 | 57.1 |
| Total | 29 | | 16 | | 21 | |

**Joint degeneration in males**

The age structure of the male samples showed no significant differences among the three groups for the subsample of individuals in both osteoarthritis (Table 5.10) ($Chi^2$ = 0.73, p-value = 0.70) and vertebral osteophytosis (Table 5.11) ($Chi^2$ = 0.23, p-value = 0.89) calculations. All samples had slightly more older adult males (≥ 30 years) present and able to be observed than the younger (< 30 years) group.

One potential problem with this data, as with the females, was the sample size of the Ban Na Di males. This was very small and caution with interpreting the result obtained from these individuals will be taken.

The osteoarthritis data for the upper limb joints of the male samples are presented in Table 5.12 and Figure 5.5. No statistically significant differences were calculated among the three samples (Table 5.13) in the prevalence of osteoarthritis in any of the upper limb joints. However, some non-statistical patterns are worth noting. The Khok Phanom Di males showed high prevalences of osteoarthritis in the hand joints compared with the other samples. Osteoarthritis in the shoulders of the Khok Phanom Di sample was also higher than any other sample. However, the difference was not as marked as it was in the female shoulder joints. Osteoarthritis in the wrist joints of the Khok Phanom Di males were not particularly different than any other sample. All joints in the upper limb were affected in the males from Ban Lum Khao. The high level of osteoarthritis in the elbow and shoulder was particularly obvious. The former was markedly higher compared with the other samples. Ban Na Di males only showed osteoarthritis in the shoulder and wrist joints. The apparent high prevalence of shoulder degeneration was, however, an artefact of the small sample size where only one individual had osteoarthritis of the shoulder.

The osteoarthritis data for the lower limb joints of the male samples are presented in Table 5.14 and Figure 5.6. No statistically significant differences were calculated among the osteoarthritis prevalences of the three samples for the lower limb joints (Table 5.15). However, non-statistical patterns were noted. All lower limb joints in Khok Phanom Di males had some evidence of osteoarthritis. However, only one individual had degeneration of the hip. Osteoarthritis of the knee in the Khok Phanom Di males was high compared with the other samples, as it was in the females. Ban Lum Khao males also showed osteoarthritis in all lower limb joints, mostly at a moderate level. The joints of the feet showed a particular vulnerability to osteoarthritis compared with the other samples. Ban Na Di males had evidence of osteoarthritis in all types of joints at slightly lower levels than Ban Lum Khao, except in the hip joints. This was an artificially high prevalence however, as it represented only one individual who had osteoarthritis in this joint.

The vertebral osteophytosis data from the male samples are presented in Table 5.16 and Figure 5.7. These results showed no significant differences in any spinal region among the three samples (Table 5.16 and 5.17). However, non-statistical patterns are discussed. All samples showed the same pattern of increasing vertebral osteophytosis from the cervical region to the lumbar region.

The Khok Phanom Di males had consistently higher levels of vertebral osteophytosis in all regions of the spine, compared with the other two samples. The Ban Lum Khao males showed consistently higher levels of vertebral osteophytosis than Ban Na Di males in all regions except the lumbar, where prevalences were similar in the two samples.

**Table 5.12:** Osteoarthritis prevalences in the upper limb joints of the male samples

| Joint | Khok Phanom Di* | | Ban Lum Khao | | Ban Na Di | |
|---|---|---|---|---|---|---|
| | A/N | % | A/N | % | A/N | % |
| Shoulders | 13/25 | 52.0 | 3/7 | 42.9 | 1/3 | 33.3 |
| Elbows | 6/25 | 24.0 | 4/8 | 50.0 | 0/1 | 0 |
| Wrists | 6/19 | 31.6 | 3/11 | 27.3 | 3/7 | 42.9 |
| Hands | 5/18 | 27.8 | 1/13 | 7.7 | 0/10 | 0 |

A = number affected, N= number of individuals with both left and right joints + number of individuals with single joints with osteoarthritis. * Tayles (1992)

**Table 5.13:** Fisher's Exact p-values for comparing osteoarthritis prevalences amongst the male samples at each upper limb joint

| Joint | Khok Phanom Di vs Ban Lum Khao | Khok Phanom Di vs Ban Na Di | Ban Lum Khao vs Ban Na Di |
|---|---|---|---|
| Shoulders | >0.99 | >0.99 | >0.99 |
| Elbows | 0.21 | >0.99 | >0.99 |
| Wrists | >0.99 | 0.66 | 0.63 |
| Hands | 0.36 | 0.13 | >0.99 |

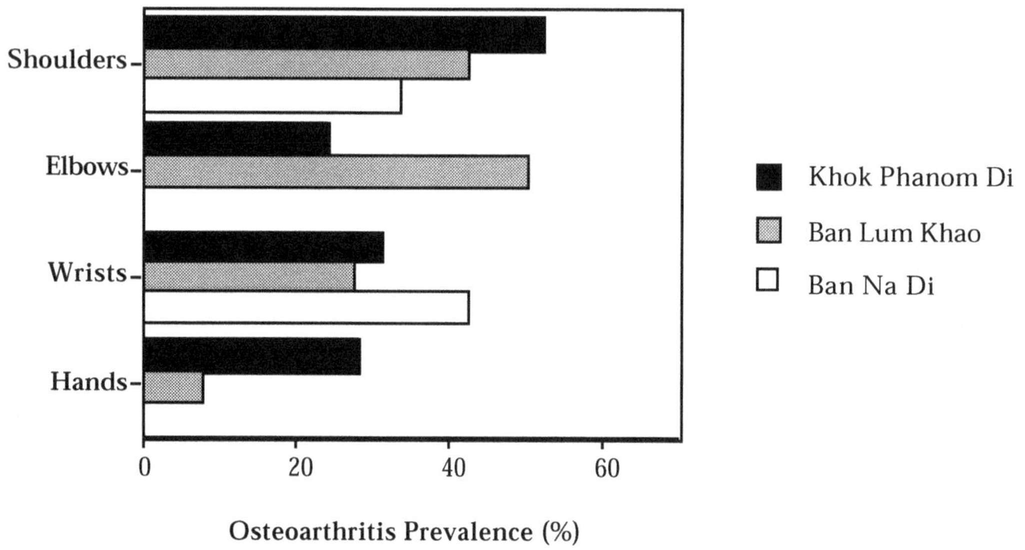


**Figure 5.5:** Osteoarthritis prevalences in the upper limb joints of the male samples

**Table 5.14:** Osteoarthritis prevalences in the lower limb joints of the male samples

| Joint | Khok Phanom Di* | | Ban Lum Khao | | Ban Na Di | |
|---|---|---|---|---|---|---|
| | A/N | % | A/N | % | A/N | % |
| Hips | 1/17 | 5.9 | 2/7 | 28.6 | 1/2 | 50.0 |
| Knees | 11/23 | 47.8 | 3/10 | 30.0 | 1/5 | 20.0 |
| Ankles | 6/24 | 25.0 | 3/13 | 23.1 | 1/10 | 10.0 |
| Feet | 3/19 | 15.8 | 5/13 | 38.5 | 2/11 | 18.2 |

A = number affected, N= number of individuals with both left and right joints + number of individuals with single joints with osteoarthritis. * Tayles (1992)

**Table 5.15:** Fisher's Exact p-values for comparing osteoarthritis prevalences amongst the male samples at each lower limb joint

| Joint | Khok Phanom Di vs Ban Lum Khao | Khok Phanom Di vs Ban Na Di | Ban Lum Khao vs Ban Na Di |
|---|---|---|---|
| Hips | 0.19 | 0.20 | >0.99 |
| Knees | 0.46 | 0.36 | >0.99 |
| Ankles | >0.99 | 0.64 | 0.60 |
| Feet | 0.22 | >0.99 | 0.39 |

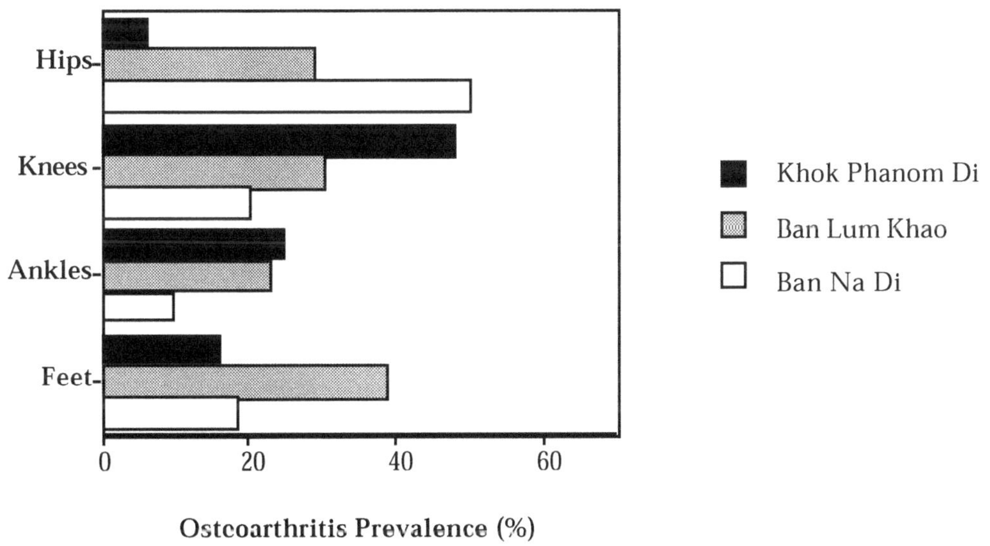


**Figure 5.6:** Osteoarthritis prevalences in the lower limb joints of the male samples

**Table 5.16:** Vertebral osteophytosis prevalences in the male samples

| Spinal Level | Khok Phanom Di* A/N | % | Ban Lum Khao A/N | % | Ban Na Di A/N | % | Comparison among samples $Chi^2$ Stat | p-value |
|---|---|---|---|---|---|---|---|---|
| Cervical | 6/27 | 22.2 | 1/14 | 7.1 | 0/8 | 0 | - | - |
| Thoracic | 10/27 | 37.0 | 5/16 | 31.3 | 2/10 | 20.0 | - | - |
| Lumbar | 16/29 | 55.2 | 5/13 | 38.5 | 4/10 | 40.0 | 1.33 | 0.51 |

A = number affected, N= number of individuals with vertebrae able to be observed. Stat = statistic; $Chi^2$ tests comparing the samples could not be performed on the cervical and thoracic data as samples sizes are too small, therefore FET are used. * Tayles (1992)

**Table 5.17:** Fisher's Exact p-values for comparing vertebral osteophytosis prevalences amongst the male samples at the cervical and thoracic level

| | Khok Phanom Di vs Ban Lum Khao | Khok Phanom Di vs Ban Na Di | Ban Lum Khao vs Ban Na Di |
|---|---|---|---|
| Cervical | 0.39 | 0.30 | >0.99 |
| Thoracic | 0.75 | 0.44 | 0.67 |

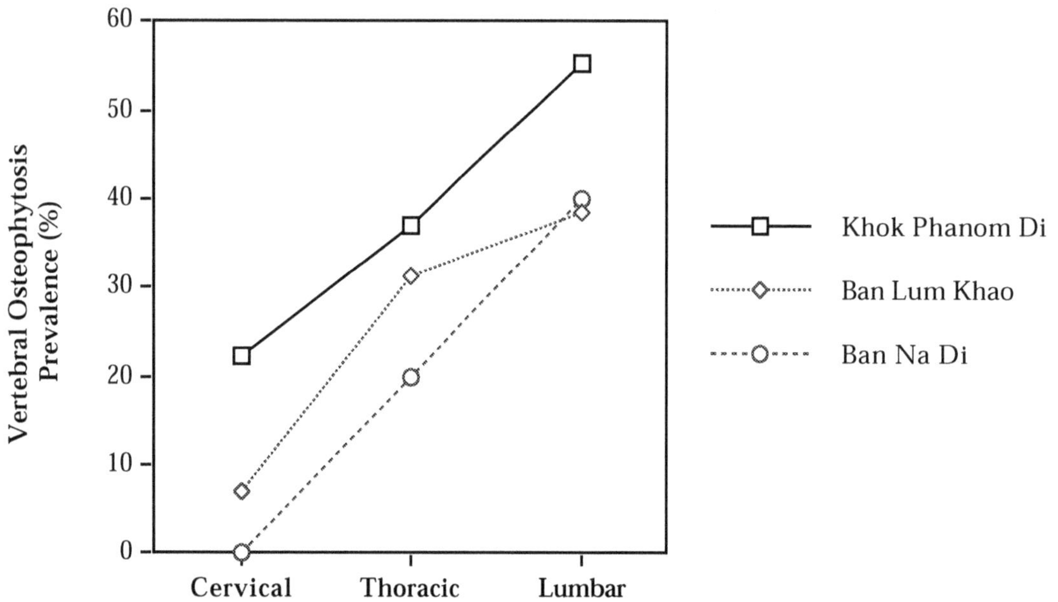


**Figure 5.7:** Vertebral osteophytosis prevalences in the male samples

**Table 5.18:** Age structures of the total sample observed for osteoarthritis

| Total | Khok Phanom Di N | % | Ban Lum Khao N | % | Ban Na Di N | % |
|---|---|---|---|---|---|---|
| < 30 years | 27 | 42.9 | 25 | 52.1 | 12 | 36.4 |
| ≥ 30 years | 36 | 57.1 | 23 | 47.9 | 21 | 63.6 |

**Table 5.19:** Age structures of the sample observed for vertebral osteophytosis

| Total | Khok Phanom Di N | % | Ban Lum Khao N | % | Ban Na Di N | % |
|---|---|---|---|---|---|---|
| < 30 years | 27 | 42.2 | 23 | 56.1 | 12 | 36.4 |
| ≥ 30 years | 37 | 57.9 | 18 | 43.9 | 21 | 63.6 |

**Joint degeneration in the total sample**

This section combines the data from male and female adults from each sample. Male and female prevalences of joint degeneration within each sample showed no significant differences for joint disease at each joint analysed (data not shown), with the single exception of the levels of vertebral osteophytosis in the thoracic region of Ban Lum Khao males compared with Ban Lum Khao females (FET <0.01). Therefore, it is valid to combine the male and female data for further analysis to obtain a general picture of joint disease among the samples.

There were no significant differences among the age structures of the three samples in those that could be observed for osteoarthritis (Table 5.18) ($Chi^2 = 2.07$, p-value = 0.36, n.s.) or vertebral osteophytosis (Table 5.19) ($Chi^2 = 3.24$, p-value = 0.20).

The osteoarthritis data for the upper limb joints of the samples are presented in Table 5.20 and Figure 5.9. Analysis of the upper limb joints in the Khok Phanom Di sample indicated a particularly high level, compared with the other samples, in the shoulder and hand joints. The level of osteoarthritis in the hand joints was statistically higher than in the Ban Lum Khao people and close to significant compared with the Ban Na Di sample (Table 5.21). Osteoarthritis was evident in the other upper limb joints, the elbow and wrist, in the Khok Phanom Di people, but not at particularly different levels from the other two samples. The Ban Lum Khao people had consistently moderate levels of osteoarthritis in all upper limb joints, with the single exception of the joints of the hand. Only one individual showed osteoarthritis in the hand joints. The Ban Na Di people showed no evidence of osteoarthritis in the joints of the hand or the elbow joints. This is in contrast to both the Khok Phanom Di and Ban Lum Khao samples. Evidence of osteoarthritis in the Ban Na Di sample was apparent only in the shoulder joints of one individual, and in the wrist joints, at a similar level as was evident in the Khok Phanom Di people.

The osteoarthritis data for the lower limb joints of the samples are presented in Table 5.22 and Figure 5.9. The Khok Phanom Di lower limb joints overall had quite similar prevalences of osteoarthritis to the other samples, with the exception of the knee joint. The Khok Phanom Di knee joints had a much higher prevalence and this was statistically significant compared with the Ban Lum Khao knee joints (Table 5.23). All lower limb joints at Ban Lum Khao were similarly affected at reasonably low to moderate prevalences in comparison with the other samples. Ban Na Di lower limb joint osteoarthritis was similar to Ban Lum Khao rates except in the joints of the feet. Ban Na Di feet showed high levels of osteoarthritis compared with both Ban Lum Khao and Khok Phanom Di foot joints, although not to a statistical level.

The vertebral osteophytosis data for each region of the spine in each sample are presented in Table 5.24 and Figure 5.10. Vertebral osteophytosis prevalences compared among the samples are not statistically different (Table 5.24 and 5.25). All samples showed the same pattern of increase in vertebral osteophytosis prevalence from the cervical region to the lumbar region. The cervical and lumbar regions of the Khok Phanom Di sample showed higher vertebral osteophytosis prevalences compared with the other samples. The prevalence of vertebral osteophytosis in the thoracic region was almost identical in the Ban Na Di individuals. The Ban Lum Khao people consistently showed the lowest vertebral osteophytosis prevalences in all regions of the spine compared with the other samples.

**Table 5.20:** Osteoarthritis prevalences in the upper limb joints of the combined sex samples

| | Khok Phanom Di* | | Ban Lum Khao | | Ban Na Di | |
|---|---|---|---|---|---|---|
| Joint | A/N | % | A/N | % | A/N | % |
| Shoulders | 22/51 | 43.1 | 4/18 | 22.2 | 1/4 | 25.0 |
| Elbows | 11/55 | 20.0 | 6/22 | 27.3 | 0/5 | 0 |
| Wrists | 12/41 | 29.3 | 3/25 | 12.0 | 5/13 | 38.5 |
| Hands | 9/33 | 27.3 | 1/34 | 2.9 | 0/18 | 0 |

A = number affected, N= number of individuals with both left and right joints + number of individuals with single joints with osteoarthritis. * Tayles (1992)

**Table 5.21:** Fisher's Exact p-values for comparing osteoarthritis prevalences amongst the combined sex samples at each upper limb joint

| Joint | Khok Phanom Di vs Ban Lum Khao | Khok Phanom Di vs Ban Na Di | Ban Lum Khao vs Ban Na Di |
|---|---|---|---|
| Shoulders | 0.16 | 0.63 | >0.99 |
| Elbows | 0.55 | 0.57 | 0.56 |
| Wrists | 0.14 | 0.73 | 0.09 |
| Hands | **<0.01** | 0.02 | >0.99 |

Bold indicates statistical significance at 1% level.

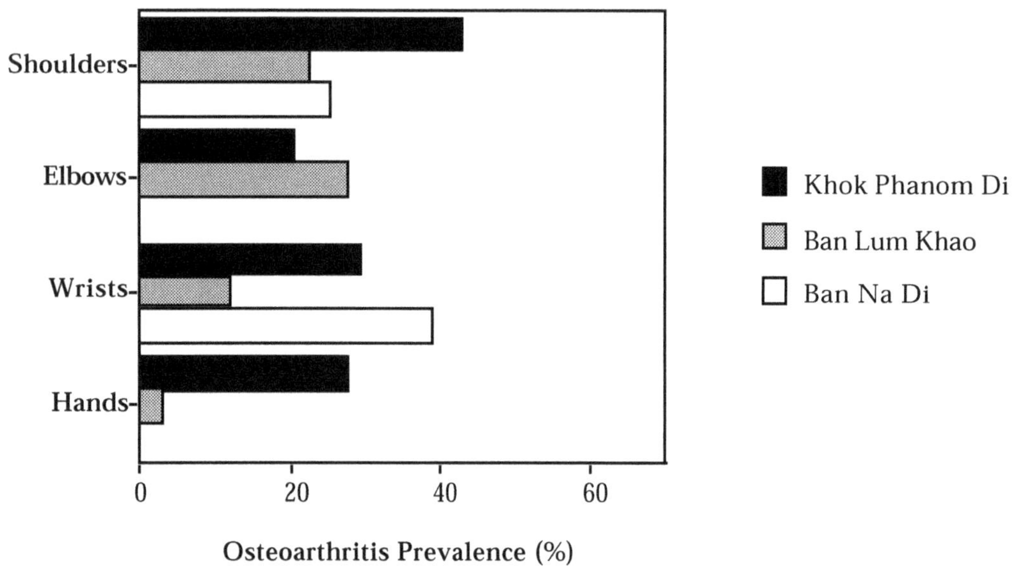


**Figure 5.8:** Osteoarthritis prevalences in the upper limb joints of the combined sex samples

**Table 5.22:** Osteoarthritis prevalences in the lower limb joints of the combined sex samples

| | Khok Phanom Di* | | Ban Lum Khao | | Ban Na Di | |
|---|---|---|---|---|---|---|
| Joint | A/N | % | A/N | % | A/N | % |
| Hips | 7/34 | 20.6 | 3/20 | 15.0 | 1/4 | 25.0 |
| Knees | 28/51 | 54.9 | 3/21 | 14.3 | 2/9 | 22.3 |
| Ankles | 11/50 | 22.0 | 3/30 | 10.0 | 2/14 | 14.3 |
| Feet | 7/44 | 15.9 | 6/31 | 19.4 | 7/20 | 35.0 |

A = number affected, N= number of individuals with both left and right joints + number of individuals with single joints with osteoarthritis. * Tayles (1992)

**Table 5.23:** Fisher's Exact p-values for comparing osteoarthritis prevalences amongst the combined sex samples at each lower limb joint

| Joint | Khok Phanom Di vs Ban Lum Khao | Khok Phanom Di vs Ban Na Di | Ban Lum Khao vs Ban Na Di |
|---|---|---|---|
| Hips | 0.73 | >0.99 | 0.54 |
| Knees | **<0.01** | 0.09 | 0.62 |
| Ankles | 0.23 | 0.72 | 0.65 |
| Feet | 0.76 | 0.11 | 0.32 |

Bold indicates statistical significance at 1% level.

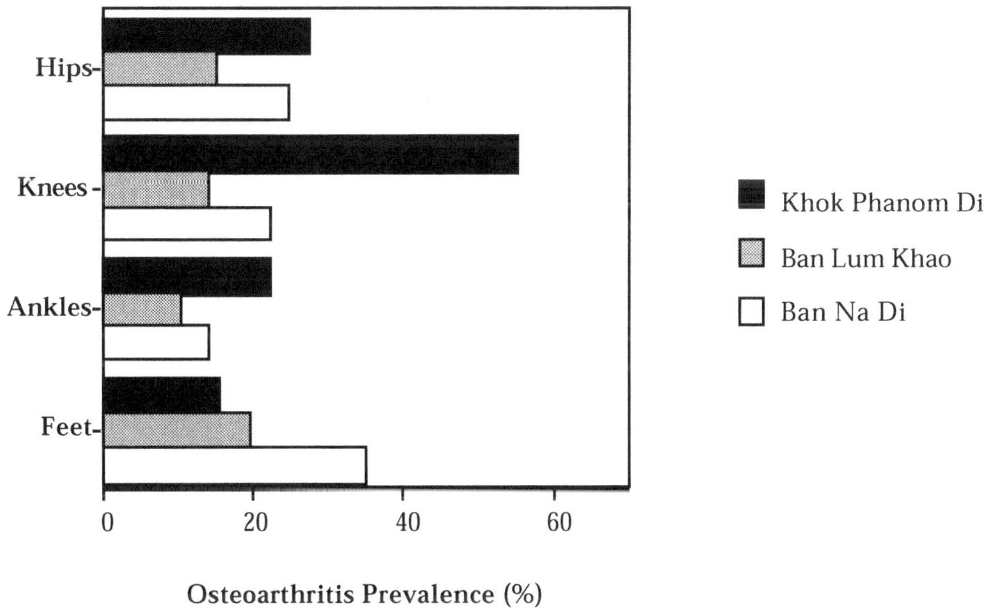


**Figure 5.9:** Osteoarthritis prevalences in the lower limb joints of the combined sex samples

**Table 5.24:** Vertebral osteophytosis prevalences in the combined sex samples

| Spinal Level | Khok Phanom Di* A/N | % | Ban Lum Khao A/N | % | Ban Na Di A/N | % | Comparison among samples Chi$^2$ Stat | p-value |
|---|---|---|---|---|---|---|---|---|
| Cervical | 11/64 | 17.2 | 1/32 | 3.1 | 2/18 | 11.1 | - | - |
| Thoracic | 15/64 | 23.4 | 5/40 | 12.5 | 5/21 | 23.8 | - | - |
| Lumbar | 34/64 | 53.1 | 11/34 | 32.4 | 9/22 | 40.9 | 4.05 | 0.13 |

N = number of individuals with vertebrae able to be observed. Chi$^2$ tests comparing the samples could not be performed on the cervical and thoracic data as samples sizes are too small, therefore FET are used. * Tayles (1992)

**Table 5.25:** Fisher's Exact p-values for comparing vertebral osteophytosis prevalences amongst the combined sex samples at the cervical and thoracic level

| | Khok Phanom Di vs Ban Lum Khao | Khok Phanom Di vs Ban Na Di | Ban Lum Khao vs Ban Na Di |
|---|---|---|---|
| Cervical | 0.06 | 0.72 | 0.29 |
| Thoracic | 0.21 | >0.99 | 0.29 |

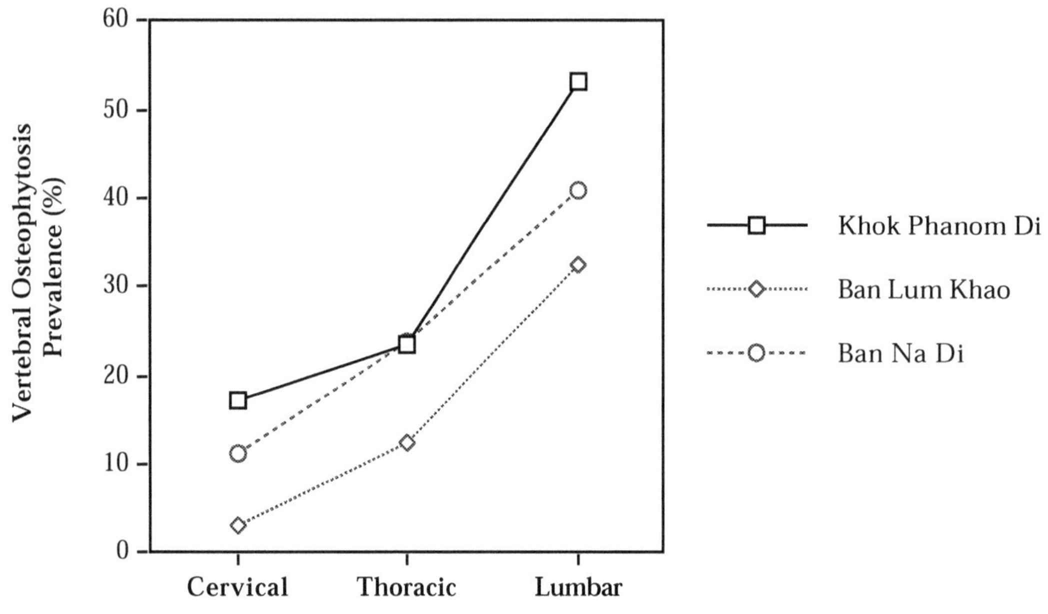


**Figure 5.10:** Vertebral osteophytosis prevalences in the combined sex samples

**Joint degeneration overall**

Osteoarthritis

To obtain an overall picture of osteoarthritis in each sample, the percentage of individuals with one or more joint affected by osteoarthritis was calculated. The results are presented in Table 5.26 and Figure 5.11.

For the total samples (males and females combined) the prevalence of osteoarthritis was highest in the Khok Phanom Di people compared with both the Ban Lum Khao and Ban Na Di samples. As the $Chi^2$ statistic and the Fisher's Exact p-values indicated (Table 5.26 and 5.27), this was a statistically higher result compared with the Ban Lum Khao sample. The level of osteoarthritis in the males of Khok Phanom Di was very similar to that of the Ban Lum Khao males and these were both higher, although not significantly so, than Ban Na Di males. In contrast, the Khok Phanom Di females were most similar to the Ban Na Di females.

Overall the Ban Lum Khao people showed the least amount of osteoarthritis, for males and females combined, but this prevalence of osteoarthritis was only slightly lower than that of the Ban Na Di sample. However, there were differences in the distribution between the sexes in these samples. The Ban Lum Khao people had a significantly higher level of osteoarthritis in the males compared with the females (FET$<$0.01). In contrast, the Ban Na Di sample showed the opposite to this, where the females had considerably higher levels of osteoarthritis than the males. However, the difference is not significant in the latter sample (FET = 0.14). These differences appear to directly reflect the age structures of each of these samples as will be discussed later. Khok Phanom Di males and females, in contrast to both of these groups, had a similar overall level of osteoarthritis between the sexes.

Two important points were evident from these results. First, a lower level of osteoarthritis was apparent between Khok Phanom Di and the two northeastern samples. Secondly, the distribution of osteoarthritis between the sexes was different in the Khok Phanom Di sample compared with the two northeastern samples.

Vertebral Osteophytosis

Summary results of joint degeneration data in the vertebral column are presented in Table 5.28 and Figure 5.12. Joint degeneration over the whole spine showed similar patterns to that of the appendicular osteoarthritis amongst the samples. Khok Phanom Di people showed the highest prevalences in both males and females. The people of Ban Lum Khao showed the least amount of vertebral osteophytosis, but within this sample the males showed a predominance of degeneration compared with the females. The vertebral osteophytosis in the Ban Na Di people was present at a moderate level with no contrast between the males and females.

The $Chi^2$ test that compared the total prevalences among the samples showed a significant difference. However, Fisher's Exact p-values failed to reach a significant level between any two samples (Table 5.29).

**Table 5.26:** Overall prevalence of osteoarthritis in each sample

| | Khok Phanom Di* | | Ban Lum Khao | | Ban Na Di | | Comparison among samples | |
|---|---|---|---|---|---|---|---|---|
| | A/N | % | A/N | % | A/N | % | $Chi^2$ stat | p-value |
| Female | 22/34 | 64.7 | 4/26 | 15.4 | 7/12 | 58.3 | 15.34 | **<0.01** |
| Male | 16/29 | 55.2 | 12/22 | 54.5 | 6/21 | 28.6 | 4.14 | 0.13 |
| Total | 38/63 | 60.3 | 16/48 | 33.3 | 13/33 | 39.4 | 8.85 | **0.01** |

A = number affected, N= number of individuals with one or more joint. Bold indicates statistical significance among the samples at 5% level; stat = statistic; * Tayles (1992)

**Table 5.27:** Fisher's Exact p-values for inter-sample comparisons of the overall prevalence of osteoarthritis

| | Khok Phanom Di vs Ban Lum Khao | Khok Phanom Di vs Ban Na Di | Ban Lum Khao vs Ban Na Di |
|---|---|---|---|
| Female | **<0.01** | 0.74 | 0.02 |
| Total | **<0.01** | 0.06 | 0.64 |

Bold indicates statistical significance at 1% level.

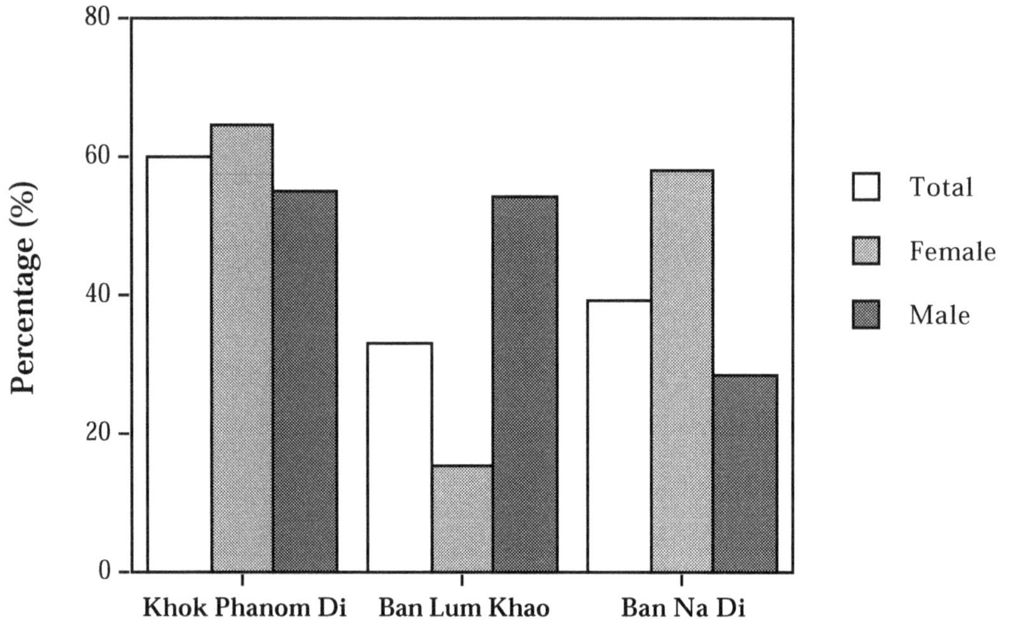


**Figure 5.11:** Overall prevalence of osteoarthritis

**Table 5.28:** Overall prevalence rates of vertebral osteophytosis in each sample

| | Khok Phanom Di* | | Ban Lum Khao | | Ban Na Di | | Comparison among samples | |
|---|---|---|---|---|---|---|---|---|
| | A/N | % | A/N | % | A/N | % | $Chi^2$ stat | p-value |
| Female | 18/35 | 51.4 | 6/25 | 24.0 | 6/14 | 42.9 | 4.59 | 0.10 |
| Male | 18/29 | 62.1 | 7/16 | 43.8 | 4/11 | 36.4 | 2.69 | 0.26 |
| Total | 36/64 | 56.3 | 13/41 | 31.7 | 10/24 | 41.7 | **6.26** | **0.04** |

A = number affected, N= number of individuals with vertebrae Bold indicates statistical significance among the samples at 5% level. stat = statistic. * Tayles (1992)

**Table 5.29:** FET results for inter-sample comparisons of overall vertebral osteophytosis prevalences

| | Khok Phanom Di vs Ban Lum Khao | Khok Phanom Di vs Ban Na Di | Ban Lum Khao vs Ban Na Di |
|---|---|---|---|
| Total | 0.02 | 0.24 | 0.43 |

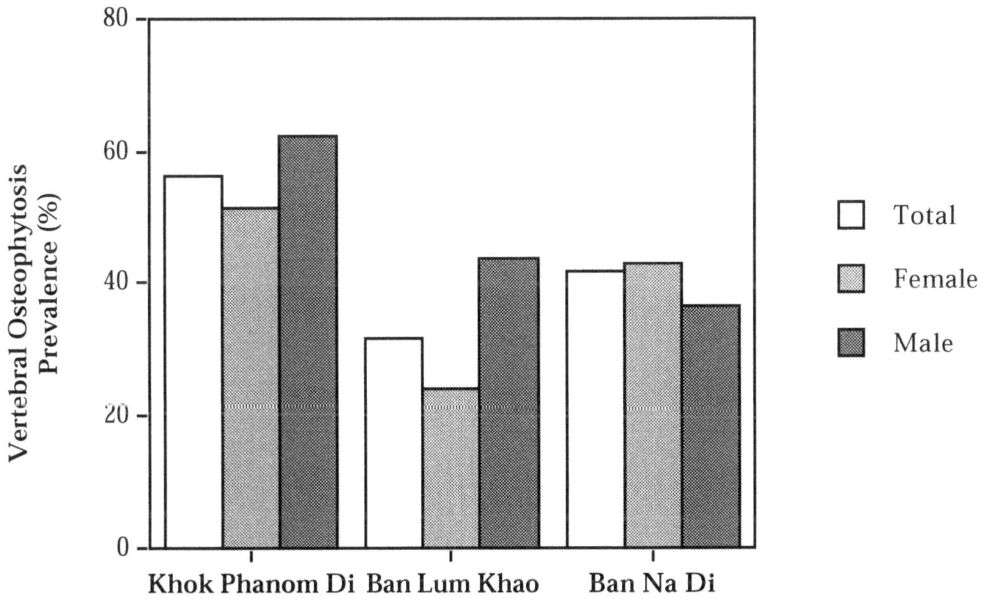


**Figure 5.12:** Vertebral osteophytosis prevalences over all regions of the spine

**Table 5.30:** Joints showing the most osteoarthritis among all the samples

| Sample | Joints with the most osteoarthritis | Sex |
|---|---|---|
| Khok Phanom Di | Shoulders<br>Hands<br>Knees | Males and Females |
| Ban Lum Khao | Feet<br>Elbows | Males only |
| Ban Na Di | Wrists<br>Feet | Males and females<br>Females only |

# Results Summary

Table 5.30 indicates the joints with the highest levels of osteoarthritis among the samples and which sex was predominantly affected. The males and females of Khok Phanom Di showed the highest prevalences of osteoarthritis in four out of the eight appendicular joints. In particular, the shoulder, hand, and knee joints were affected more than the northeastern samples. There was very little difference in the prevalence of osteoarthritis between males and females from Khok Phanom Di overall.

Ban Lum Khao males were predominantly affected by osteoarthritis in the joints of the feet and the elbows. These joints showed higher prevalences of osteoarthritis than the other samples. The differences between the males and females from Ban Lum Khao were quite significant, and in all joints males consistently showed higher prevalences of osteoarthritis. This difference between the sexes is most likely directly related to the age and sex structure of this sample which was comprised of predominantly young females and older males.

The results from the Ban Na Di people are taken with caution considering the small sample sizes available. However, two joints with adequate sample sizes had somewhat higher prevalences of osteoarthritis compared with the other samples. These were the wrist joints in both the males and females, as well as the joints of the feet in females only. Overall, females showed a predominance of osteoarthritis. However, this was not consistent over all individual joints. Females tended to have a higher prevalence of osteoarthritis than males in the lower limb joints, while males were more predominantly affected in the upper limb joints. The age and sex structure of this sample has resulted in some of the differences in joint degeneration prevalences observed. The sample of Ban Na Di was comprised of more younger males and older females explaining the higher prevalence of osteoarthritis in women. However, the higher prevalences in the lower limb joints of the male subsample is not an artefact of the age structure.

The vertebral osteophytosis analysis indicated all groups showed the same general pattern of degeneration among the different regions of the spine. An increase in the prevalence of vertebral osteophytes was evident from the cervical region through to the lumbar region, with the higher prevalence of vertebral osteophytosis evident in the lumbar vertebrae. The people of Khok Phanom Di consistently showed the highest levels of vertebral osteophytosis in the lumbar region.

# Discussion

The importance of assessing joint disease in any skeletal population is not just considering whether or not it is present. Osteoarthritis and vertebral osteophytosis are extremely common and are almost always present at some level in all archaeological populations. The focus instead is placed on the patterns of joint involvement.

Although joint degeneration can begin as early as the second decade (Moskowitz 1984), it is rarely of a debilitating nature at this stage, as far as is evident in modern populations. More severe joint degeneration is not commonly seen until greater than 40 years of age. However, as the majority of individuals in this study did not live much beyond 40-45 years, age is probably not the only reason for the expression of joint disease in these skeletal samples. Given the reduced life expectancy of these prehistoric groups and the difference in the distribution of osteoarthritis, mechanical factors may well be a factor in their aetiology. However, differential genetic susceptibility cannot be ruled out and for the purposes of this discussion genetic homogeneity is assumed. The aim in this section is to determine whether or not the pattern of joint degeneration in each sample holds some significance to their cultural and/or natural environment.

There is a generally higher level of osteoarthritis and vertebral osteophytosis in the males and females from Khok Phanom Di, compared with both of the northeastern groups. Unfortunately, due to the lack of comparative data from the Nong Nor people, it is difficult to suggest whether this difference may be associated with Khok Phanom Di's economic organisation or their natural environment. For example, cultural practices can influence the type and degree of joint use (Ortner 1968) through determining what type of subsistence technology is used. In comparison with the people of the northeast, the people of Khok Phanom Di had a subsistence economy that included minimal rice agriculture and more reliance on fishing and gathering coastal and estuarine resources. These differences in activities between the two areas may be a factor influencing the different pattern and severity of osteoarthritis. Perhaps the general lifestyle at Khok Phanom Di placed more mechanical stress on individuals, as they had to partake in a generally physically harder workload than in the northeast. The intensification of rice agriculture in the later northeastern samples may have led to a less stressful workload. Alternatively, the problems with age estimation, discussed later may also be influencing these differences.

Roosevelt (1984) concluded in her synthesis of a number of studies presented in Cohen and Armelagos (1984a) that there was, in general, a reduction in the amount of physical activity after the development of agriculture. This appears to be the case in the Midwest and Southeast United States, Western Europe, Eastern Mediterranean, the Near East and South Asia. However, this is not a universally consistent finding. For example, the study by Goodman *et al.* (1984a) of the Dickson Mound skeletal samples showed a significant increase in degenerative pathology with time, claiming the increase in agriculture saw an increase in physical workload. In their summation of these studies, Cohen and Armelagos (1984b) acknowledged these mixed results but still they consider, on balance, that there was a reduction in physical activity with the adoption of agriculture. However, it is stated that the hunter-gatherers in these studies had a higher average age at death compared with the farmers, therefore it would be expected that the hunter gatherers would have more arthritis through having lived longer. This indicates the importance of considering the age structures of the samples and how this may influence the results. In a more recent review of the effects of the development of agriculture on joint degeneration, Larsen (1997) concurs with the conclusions of Cohen and Armelagos (1984b), although again a few exceptions are acknowledged.

The results from prehistoric Thailand support the general trend of decreasing workload with the intensification of agriculture. However, there is the possibility that this generally higher level of osteoarthritis in earlier times, is not solely due to a harder life. It is possible that the aging methods do not accurately discriminate among the older adults. Given the higher prevalences of joint degeneration found in the Khok Phanom Di sample, it is possible that this sample was comprised of more older adults that the age methods have allowed for. It is a limitation in all skeletal assessments of age that the present methods available are more difficult to apply to older adults, with few, if any methods that rely on factors other than those of a degenerative nature. However, considering the aging methodologies in this study were all as similar as possible over the three samples, it should be that any bias present has been present in all three samples.

There are a number of known aetiological factors for the development of osteoarthritis, but often the initiating cause is unknown. Even in modern clinical situations, determining the initiating factor for each case of osteoarthritis is far from straightforward and sometimes impossible. Because of the various factors that could potentially be involved, rarely can a single main determinant, for example mechanical factors, be singled out with any confidence (Jurmain 1991a; Rogers and Waldron 1995).

It is apparent, from archaeological evidence, that the subsistence activities were different in some respects between the northeast and the southeast of prehistoric Thailand. However, within the northeast it cannot be as confidently stated that the people of Ban Na Di and Ban Lum Khao, only separated by a few hundred years, were partaking in greatly different subsistence strategies. The Ban Na Di people may have been undertaking more rice agriculture with more advanced tools, but this probably involved the same type and level of activity. This is supported by the similarity in overall prevalences of osteoarthritis (Table 5.26). The most likely explanation for the differences between the sexes is the different age structures of these two northeastern samples.

Fishing from boats would probably have been a common occurrence for a coastal community like Khok Phanom Di, more so than in the landlocked northeast. The sea may also have been more demanding in muscular effort to paddle in compared with the rivers in which the people of the northeast may have fished. It has previously been suggested that one possible reason for the high level of shoulder osteoarthritis, as well as the development of particular upper arm and shoulder girdle entheses, particularly in the males, and the degeneration of the spine, may be consistent with the movements involved in paddling a canoe or small boat. It is acknowledged that this is only one possible explanation (Tayles 1999).

A further explanation concerning why the prevalence of shoulder osteoarthritis and other joints were higher in the Khok Phanom Di sample is related to the preservation of this sample as mentioned in the methods section. The Khok Phanom Di skeletal collection was very well preserved and therefore it was more likely that observations could be made for these joints, and all other joints, as they were more likely to be present. This may be particularly relevant in the assessment of osteoarthritis in the hands and feet. The lower sample sizes due to poorer preservation in the other samples may have lead to an underestimation of the true prevalence of osteoarthritis.

There is a significantly higher level of osteoarthritis in the hands of the Khok Phanom Di people compared with the northeastern groups which has been suggested to have been influenced by the manufacturing of pots (Tayles 1999). This possibility was further investigated by Domett (1994, 1998) because, from archaeological evidence, "It is evident that high energy expenditure and artistic skill..." were involved in the manufacture of the burial pots by the people of Khok Phanom Di (Hall 1993: 273). The study involved a focused assessment of osteoarthritis in all the joints of the hand similar to the approach presented by Hadler *et al.* (1978). Khok Phanom Di males had a similar pattern of joints affected in both left and right hands while females, in contrast, had quite different patterns of osteoarthritis between their left and right hands. These results imply different patterns of usage between males and females which could suggest that there were some sex-specific activities carried out in this population. It has been suggested that the women carried out most of the pottery work based on the distribution of anvils and burnishing pebbles commonly in the female burials (Moore 1993). The use of an anvil and paddle in making pottery involves each hand performing a different task which may well relate to the difference in joint degeneration between the left and right hands of the women.

The knee joints are also predominantly affected in the Khok Phanom Di people and while this is commonly associated with obesity in modern populations, this is unlikely to be the explanation in prehistory. Osteoarthritis of the knee has been linked with that of the hand in clinical studies (Cushnagen and Dieppe 1991, Hirsch *et al.* 1996 cited in

Waldron 1997). A person who has developed osteoarthritis of the hand is also more likely to develop osteoarthritis in other joints leading to generalised osteoarthritis (Doherty *et al.* 1998), although the factors behind this association are not clear. Given the high prevalence of hand and knee osteoarthritis in the people of Khok Phanom Di, it is possible there were a number of individuals suffering from generalised osteoarthritis and/or this sample was comprised of older individuals as mentioned earlier.

The predominant joints affected in the northeast are the joints of the feet. The sex differences in osteoarthritis between the two northeast samples is an artefact of the age and sex structure of these samples. However, the total prevalence in each sample can be considered relatively reliable as the effects of age on osteoarthritis prevalence within each sex, within each sample, are reduced when the data for each sex are combined. The feet are more likely to be exposed to repeated microtrauma than many other joints and this may explain the high prevalences of osteoarthritis seen in these joints. There is also the possibility that these people had to walk longer distances to obtain the necessary variety of resources. However, there is no reason to expect this was the case as both the sites of Ban Lum Khao and Ban Na Di appeared to have a good supply of local resources. However, the dry season can be harsher in the northeast compared to coastal regions. This may have lead to a reduction in immediate resources for periods of each year resulting in an expansion of the area exploited.

The predominance of osteoarthritis in the elbow joints of males of Ban Lum Khao is at least partially related to the relatively higher incidence of forearm fractures experienced by these people in comparison to the other samples. Of the four males with forearm fractures, two had articular surfaces of the elbow able to be assessed and both showed evidence of osteoarthritis. Chapter 6 discusses the significance of this more fully.

The involvement of the wrist in osteoarthritis at a relatively high frequency in the people of Ban Na Di is a rather enigmatic finding, especially given the absence of osteoarthritis in the hands of these people. The wrist and hand commonly act as a functional unit, therefore it is surprising to see joint changes in one without the other.

Two other skeletal samples from the northeast of Thailand, those from Non Nok Tha and Ban Chiang, span the time that includes the samples from Khok Phanom Di through to early Ban Na Di. Both of these samples show osteoarthritis more commonly in the elbow, wrist, ankle, and feet (Douglas 1996). Except for the involvement of the ankle, this is a strikingly similar pattern to that seen in the people of Ban Lum Khao and Ban Na Di. This evidence would appear to support that the osteoarthritic patterns seen in the northeast and the southeast are geographical patterns, rather than one influenced by the cultural environment.

The second main point to be considered is the increase, through time, in the differentiation of males and females affected by osteoarthritis. There is no such sex difference in the people from the earlier site of Khok Phanom Di. However, there is a marked sex difference in both the Ban Lum Khao sample, where the males are predominantly affected, and in the Ban Na Di group, where the females are more often affected overall. As will be shown, these sex differences in the northeastern samples are in fact most likely directly related to the age and sex structures of these samples.

Through time, populations may become more complex, involving a greater organisation of labour which may lead to sexual divisions. Therefore, different joints may be more mechanically stressed in each sex. The males from Ban Lum Khao developed more osteoarthritis in all joints and more vertebral osteophytosis in all spinal regions compared with females. Therefore, at this site it may have been the males who performed the heavy manual labour. However, a simpler explanation is at hand. The presence of a higher number of older males compared with older females at Ban Lum Khao may have been aberrantly creating this result. The significantly lower level of osteoarthritis in females may have to do with the age at which they began their involvement in physical labour. The later they began, the less exposure to mechanical stress. While it may be more interesting to suggest males began earlier than the females thus developing more joint degeneration from longer exposure to mechanical stress, the more likely explanation again concerns the age structure. The higher proportion of young females in this sample has contributed to the lower prevalence of osteoarthritis, as these young women have not had time to develop degenerative changes.

A similar situation is present in the Ban Na Di sample, although here females were affected more frequently by osteoarthritis than males. The higher prevalence of females with osteoarthritis in this late prehistoric sample approaches the pattern observed in modern Thailand (Chaiamnuay *et al.* 1998). However, the following discussion suggests some of the osteoarthritis patterns may not be truly representative of this sample. In the Ban Na Di sample, the higher prevalence of osteoarthritis in females, compared with males, is restricted to three lower limb joints, the knees, ankles, and feet. The males, in fact, show higher rates than females in three upper limb joints as well as the hip. The fact that females overall are showing more osteoarthritis could be a problem with the sample size creating this result. Table 5.26 indicates that while a similar number of males and females had osteoarthritis in one or more joints, the male sample size was nearly twice as large therefore creating a prevalence half that of the females. However, the individual joints showing high prevalences, the wrists and the feet, in comparison with the other samples do have reasonable sample sizes and are considered reliable results.

The individual underlying genetic susceptibility to developing osteoarthritis may also have been different in males and females in both of these northeastern samples. Cultural customs, such as males who move into the village of his wife on marrying, as occurs in some parts of Thailand today, could partly explain the presence of different genetic susceptibilities between the sexes.

During the assessment of joint changes in the Thai skeletal sample few cases of very severe osteoarthritis were observed. Therefore, the level of disability suffered by those with osteoarthritis is unlikely to have been overly debilitating. Symptoms may even be less severe in a warmer climate (Moskowitz 1993) as would have been present in prehistoric Thailand.

When comparing this study of joint degeneration with that from other parts of the world, the focus is on the patterns of joint involvement rather than the actual level of involvement. This is carried out in order to reduce interobserver differences in the classification of osteoarthritis and also to gain an impression of how the prehistoric Thai people compare to other regions. In a comparative study of a hunter-gatherer group and an agricultural group from prehistoric northwest Alabama very few significant differences were found in the pattern of joints affected (Bridges 1991). Osteoarthritis in the shoulder, elbow, and knee joints was the highest in both groups and low in the hip and ankle. The joints of the hands and feet were not assessed. These general trends correlate, to some degree, with that observed in prehistoric Thailand. However, Bridges (1991) found few sex differences between the two samples she studied. While the hunter-gatherer group overall had the higher frequencies of osteoarthritis the differences were not always statistically significant. In general, Bridges concluded from her study and a review of others, that the patterning of osteoarthritis was not clearly related to the subsistence economy with both hunter-gatherer and agriculturalists having similar patterns of osteoarthritis (Bridges 1991).

In another part of the world an analysis of joint degeneration was carried out on three samples varying in time, one each from the premedieval, medieval, and the postmedieval periods in England (Waldron 1995). The shoulder and spine were the most commonly affected by osteoarthritis, although levels proportionally decreased with time. There was also a general decrease in the involvement of osteoarthritis in the hip and wrist joints, while the knee and hand joints increased in their prevalence of osteoarthritis. However, only some of the differences were significant. As in the Thai samples, similarities were found in spinal degeneration in all samples in historic England. It is suggested this is a function of locomotion and weight bearing rather than specific activity (Waldron 1995).

Knüsel *et al.* (1997) also picked up on this theory as the three subsamples of different status skeletons they studied from medieval England also showed very similar trends in spinal degeneration along the spine, in both vertebral osteophytosis, and osteoarthritis of the zygapophyseal joints. The spine is more under the influence of biomechanical factors in keeping the body upright and transmitting the body weight during locomotion. Knüsel *et al.* (1997) concluded that given this inherent biomechanical function of the spine and the similarity in the pattern of joints affected in the spine among different populations, analysis of the joints of the spine are not very useful in the study of physical activity. However, this is a strong statement given the positive association of vertebral osteophytosis with physical activity seen in some studies (for example Videman *et al.* 1990; O'Neill *et al.* 1999). Additionally, the question is raised of why the joints of the vertebral column are more 'functionally constrained' than other joints such as the lower limb joints that must also be both mobile and stable during locomotion.

In the Thai samples a similar pattern of joints affected by vertebral osteophytosis is evident and thus, given the preceding discussion, precautions against inferring the level of physical activity among these samples. However, the actual frequency of vertebral osteophytosis does differ among some of the samples. The higher prevalence of vertebral osteophytosis in the lumbar vertebrae of the Khok Phanom Di people may imply more weight bearing activities were carried out by them. On the other hand, this may also be explained by the better quality of preservation of the material. It may also have been observed, as discussed, because of the lack of discrimination of the age at death estimation of older adults where, in fact, the Khok Phanom Di sample, may have comprised more adults older than estimated compared with the northeastern samples that available aging methods are not able to identify.

The pattern of joint involvement in osteoarthritis in prehistoric Thailand does not correlate in all respects with the epidemiology of osteoarthritis in modern populations. The shoulder, elbow, and ankle joints are some of the least commonly affected in modern groups and are principally affected in the very elderly only (Doherty *et al.* 1998). The ankle was rarely involved in Thai prehistory, although the shoulder was one of the most commonly affected joints. In general modern populations show the interphalangeal joints (hands and feet), and the hip and knee joints as the most commonly affected by osteoarthritis (Doherty *et al.* 1998). While the hand, feet, and knee joints were commonly affected in some of the prehistoric Thai samples, the hips were rarely involved. Osteoarthritis of the knee and hand joints were the most common in Chaiamnuay *et al.* (1998) study of modern Thai which correlates with the epidemiology seen in the Khok Phanom Di sample.

The hip joint in Oriental (from India to southeastern Asia, Borneo and the Philippines) and Chinese populations appears to be protected in comparison with other parts of the world (Hoaglund *et al.* 1973; Chaiamnuay *et al.* 1998; Doherty *et al.* 1998). Given the low frequency of hip osteoarthritis in Thai prehistory this may be a pattern that has long been established perhaps in relation to the squatting position adopted by so many Asians. This position puts the hip joint through its full range or movement and this mobility may be protective against developing osteoarthritis (Harrison *et al.* 1953 cited in Hoaglund *et al.* 1973).

Further studies that may shed more light on these patterns of joint degeneration in prehistoric Thailand include the analysis of osteoarthritis prevalences by side (left and right) or a more detailed analysis of which parts of each joint show the degenerative changes. Combine this with the study of muscle markings (enthesopathies) and a more detailed portrayal of physical activity and how it may or may not be related to joint degeneration may be found. Some researchers (for example Dutour 1986; Kennedy 1998) believe the study of musculoskeletal stress markers may be a more useful indicator of physical activity because of their association with prolonged hyperactivity of a particular muscle. However, recent work suggests some people may be more likely to form bone than others to begin with (Rogers *et al.* 1997). This may provide an alternative explanation for the heterogeneity of the outcome in the response by the musculoskeletal system to stress and injury, including the development of osteophytes and osteoarthritis (Rogers *et al.* 1997).

## Summary

The main points identified from the analysis of osteoarthritis was firstly that the pattern of joints most predominantly affected within each sample was different among the samples. Khok Phanom Di adults were particularly affected in the shoulder and knee joints, and the joints of the hands. This pattern is evident in both males and females. Both males and females also showed the same predominance of vertebral osteophytosis in the lumbar region. However, males consistently had higher rates than females in the cervical and lumbar regions.

In the males from Ban Lum Khao osteoarthritis in the elbows and joints of the feet were high compared to the other samples. This sample also showed males with consistently higher prevalences of both osteoarthritis and vertebral osteophytosis than females. Given the association of joint degeneration with age, the fact that this sample had a high proportion of young females may be affecting these results.

Small sample sizes for the Ban Na Di sample probably prevented a true picture of the presence of joint degeneration being obtained. However, some joints were reasonably represented and of those, the wrist joints in both males and females as well as the joints of the feet in females had quite high prevalences of osteoarthritis in comparison with the other samples. Ban Na Di females show higher levels of vertebral osteophytosis compared with the other samples, and compared to their male counterparts, particularly in the cervical and thoracic regions. The predominance of females affected by these joint diseases may at least be partially explained by the higher numbers of older females than males.

The second main point is that there is a trend for decreasing osteoarthritis with time. Khok Phanom Di adults had a generally higher prevalence of this pathology compared with the other samples that could be assessed, Ban Lum Khao and Ban Na Di. Vertebral osteophytosis prevalences do not show as clear patterns with time. In general the Khok Phanom Di and Ban Na Di samples show the higher prevalences.

Confounding problems in the interpretation of such patterns of osteoarthritis and vertebral osteophytosis include the age and sex structures of some of the samples. Additionally, while some explanations have been suggested concerning the types of activities influencing these patterns, it is emphasised that these are purely suggestions. There are possibly other alternative explanations that include genetic factors.

# 6 Trauma

## Introduction

> The record of trauma imprinted upon an individual skeleton may contain fascinating information, not always decipherable, about a lifetime of encounters with the environment and fellow humans. (Merbs 1989: 184)

The assessment of traumatic injuries can not only provide information about an individual's relationship with his/her environment, but observations over a whole skeletal sample can provide information in a more general way about the sample's general lifestyle conditions. "Prevalence of skeletal injuries by element type as well as regional temporal patterns give insight into the influence of different lifestyles" (Larsen 1997: 110).

Trauma indicates an extrinsic influence has affected the skeleton (Lovell 1997). This might include the environment and physical activity associated with subsistence patterns. However, the nutritional status and genetic composition of an individual can predispose them to fracture, as mentioned in connection with loss of bone mass in osteoporosis (Mazess 1982) (see Chapter 4).

The analysis of trauma in the Thai skeletal collections has been restricted to the description of two types of bony fractures. The first type includes all those that are non-vertebral. This includes long bone fractures as well as craniofacial fractures and other fractures of the axial skeleton other than the vertebrae. The second type is restricted to a particular form of vertebral fractures called spondylolysis. These two general types of fractures are common and were chosen for systematic analysis. Vertebral compression fractures are not included in this analysis as they can often be linked with pathological processes, such as osteoporosis. Other isolated skeletal lesions possibly indicative of acute trauma are not included as these cannot be systematically compared among samples.

### Non-vertebral fractures

Definition

A fracture may be an incomplete or complete break of a bone with the bony ends piercing the skin ('compound') or not ('simple'). When a break occurs at the point of impact this is the result of a direct force which may cause a transverse, penetrating, comminuted, or crush type fracture. Indirect trauma is where a fracture occurs in a place other than at the point of impact and may be an oblique, spiral, greenstick, impacted, burst, or an avulsion type fracture (Lovell 1997). By determining the type of break, the type of force may be discovered and consequently the cause of the fracture.

Aetiology

A fracture is often the result of abnormal stress applied to bone with the stress being dynamic, static, or chronic. However, pathological conditions can weaken bone so that even minimal stress results in a break (Ortner and Putschar 1985). The force creating the stress may be tension, compression, torsion or twisting, flexion or bending, or shearing (Ortner and Putschar 1985). These can generally be termed either direct or indirect force. These aspects may provide insights regarding the most common types of traumatic stresses in an archaeological population as discussed later.

The clinical literature details the extensive array of different types of fractures possible, their cause, and potential complications. The majority of fractures seen clinically are the result of an accident (Lovell 1997; Smith 1996). Interpersonal violence is another possible cause. However, some individuals may be more predisposed to fracture through an underlying pathological condition. Certain diseases and disorders can result in abnormal bone tissue that leaves such bones more vulnerable to fracture than normal bone, as mentioned above. The force required to fracture abnormal bone is much less than that required to fracture normal bone and can occur through daily activity or with only minor trauma (Springfield and Jennings 1991). Osteoporosis is one of the more common predisposing conditions, but infection, developmental abnormalities, and Paget's disease (increase in bone resorption and formation) (Springfield and Jennings 1991; Aufderheide and Rodriquez-Martin 1998) may also predispose an individual to fracture of a bone.

The accumulation of microfractures from repetitive forces can also eventually lead to a complete fracture. This is termed a stress or fatigue fracture (Aufderheide and Rodriquez-Martin 1998). These types of fractures are frequently seen in those undergoing vigorous activity such as highly trained athletes (Harkess *et al.* 1991). They are typically seen in metatarsals, the calcaneus, and the tibia (Lovell 1997) as these bones bear much of the body weight and must absorb the force as the feet hit the ground. Muscle fatigue and the resultant abnormal loading on bones is thought to be one of the mechanisms involved (Harkess *et al.* 1991). Since bone has the ability to repair, rest and protection from the stress will allow microfractures to heal (Harkess *et al.* 1991).

Pathology - repair and complications

A fracture begins to heal immediately after the bone has broken (Lovell 1997). The healing process can be divided into three main stages, namely inflammation, repair, and remodelling. These process form a continuum and the rate of healing may be influenced by both local and systemic factors (Buckwalter and Cruess 1991; Resnick *et al.* 1995b).

A fracture not only damages bone, cells, and blood vessels but also the overlying soft tissues including the periosteum and muscle (Buckwalter and Cruess 1991). In the first stages immediately after a fracture, a haematoma (a mass of clotted blood) forms from the broken blood vessels within the medullary cavity and beneath the periosteum. Osteocytes (bone cells) and other cells are no longer supplied with nutrients and the resulting necrotic tissue induces an intense, acute inflammatory response (Resnick *et al.* 1995b). This in turn begins the migration of the cells involved with repair and the removal of necrotic tissue (Buckwalter and Cruess 1991).

Repair cells have the capacity to form all different tissues such as bone, cartilage, and blood vessels. Initially the cells are undifferentiated and are involved with the synthesis of a new matrix (Buckwalter and Cruess 1991). Fibrovascular tissue invades the haematoma (Resnick *et al.* 1995b) and this initiates its conversion into a fibrocartilaginous callus. The callus envelops the bone ends and produces stability at the fracture site (Resnick *et al.* 1995b). The cartilage in the callus is eventually ossified and the resulting bony callus provides increased stabilisation of the fracture by acting as a splint between the two bone ends (Resnick *et al.* 1995b).

The remodelling stage of fracture healing is the longest process. It can continue for many years but is usually near completion in a matter of months. Remodelling reshapes the bony callus by removing and replacing cells and matrix (Buckwalter and Cruess 1991). Trabeculae are formed along the lines of stress and are subsequently remodelled to perform the functions demanded of the bone as it was originally (Resnick *et al.* 1995b).

The healing process is influenced by a number of factors including the bone fractured, the type of fracture, the severity of the injury, what tissues other than bone are affected, and the age and nutritional status of the patient (Buckwalter and Cruess 1991; Roberts and Manchester 1995). For example, more force is required to induce a fracture of the femur than the ulna, given the difference in shaft thickness. The fracture of the bones of the hand or feet are less likely to be of a significant disability than that of the larger limb bones.

Complications arising from fractures in ancient populations are more likely to occur than in modern populations because of the lack of medical knowledge (Ortner and Putschar 1985). Nevertheless, more violent fractures occur in modern times and have complications of their own, most requiring surgery. Common complications in prehistory include infection. This would likely develop secondary to a compound fracture. Infection can result in delayed healing, especially if living conditions are less than hygienic. In skeletal remains signs of infection may be indicated by the presence of a localised periosteal reaction, evident as an area of new bone on the subperiosteal surface, or osteomyelitis (inflammation of the bone marrow and adjacent bone) (Lovell 1997).

Bone deformity, inadequate or non-fusion of bones, or joint fusion can all result from the malalignment of the fracture ends caused by inadequate reduction or immobilisation. Secondary arthritis may also result through abnormal stress placed on a joint adjacent to the fracture site or as a result of a shift of biomechanical stress on to part of the joint not use to such stress. The fracture does not necessarily need to include the joint for this to occur.

Bones from younger individuals will generally heal more quickly than those of older people because they generally have a better remodelling rate. In the elderly, the outcome will be less likely to restore normal structure and function (Buckwalter and Cruess 1991). The repair process also requires substantial energy input so that the metabolic state of the person can affect the outcome of the repair process. For example, a fracture may fail to heal in severely undernourished individuals (Buckwalter and Cruess 1991). The fact that an individual has been able to repair a fracture with few complications is generally indicative of a good level of health and their immune response has been adequate in response to infection.

There are other complications that can result from a fracture and lead to poor healing. These include tissue necrosis, nerve damage, haemorrhage, myosotis ossificans (muscle ossification), and trauma to overlying muscle, tendon or ligament resulting in mobility problems (Ortner and Putschar 1985).

Biocultural interpretation

Lovell (1997) recommends classifying fractures according to their predominant characteristics. In this way a single cause is not immediately inferred before all possible evidence of other causes are considered. For example, labelling a forearm fracture as a 'parry' type fracture implies a violent causation that is not necessarily true of all forearm fractures. Forearm fractures can in fact occur from many different causes including, most frequently, falls.

By systematically analysing fractures in a skeletal sample much information may be derived concerning the cause of the fracture and the subsequent effect of the injury on the individual. Lovell (1997) provides a descriptive protocol that involves recording four aspects of a fracture for each fracture analysed. They are the maximum length of the bone, the apposition of the fracture ends, the degree of rotation, and the degree of angulation of bone at the fracture site.

By comparing the maximum length of the fractured bone with the unfractured bone from the opposite side, the degree of shortening can be determined (Lovell 1997). This may indicate whether the fracture was effectively reduced (Roberts 1991) and would have implications for determining the level of medical knowledge and care provided within a population.

Apposition is the percentage of bony contact between fragment ends and is measurable by x-ray (Lovell 1997). However, in the absence of radiographs the direction of shifting of the distal fragment from the proximal end can be recorded. This also provides information on how well the fracture was splinted and reduced (Roberts 1991).

When the distal fragment has turned relative to the proximal end this is termed rotation. It may be described as internal (medial) or external (lateral) depending on the direction of the rotation. Adjacent joint problems may be apparent if rotation is observed. The angulation of a fracture may also be recorded in degrees when possible (Lovell 1997).

The quality of healing of a fracture in a skeleton can be indicative of the knowledge or lack of knowledge regarding medical procedures in prehistory. The lack of marked shortening or angulation indicates an excellent healing process which implies both the knowledge and ability to care for people with traumatic lesions (Lovejoy and Heiple 1981). The type and location of a fracture may also indicate the cause was an accident rather than a deliberate act, such as interpersonal violence.

In interpreting the patterns of trauma in a prehistoric population three aspects of the injury are taken into consideration. First the characteristics of each fracture are determined. This includes describing the fracture itself in the terms defined above, subsequent complications, and determining the likely cause. Secondly, the pattern of trauma in the population can be analysed, including age and sex differences within and among samples. Thirdly, the social and cultural history, and the environmental context of the population can be considered (Lovell 1997).

The incidence and location of traumatic events is influenced by culture. For example, hunters will probably have had a different pattern than sedentary farmers. Patterns of fractures in females often differ from those of males, and adults from children (Ortner and Putschar 1985) if there is a difference in the level of risk, perhaps as a result of divisions of labour.

There are some limitations in the analysis of fractures in skeletal remains. These have been reviewed in (Roberts 1991) and are outlined briefly here. First, there is no way of knowing at what age the fracture in an individual occurred. Second, some fractures, especially in subadults, can heal extremely well, leaving no sign of their occurrence. Small, but significant, stress fractures may not be easily identified unless every bone is radiographed. Fractures occurring just before death may be indistinguishable from postmortem fractures because of burial or excavation practices.

Fracture rates can be difficult to compare between populations as often the level of completeness of the sample is not taken into account. The presence of each element should be counted and used as the denominator, not the number of individuals.

### Spondylolysis

Spondylolysis is the osseous separation of a vertebra into anterior (body) and posterior parts (arch). The separation is usually, but not necessarily, complete (Merbs 1995). The two parts are normally held together by a soft tissue connection called the spondylolysis ligament on either side (Merbs 1996). The exact location of the separation may vary and it is most commonly seen through the pars interarticularis. However, the pedicle, the transverse process, and the lamina, are also possible sites of separation (Merbs 1996). Spondylolysis most commonly affects the lumbar vertebrae and the fifth is particularly vulnerable (Resnick *et al.* 1995b).

The aetiology of this condition is under dispute. One theory states spondylolysis is a congenital abnormality with failure of the arch to ossify with the body. The alternative is the stress fracture theory. This suggests that it is repetitive microtrauma that results in the complete separation through the arch. The reality is probably a combination of both these theories, with some individuals having a genetic predisposition to a weakness in the pars interarticularis with fracturing occurring after chronic stress to the back (Resnick *et al.* 1995b; Aufderheide and Rodriquez-Martin 1998).

Spondylolysis results most frequently from a chronic fatigue fracture, although acute trauma can also induce this lesion (Resnick *et al.* 1995b). Considerable force would be required to fracture the arch in a single traumatic event and can usually be distinguished from a fatigue fracture by the presence of damage to adjacent vertebrae (Merbs 1996).

A spondylolytic fracture is commonly acquired during childhood after a traumatic episode or series of episodes. The incidence of this condition does not increase after 20 years of age (Resnick *et al.* 1995b). However, archaeological studies on skeletal samples have rarely found cases of spondylolysis in children; spondylolysis is more commonly found in older individuals (Merbs 1996).

The prevalence of spondylolysis in the general population ranges from three to seven percent and there is a slight predominance in males compared with females (Resnick *et al.* 1995b). An association with activities that concentrate abnormal stress in the lumbosacral spine have been reported with the presence of spondylolysis. For example, those involved in heavy labour, particularly involving repetitive and vigorous movements of the back, especially hyperextension, have frequently suffered from spondylolysis (reviewed in Merbs 1996).

In some cases there may be no symptoms or signs associated with spondylolysis. Others may experience back pain, tenderness, gait abnormality, or neurological problems (Resnick *et al.* 1995b). Generally pain is of a mild, constant level, but not acute enough to require specific attention (Roberts and Manchester 1995; Merbs 1996). When the separation of the arch from the vertebral body occurs bilaterally (as it often does) the inferior articular process is separate and can allow the body to slip forward relative to the vertebra below. This is called spondylolisthesis and can be associated with spondylolysis but it can also occur alone (Merbs 1996). This condition can have neurological complications.

The significance of studying spondylolysis in a prehistoric sample is in its cause. While an individual may not be significantly debilitated, the presence of spondylolysis does imply they underwent either chronic loading on the back or suffered from an accident.

**Table 6.1:** Fracture rates of the major appendicular bones in each prehistoric Thai sample

| Bone | Khok Phanom Di | | Nong Nor | | Ban Lum Khao | | Ban Na Di | |
|---|---|---|---|---|---|---|---|---|
| | A/N | % | A/N | % | A/N | % | A/N | % |
| Clavicle | 2/106 | 1.9 | 1/18 | 5.6 | 1/39 | 2.6 | 0/15 | 0 |
| Radius | 0/102 | 0 | 0/13 | 0 | 3/48 | 6.3 | 0/18 | 0 |
| Ulna | 0/74 | 0 | 2/9 | 22.2 | 3/40 | 7.5 | 0/10 | 0 |
| Femur | 0/85 | 0 | 0/17 | 0 | 0/37 | 0 | 1/13 | 7.7 |
| Fibula | 0/48 | 0 | 0/0 | 0 | 1/29 | 3.4 | 0/4 | 0 |

A= number of bones affected; N= number of bones observed.

# Methods

Every bone present was assessed visually for signs of fracture. When identified they were described in detail in terms of the degree of shortening, apposition, angulation, and rotation. Any signs of complications were also noted, for example, an association with osteoarthritis in an adjacent joint or periosteal infection. Long bone length measurements were recorded for both the bone affected and the opposite bone to determine the degree of shortening. All other details are qualitatively described. The presence of all complete long bones irrespective of pathology were counted in order to calculate the fracture rate for each bone element. The fracture rate is calculated by dividing the number of bones with fractures by the number of complete bones present in the sample and is presented as a percentage. This is carried out for each bone type. This calculation is limited to fractures in the larger limb bones in this skeletal samples, and not in the small bones of the hand and feet where the presence of a fracture was noted but not calculated as a rate.

Spondylolysis is identified by the presence of a separate arch and vertebral body. This is distinguished from postmortem fracture by the remodelling of both surfaces of the fractured parts, namely in the region of the pedicle, pars interarticularis, or the transverse process. Results are described for each sample and numbers of each sex affected are provided. No fracture rates are provided as the total number of vertebrae observed were not recorded for all the samples.

# Results

### Non-vertebral Fractures

A description of each fracture in all of the skeletal samples is provided in Table 6.2. Based on this evidence, possible causes are also suggested. The data is summarised as fracture rates for each of the major bones affected (Table 6.1). Sample sizes for the Nong Nor and Ban Na Di people were relatively small which created elevated fracture rates that are probably not truly representative of these samples. Therefore, in addition to the fracture rate, the actual number of bones affected and the number observed are also provided for each sample in Table 6.1. All fractures show evidence of remodelling.

A total of 20 fractures in 18 individuals were observed among the four samples. The sample from Khok Phanom Di showed a total of five fractures. They were restricted to the small bones of the hands and feet (Figures 6.1) and the clavicle (Figures 6.2). The Nong Nor samples showed a variety of bones with healed fractures. The fractures of the ulnae (Figures 6.3) and clavicle (Figure 6.4) may have been the result of a fall in these individuals. However, the fracture of the foot phalanx (no photo) and the mandibular body (Figure 6.5) were more likely to have been the result of a direct force.

The sample from Ban Lum Khao showed a predominance of forearm fractures (Figure 6.6), but also one fibula (Figure 6.7) and one clavicle (Figure 6.8) were fractured. The forearm fractures varied considerably in their level of healing as described in Table 6.2, with some having debilitating complications, as in Burial 30 (Figure 6.6E) and Burial 98 (Figure 6.6C). Others had generally healed well. The aetiology of the forearm fractures may have been the result of falls in these individuals, although direct force is also a possibility. One individual, Burial 28, also had bony changes at the inferior end of the nasal bones indicative of a possible broken nose. Males were more predominantly affected by forearm fractures than females in this sample, with five of the six radius and ulna bones fractured in male individuals.

Only two individuals were identified in the Ban Na Di sample as having fractures. One involved the third metatarsal (Figure 6.9), most likely the result of a direct force such as dropping a heavy object on the foot. The other was a fracture of the femoral shaft (Figure 6.10). This fracture has healed well considering the debilitating implications a fracture of this type can have. This fact and the form of the callus being restricted to the posterior shaft, implies a partial fracture, possibly a greenstick fracture that occurred during childhood.

**Table 6.2:** Non-vertebral fractures in prehistoric Thailand: description and possible cause of each case

*Khok Phanom Di* *

| Burial | Sex | Age | Bone Fractured | Description | Possible Cause |
|---|---|---|---|---|---|
| 44 | Male | 18 y | Right fifth Metatarsal (Figure 6.1A) | A partial transverse, healed fracture of the distal shaft. Medially thickened shaft with lateral bowing of the fourth metatarsal. | Direct force to side of foot or weight dropped on foot. |
| 61 | Female | 48 y | Left Clavicle (Figure 6.2A) | An oblique, healed fracture of the midshaft with c.15 mm of shortening and a shorter posterior curve than the right clavicle. | Fall onto shoulder or outstretched hand or from a direct blow (Craig 1991) |
| 91 | Male | 45 y | Left fifth Metacarpal (Figure 6.1C) | Oblique, healed, midshaft fracture with some shortening, apposition laterally, and slight rotation. Concave bowing toward the palmar surface. No pathology of adjacent bones. | A torque force (force that can cause rotation) can produce this type of fracture (Green and Rowland 1991), also called a 'Boxer's fracture' |
| 110 | Female | 36 y | Left Clavicle (Figure 6.2B) | An oblique, healed fracture of the midshaft with some shortening. Curvatures are increased. A small perforation on the posterior midshaft may be a remodelled cloaca. | Fall onto shoulder or outstretched hand or from a direct blow (Craig 1991). Bone may have been weakened by infection or infection has occurred secondary to fracture, implying a compound fracture. |
| 142 | Female | 31 y | Right fourth Metacarpal (Figure 6.1B) | Healed, oblique midshaft fracture. Shortening of c.5-10 mm with apposition dorsally and concave angulation in palmar direction. No pathology of adjacent bones. | Torque force (force that can cause rotation) with the finger acting as the long lever (Green and Rowland 1991). |

*(Tayles 1992)

**Table 6.2 (cont.):** Non-vertebral fractures in prehistoric Thailand: description and possible cause of each case

*Nong Nor*

| Burial | Sex | Age | Bone Fractured | Description | Possible Cause |
|---|---|---|---|---|---|
| 6 | Female | Mid | Left Ulna (Figure 6.3A) | Well-healed fracture in the distal third of the shaft with a small callus. | Fall onto outstretched hand (indirect) or the result of a direct blow |
| 27 | ? | Adult | Proximal Foot Phalanx | Oblique, healed fracture just proximal to head. Angulation of head medio-laterally. | Direct blow such as striking toe against a hard object. |
| 32 | Male | Young | Mandibular body (Figure 6.5) | Healed fracture in region of left molars with some angular deformity. Antemortem loss of left first molar may be associated. | Direct force (?) |
| 47 | Male | Mid | Right Clavicle (Figure 6.4) | Healed midshaft fracture with thickened shaft. Osteoarthritis of shoulder and elbow. | Fall onto shoulder or outstretched hand or from a direct blow (Craig 1991) |
| 53 | ? | Mid | Left Ulna (Figure 6.3B) | Oblique or spiral, healed fracture of mid shaft with some shortening and slight apposition and angulation laterally. | Fall or a direct blow. |

**Table 6.2 (cont.):** Non-vertebral fractures in prehistoric Thailand: description and possible cause of each case

*Ban Lum Khao*

| Burial | Sex | Age | Bone Fractured | Description | Possible Cause |
|---|---|---|---|---|---|
| 28 | Male | 35-39 y | Right Ulna (Figure 6.6A) | Oblique, well-healed fracture in distal third of shaft. Small callus, slight medial apposition. Possibly a healed fracture of the nasal bones also. | Fall onto an outstretched hand with angulated and rotational force (indirect) (Lovell 1997) or the result of a direct blow to the arm (Anderson and Meyer 1991). |
| 30 | Male | 30-34 y | Left Ulna and Radius (Figure 6.6E) | Transverse fracture in proximal third of both bones. Ulna shows slight anterior apposition and angulation, and shortening. Junctional pseudoarthrosis between shafts of radius and ulna. No supination or pronation possible? | Required a large force. Direct blow to the arm or indirect via a fall from a height? (Anderson and Meyer 1991). |
| | | | Right Fibula (Figure 6.7) | Well healed fracture within proximal third. Bony spurs present on postero-lateral, antero-lateral and medial aspects. Tibia acted as a splint and assisted healing | Direct blow to the lower leg or a fall |
| 64 | Female | 40-44 y | Left Clavicle (Figure 6.8) | Well healed fracture with callus in region of conoid tubercle. A compression fracture with c. 7 mm shortening and anterior angulation of lateral third. | Fall onto the lateral shoulder, driving the shoulder and scapular down or fall onto outstretched hand or from a direct blow (Craig 1991). |
| 65 | Male | 40+ y | Right Radius (Figure 6.6B) | Impacted healed Colles' fracture of distal segment with posterior apposition and some shortening. Osteoarthritis of wrist. Low bone mass index in second metacarpal (47%). | Result of a fall onto hand. Commonly associated with underlying pathology such as osteoporosis |
| 67 | Male | 25-29 y | Left Radius (Figure 6.6D) | Midshaft, non union fracture? Evidence of possible remodelling over one fracture end with some tapering off. There is some postmortem damage and calcareous deposits limiting this diagnosis. No proximal radius was recovered. | |
| 98 | Female | 35-39 y | Right Ulna (Figure 6.6C) | Monteggia fracture in proximal third of shaft with associated dislocation of radial head. Slight anterior apposition and extreme medial angulation (c. 15°). Formation of new humero-radial joint below the lateral supracondylar ridge. Permanent dislocation of distal radio-ulnar joint? Severe elbow osteoarthritis. Severe limitation of movement? | Fall onto hand with forced pronation. Great amount of force required, fall from a height? |

**Table 6.2 (cont.):** Non-vertebral fractures in prehistoric Thailand: description and possible cause of each case

*Ban Na Di*

| Burial | Sex | Age | Bone Fractured | Description | Possible Cause |
|---|---|---|---|---|---|
| M. 17 | Male | 40-44 y | Left Femur (Figure 6.10) | Possible greenstick fracture of proximal shaft. Occurred during childhood. Slight apposition anteriorly and angulation posteriorly. Healed well. | Usually result from an indirect angular force (Resnick *et al.* 1995; Lovell 1997). |
| M. 35 | Male | 35-39 y | Right 3rd Metatarsal (Figure 6.9) | Oblique midshaft fracture with some apposition dorsally. No pathology of adjacent bones. | Probably due to a direct force such as a weight dropped on foot. Twisting of body with foot fixed also a possibility (Heckman 1991). |

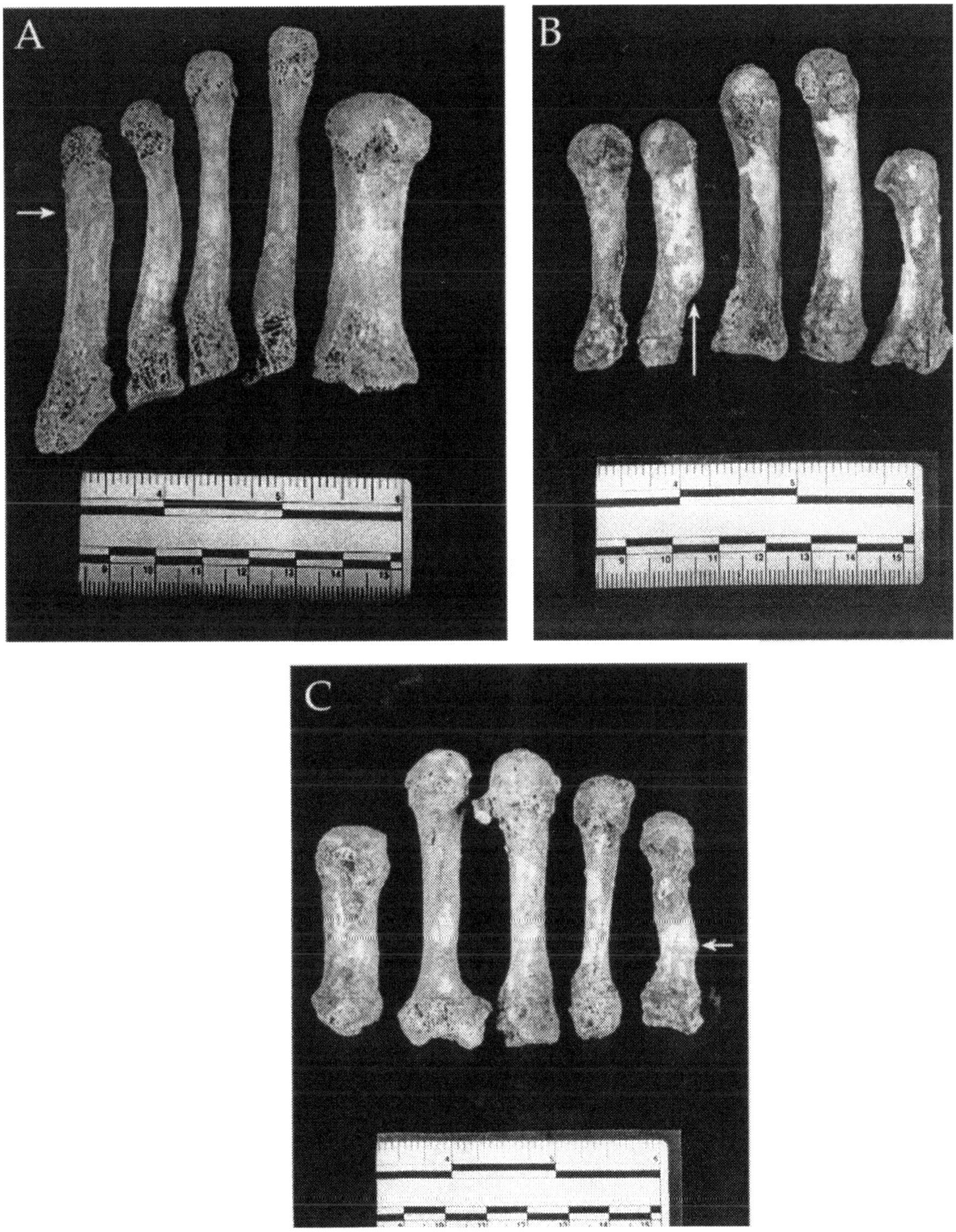

**Figure 6.1:** Khok Phanom Di fractures. (A) Right metatarsals from Burial 44, male, 18 years (plantar view). Fifth metatarsal has a healed fracture of the distal shaft, note associated bowing to fourth metatarsal. (B) Right metacarpals from Burial 142, female, 31 years (lateral view). Fourth metacarpal has a healed midshaft fracture. (C) Left metacarpals from Burial 91, male, 45 years (palmar view). The fifth metacarpal has a healed midshaft fracture

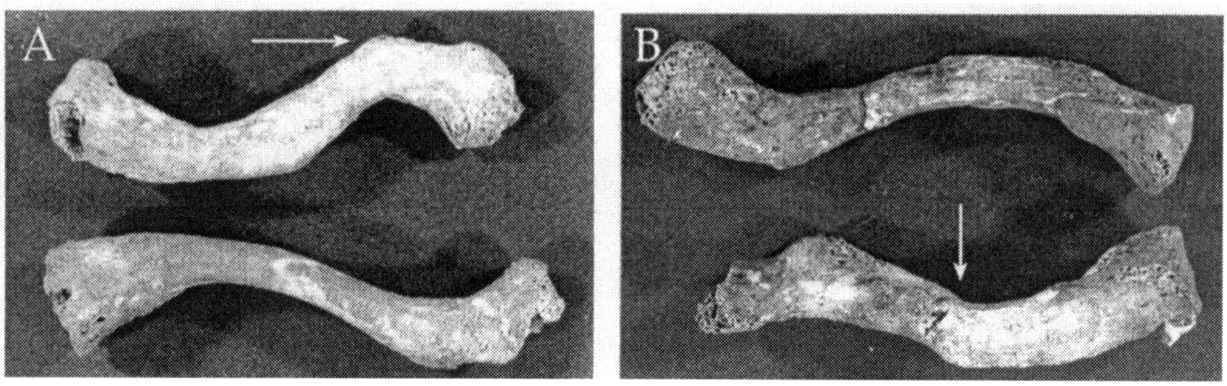


**Figure 6.2:** Khok Phanom Di fractures. (A) Clavicles from Burial 61, female, 48 years (superior view). Left clavicle (top) has a healed midshaft fracture. (B) Clavicles from Burial 110, female, 36 years (inferior view). Left clavicle (top) has a healed midshaft fracture

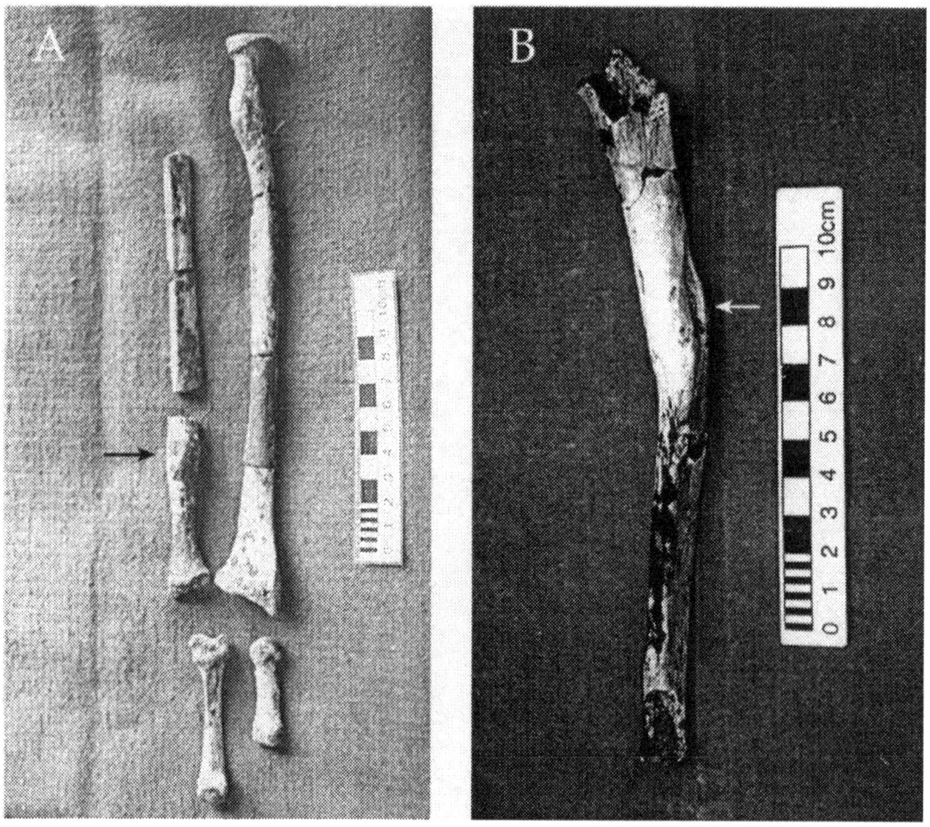


**Figure 6.3:** Nong Nor fractures. (A) Left forearm bones from Burial 6, female, 30-39 years (anterior view). Ulna has a healed fracture in distal third of shaft. (B) Left ulna from Burial 53, unknown sex, 30-39 years with healed midshaft fracture (anterior view)

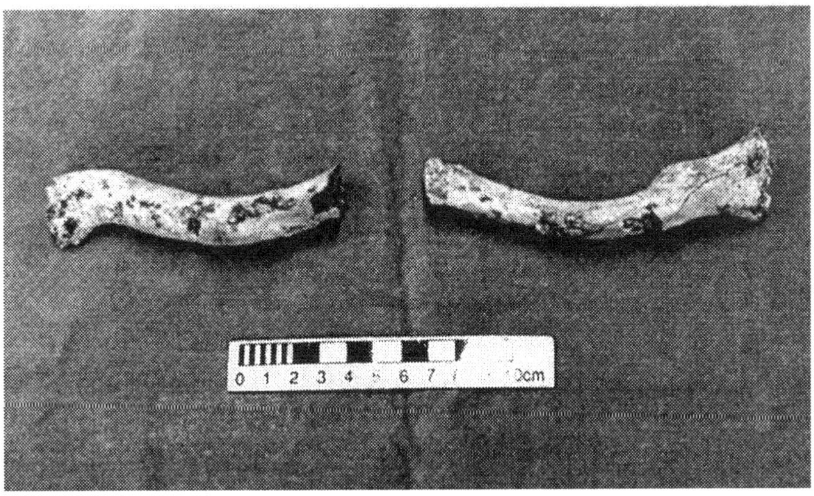


**Figure 6.4:** Nong Nor fractures. Clavicles from Burial 47, male 30-39 years (posterior view). Right clavicle (left) has a healed midshaft fracture; note thickened shaft of right clavicle

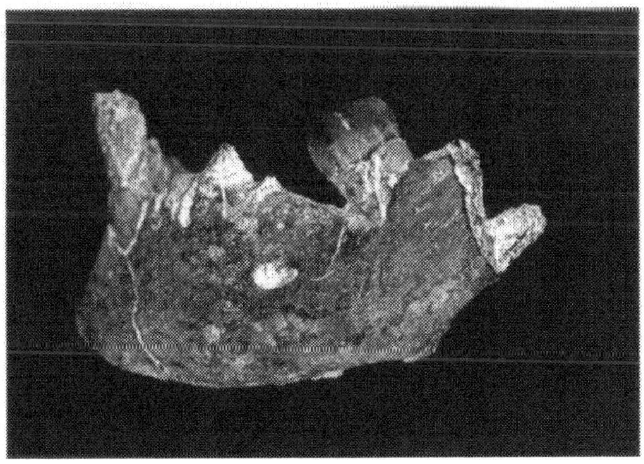

**Figure 6.5:** Nong Nor fractures. Left mandibular body from Burial 32, male, 20-29 years, with healed fracture in region of molars (lateral view)

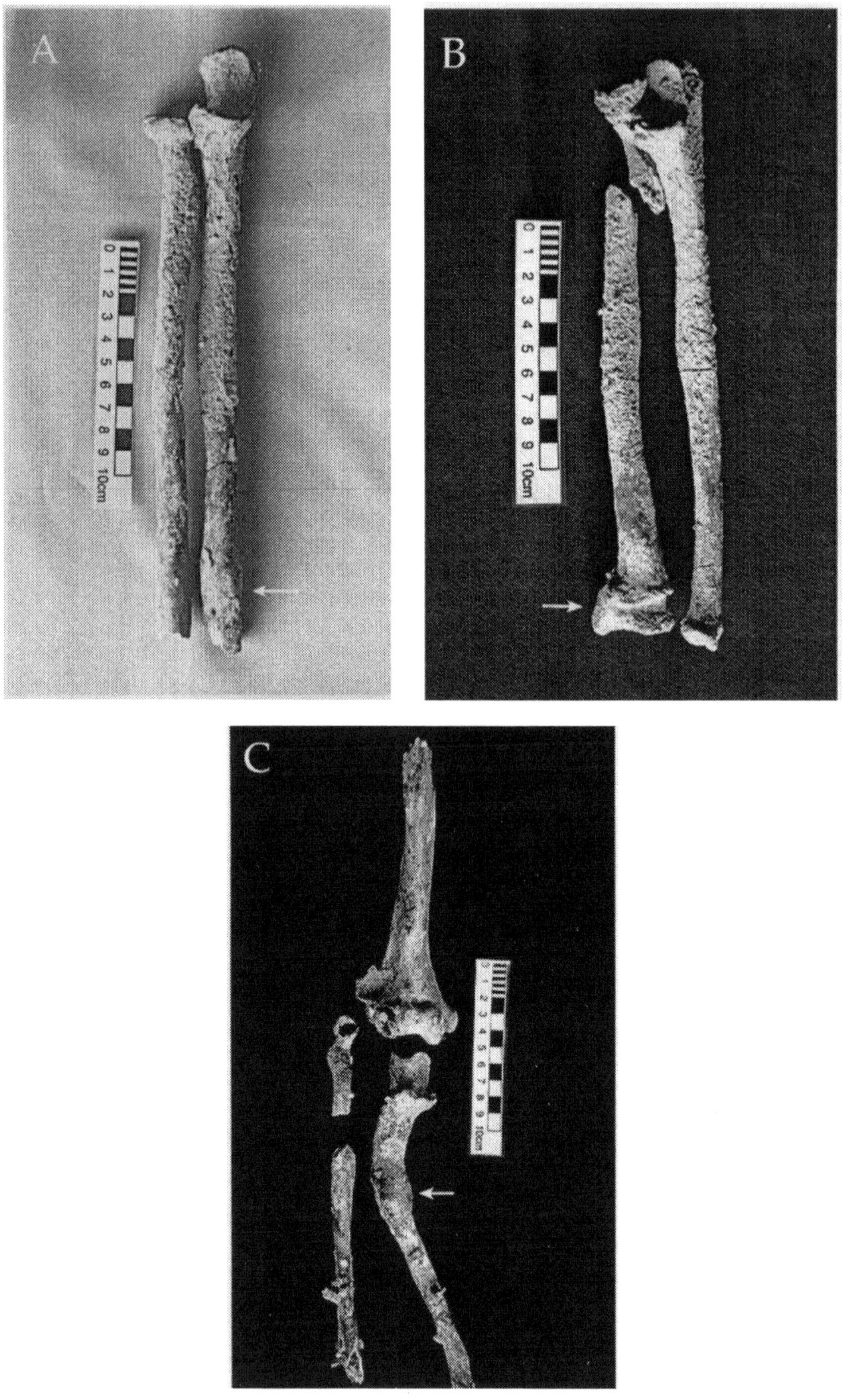


**Figure 6.6:** Ban Lum Khao fractures, see description over page

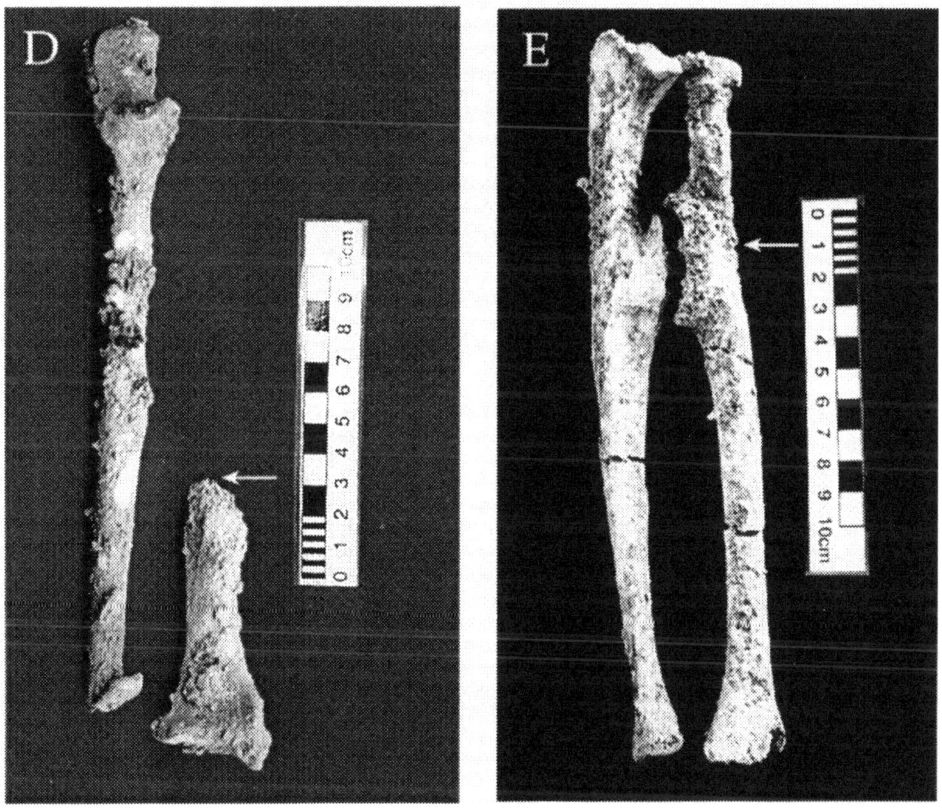


**Figure 6.6:** Ban Lum Khao forearm fractures. (A) Right forearm bones from Burial 28, male, 35-39 years (anterior view). Ulna has a healed fracture in distal third of shaft. (B) Right forearm bones from Burial 65, male, 40+ years (anterior view). Radius has a healed Colles' fracture of distal shaft. (C) Right arm bones with from Burial 98, female, 35-39 years (anterior view). Ulna has a healed Monteggia fracture of proximal third of shaft; note formation of new joint for radial head inferior to lateral supracondylar ridge of humerus. (D) Left forearm bones from Burial 67, male, 25-29 years (anterior view). Radius has a possible non-union fracture of radius midshaft. (E) Left forearm bones from Burial 30, male, 30-34 years (anterior view). Ulna and radius both have fractures in proximal third of shaft; note junctional pseudoarthrosis. N.B. Most of these bones are covered in a calcareous deposits

**Figure 6.7:** Ban Lum Khao fractures. Right tibia and fibula from Burial 30, male, 30-34 years (anterior view). Fibula has a healed fracture in proximal third of shaft. N.B. These bones are covered with a calcareous deposit

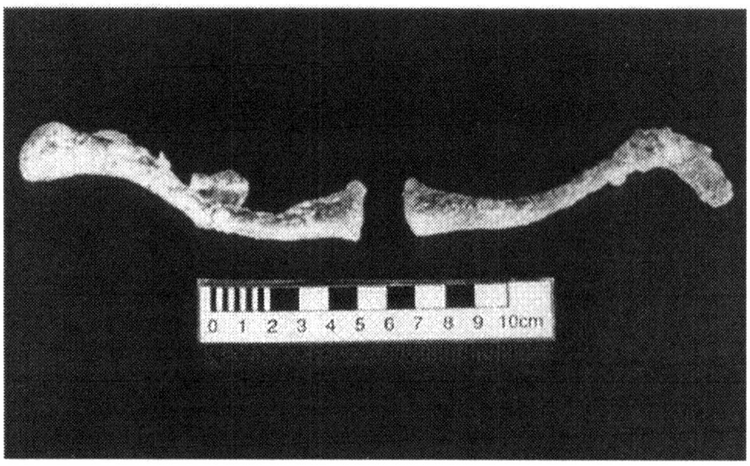


**Figure 6.8:** Ban Lum Khao fractures. Clavicles from Burial 64, female, 40-44 years (inferior view). Left clavicle (right) has a healed fracture in region of conoid tubercle. N.B. These bones are covered with a calcareous deposit

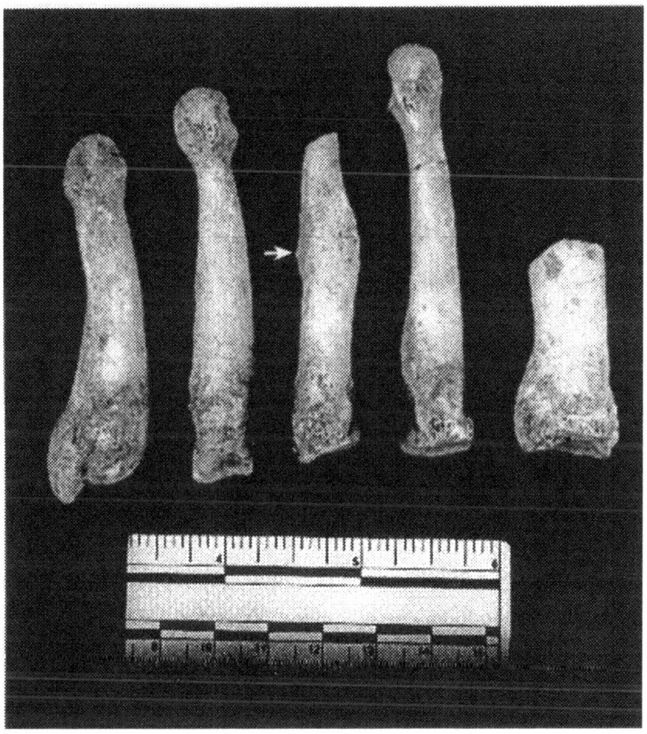

**Figure 6.9:** Ban Na Di fracture. Right metatarsals from Burial M.35, male, 35-39 years (plantar view). Right third metatarsal has a healed midshaft fracture

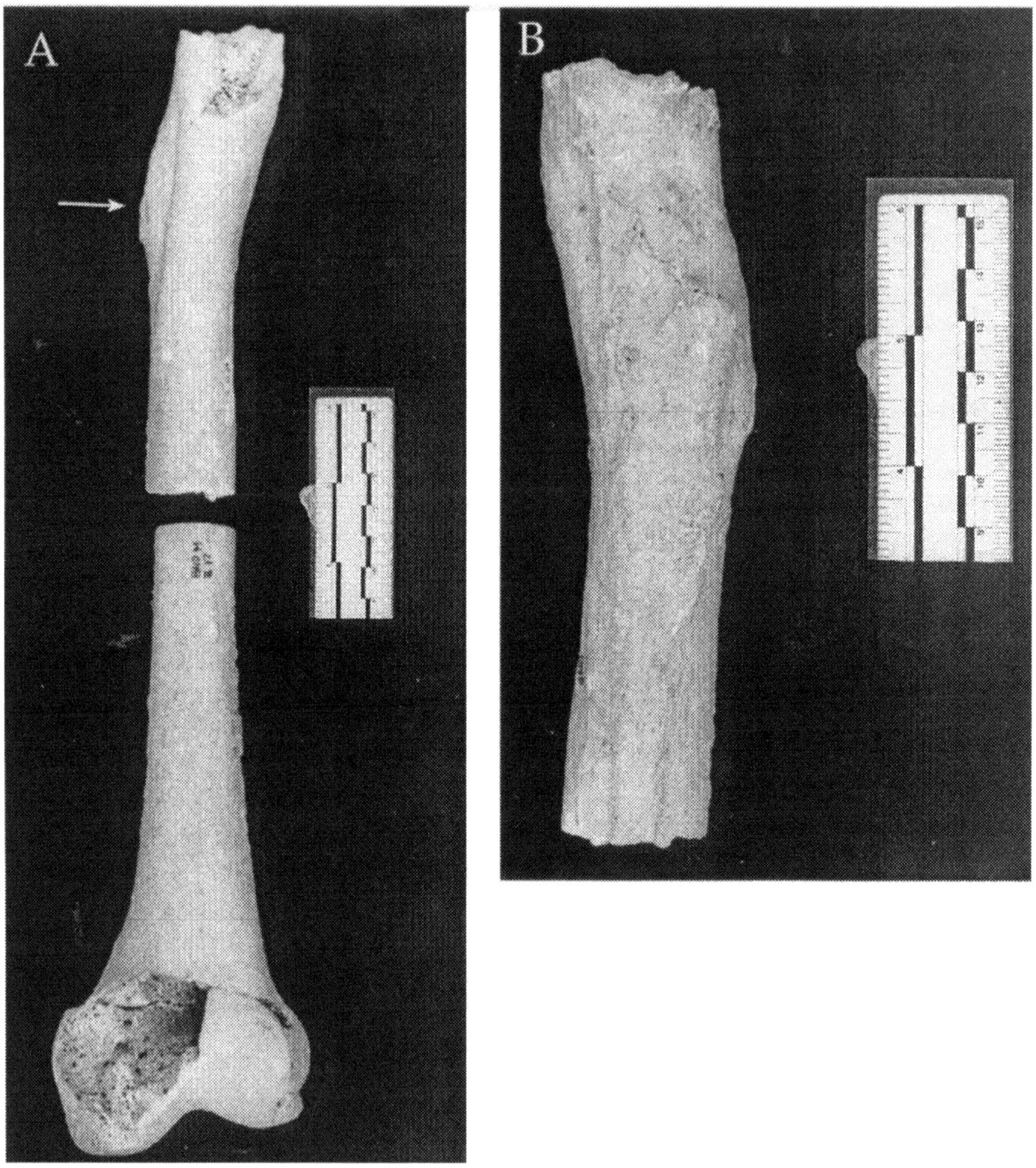


**Figure 6.10:** Ban Na Di fracture. Left femur with healed greenstick fracture of proximal shaft, from Burial M.17, male, 40-44 years. (A) anterior view (B) posterior view

**Table 6.3:** Spondylolysis in prehistoric Thailand: description and associated vertebral pathology of each case

| Burial | Sex | Age | Vertebrae Fractured | Other vertebral pathology |
|---|---|---|---|---|
| *Khok Phanom Di* | | | | |
| 28 | Male | 21 y | L 5 | Vertebral body collapse of all lumbar vertebrae & extreme osteophytosis. |
| 87 | Female | 24 y | L 3 and L 4 | L 4 extreme osteophytosis |
| 93 | Male | 42 y | L 5 | Compression of L 5 body. Moderate to severe degeneration over all spine. Some Schmorl's nodes. |
| 132 | Male | 34 y | L 5 | On left separated at the pars inter-articularis, on right side at the pedicle. Moderate to severe degeneration overall spine. |
| *Ban Lum Khao* | | | | |
| 17 | Female | 20-24 y | L 4 | No degeneration |
| 19 | Female | 20-24 y | L 4 | Slight osteophytes in lumbar region |
| 28 | Male | 35-39 y | L 4 | Moderate to severe osteophytosis in all spinal regions. Cervical Schmorl's nodes. |
| 47 | Male | 30-39 y | L 5 | Slight osteophytes in lumbar region |
| 65 | Male | 40+ y | L 5 | Severe osteophytosis and Schmorl's nodes in lumbar spine. L 1 and L 4 bodies collapsed. |
| 77 | Male | 25-29 y | L 5 | Slight osteophytes in lumbar region |
| 85 | Female | 25-29 y | L 4 and L 5 | Severe osteophytosis on lumbar spine. L 2 to L 5 bodies collapsed with increase in severity inferiorly. |
| 86 | Male | 40-44 y | L 5 | Severe osteophytosis on lumbar spine. L 5 collapsed. |
| *Ban Na Di* | | | | |
| M.36 | Female | 25-29 y | L 4 and L 5 | L5 has a diarthrodial pseudoarthrosis on left side between the superior and inferior articular surfaces. |

L= lumbar

In comparing the fracture rates among the samples (Table 6.1), the Nong Nor sample had the higher rate in the clavicle and ulna. However, sample sizes, due to poor preservation of the Nong Nor material, were very low. No other bones had fracture rates that could be compared among the samples. However, the *pattern* of bones affected differs among the samples and the significance of this will be discussed further.

Overall the Ban Lum Khao people had the most evidence of long bone fractures. No other sample was as commonly affected. The Ban Lum Khao sample showed a predominance in the males and Ban Na Di showed only males affected, while the other two samples showed a mix of males and females affected. Over all the individuals with non-vertebral fractures in all the samples only three individuals out of the total of 18 were aged less than 30 years. Over all samples of the 18 bones that could be sided, the left was fractured in 56% of the time while the right was affected in 44% of cases. Within each sample neither side was more predominantly affected.

## Spondylolysis

The results of the spondylolysis in each individual are provided in Table 6.3. No data was available for the Nong Nor skeletal material as vertebrae were seldom preserved sufficiently to be observed. Data are summarised for each sex in each sample in Table 6.4. The number of complete vertebrae able to be observed was not recorded for all samples, therefore no sample sizes are available. The Khok Phanom Di sample had excellent preservation of vertebrae, Ban Lum Khao vertebrae were relatively moderately preserved, and the Ban Na Di vertebrae the least well preserved of the three samples. An example of spondylolysis in a fifth lumbar vertebrae is presented in Figure 6.11.

The third, fourth, and fifth lumbar vertebrae were the only affected vertebrae among the three samples. The fifth lumbar vertebra was most commonly affected. A total of 16 vertebrae from 13 individuals had spondylolysis. The people of Ban Lum Khao had the highest number of individuals affected, with three females and five males with spondylolysis. That was twice that of the Khok Phanom Di sample, who had four individuals with spondylolysis. The Ban Na Di sample had only one individual affected, with both L4 and L5 were involved. The latter vertebrae in this individual had a diarthrodial pseudoarthrosis on the left side. Overall males showed a predominance of spondylolysis. Just over half of those affected by spondylolysis were aged less than 30 years.

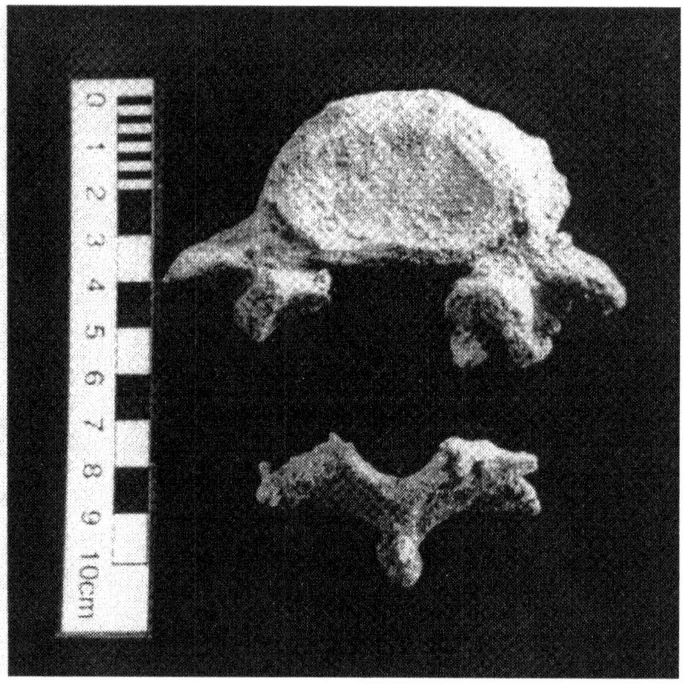

**Figure 6.11:** An example of spondylolysis in the fifth lumbar vertebra of Ban Lum Khao Burial 86, 40-44 year old male (superior view)

**Table 6.4:** Number of individuals with spondylolysis in each sample from prehistoric Thailand

| Site | Female | Male | Total |
|---|---|---|---|
| Khok Phanom Di | 1 | 3 | 4 |
| Ban Lum Khao | 3 | 5 | 8 |
| Ban Na Di | 1 | 0 | 1 |
| Total | 5 | 8 | 13 |

# Discussion

In both the Khok Phanom Di and Nong Nor samples, neither males nor females were more affected by non-vertebral fractures. From this it can be assumed that both males and females were exposed to the same degree of risk of trauma in this region of the southeast of Thailand during prehistory. However, in the northeast, particularly within the Ban Lum Khao sample, males are considerably more frequently affected with long bone fractures. Within the Ban Lum Khao sample, six of the eight bones fractured were in male individuals. Only two fractures were observed in the Ban Na Di skeletal material and both were in males. This evidence implies there was a differential risk of trauma between the sexes in the northeast. Perhaps the division of tasks exposed males to an environment rough in terrain making falls more likely.

Two of the three ulnae fractures in the Ban Lum Khao males could, alternatively, represent 'parry' fractures. The two ulnae fractures from the Nong Nor sample may also have been 'parry' fractures. However, there is no archaeological evidence of intercommunity warfare during the Bronze Age of Thai prehistory in the northeast (O'Reilly 1999) , although conflict may have arisen within the community on an individual level. Alternatively, some activities, such as boxing, could result in similar injuries.

Craniofacial injury, such as facial or cranial fractures, may support a violent aetiology for midshaft fractures (Smith 1996) . The individual either successfully protects their face and head from a blow and receives a forearm fracture, or, alternatively, unsuccessfully protects their head and face and receives a craniofacial injury. There was one case of craniofacial injury among the Ban Lum Khao sample. Burial 28, a 35-39 year old male, had a possible healed fracture of the nasal bones. This individual also had an ulna midshaft fracture. This combination of injuries may reflect successfully and unsuccessfully warding off a number of blows to the head and face. The mandibular body fracture in the young adult male from Nong Nor, Burial 32, may also have been the result of interpersonal violence. The ulna fractures in two other individuals could also be consistent with defending themselves from blows to the face and head, 'parry' fractures.

Although it cannot be determined at what age a fracture occurred in this skeletal material, the age structure of those individuals with evidence of trauma is worthy of comment as interesting patterns were evident. In the skeletal samples from Khok Phanom Di, Nong Nor, and Ban Lum Khao, of those individuals with evidence of a fracture, only one individual in each of these samples was aged less than 30 years. Neither of the two individuals affected in the Ban Na Di sample were less than 30 years old. In contrast, the age structures of those affected by spondylolysis is quite different, with more than 50% of individuals affected, over all samples, aged less than 30 years. Perhaps these differences can be explained by the fact that the older the individual is, the more likely they had experienced trauma, while the spondylolysis age structure fits with the clinical epidemiology of this pathology, which suggests an onset in young individuals is most common.

No obvious pattern is apparent for one side to be more affected by fracture than the other in the appendicular skeleton in these results. One possible exception may be in the clavicular fractures. The three females with clavicular fractures are all affected on the left side, while the one male with a clavicle fracture was affected on the right side. Sample sizes, however, are too small to draw any inferences from this evidence.

Overall it would appear a relatively random pattern of bones were affected throughout these samples from prehistoric Thailand. The common occurrence of fractures of the clavicle, forearm bones, and the small bones of the hands and feet in these samples is probably a reflection of the vulnerability to fracture of these bones. They represent the bones of smaller diameter that would require less force to fracture compared with the humerus or femur, for example.

However some patterns may not be so random but a reflection of the environment and/or activities. For example, the people of Khok Phanom Di appear to be more frequently affected in the bones of the hands and feet than other samples. They also lack forearm fractures. In contrast, the Ban Lum Khao sample has a predominance of forearm fractures. While the better preservation of small bones at Khok Phanom Di may explain the higher number of fractures observed in these bones, it does not explain the difference in forearm fractures. The difference in the pattern of fractures may indicate that these two groups had different environments and physical activities that placed on them a different type of risk of injury.

Most of the fractures had healed well in all of the samples indicating some level of knowledge and/or care of such injuries. However, injuries resulting in death are not observed, as fractures occurring immediately prior to death are almost impossible to differentiate from postmortem fractures. Therefore, the individuals who were not successfully cared for are not observed, which can bias interpretations of the level of medical knowledge.

The fractures that have healed particularly well include the radius and fibula fractures. Both of these bones have a stable adjacent bone, the ulna and tibia respectively, that can act as a splint. This supports the bone keeping it relatively immobile while it heals.

The hand and foot bones fractured in the Khok Phanom Di samples would not have been able to be given much care due to their anatomical position. Even in modern cases, little is done for simple fractures of the metacarpals and metatarsals as they have adjacent bones to act as splints. One clavicle fracture in a Khok Phanom Di female appears to have been infected which implies this was a compound fracture with an open wound. Alternatively, the infection may have been the primary problem which weakened the bone leaving it more susceptible to fracture.

The Nong Nor fractures were all well healed with few severe complications. The fractured clavicle of a middle aged male may have resulted in osteoarthritis observed in the shoulder. A fracture of the mandibular body is commonly the result of a direct force applied to the face and can be complicated by temporomandibular joint derangement and loss of teeth (Murphy and Kaplan 1995) . In such a fracture in a young male from Nong Nor the first molar is suspected to have been lost as a result of the fracture, but the remainder of the mandible and maxilla were not present for analysis of any occlusal or arthritic problems.

Just over half, four of the seven, fractures at Ban Lum Khao have healed well indicating some degree of care within the community and/or good general health and nutrition of the individual. There was no pattern of better healing in either sex. Complications in the other cases however, were severe and possibly debilitating. This seems to be the case in the Ban Lum Khao sample more so than in the other samples. This implies either a lack of knowledge concerning the reducing and setting of fractures and/or the individual was not sufficiently healthy to enable adequate remodelling and repair of the fracture site or to overcome infection (Lovejoy and Heiple 1981) . However, these fractures may not have healed better without surgery which is the common treatment in modern fracture cases

The fracture of the right ulna in Burial 98 from Ban Lum Khao, a 35-39 year old female, is termed a Monteggia type I fracture and shows characteristic complications. These include associated dislocation of the radial head, new radio-humeral joint formation, and osteoarthritis as a result of the malalignment. Quite a lot of force would have been required for this to have occurred compared with the other forearm fractures. Possibilities include a fall from a height (with forearm in full pronation) or a direct blow over the posterior ulna (Anderson and Meyer 1991) . Movement at the elbow in this individual may have been possible to some degree considering a new joint socket has formed for the radius; without biomechanical input such bony remodelling would not occur.

The formation of a junctional pseudoarthrosis between the shafts of the radius and ulna in Burial 30 from Ban Lum Khao would have prevented the movement of supination and pronation. The forearm would have been 'locked' in the supinated position. Therefore this fracture would have been quite restrictive to movement.

The diagnosis of the non-union fracture in Burial 67 from Ban Lum Khao was made more difficult by the lack of the proximal segment and is presented as a possibility only.

Malunion is a common complication of a radial shaft fracture (Lovell 1997) but it cannot be confirmed in this case.

Burial 65 from Ban Lum Khao with the Colles' fracture was suspected to be suffering from osteoporosis. He had the lowest cortical index score (47.0) in this sample (see Chapter 4) and suffered from other problems known to be potential signs of osteoporosis. These included vertebral body collapse and vertebral arch fracture. A Colles' fracture is typical in cases of osteoporosis and a simple fall onto the hand can induce it.

The femoral fracture in Burial M.17 from Ban Na Di was most likely a greenstick fracture. This diagnosis is based on the callus being well healed and restricted to the posterior aspect of the shaft; no evidence of remodelling is evident anteriorly. Greenstick fractures are the most frequent type to occur in children. They are characterised by the maintenance of bony continuity, such that the bone does not completely break (Dandy 1993) . However, "...fractures in children remodel well enough to restore the normal anatomy and the deformity of most greenstick fractures can be accepted" (Dandy 1993: 95). Burial M.17's fracture has healed well and the angulation that resulted is not significant.

Two further northeastern Thai prehistoric skeletal samples have been assessed for evidence of fractures by Douglas (1996) . In the Ban Chiang sample the most common site of fracture was the rib, with five individuals having one or, more commonly, more than one rib fractured. Considerably more males over all were affected by fractures than females in this sample, similar to the pattern in the Ban Lum Khao sample. However, there is no obvious pattern of fractures at Ban Chiang, with a random assortment of bones affected including one femur, two radii, three hand bones, two clavicles and two frontal bones, one possibly perimortem (Douglas 1996) . The femoral shaft fracture is similarly well healed to that in Burial M.17 from Ban Na Di.

Fracture patterns in the Non Nok Tha sample are somewhat different again, with the clavicle most commonly affected, in four individuals (Douglas 1996) . Rib fractures are less commonly observed than at Ban Chiang, with three individuals observed with this fracture. Again, an apparently random distribution of fractures among bones was generally found. Other bones fractured included only one radius, two tibiae, two hand bones, and one humerus (Douglas 1996) . No difference was evident between males and females in this sample. Overall it was suggested that because of the presence of significant angulation in many of the fractures in the Non Nok Tha sample, these people lacked a sophisticated knowledge of fracture reduction (Douglas 1996) .

In reports in the clinical literature, the occurrence of spondylolysis shows a predominance in males. This pattern was also found in prehistoric Thailand as these results indicated 62% of those with spondylolysis were males. The common aetiology of spondylolysis is as a chronic fatigue fracture, particularly in those undertaking heavy labour and vigorous back movements (Merbs 1996) . It could, therefore, be suggested that the males in both the Ban Lum Khao and Khok Phanom Di samples were commonly undertaking the heavy labour. However, there is also the question of genetic susceptibility which cannot be determined in these prehistoric samples.

The sample of Ban Lum Khao vertebrae does appear to have a higher predilection for spondylolysis compared with the other two samples although the differences are not great. This may, however, indicate the higher prevalence of a genetic propensity for this defect at Ban Lum Khao and/or a higher level of heavy labour in some individuals. In general the results follow the typical modern epidemiology with males showing higher levels than females and commonly involving the fifth lumbar vertebra.

In comparison with two other northeastern Thai prehistoric skeletal samples, the presence of spondylolysis is present at varying levels. Six individuals from the Ban Chiang sample were observed with either L4 or L5 spondylolysis (Douglas 1996) . However, in contrast with Ban Lum Khao and Khok Phanom Di samples, the Ban Chiang spondylolysis was predominantly found in females, with only one male observed with this vertebral fracture. Spondylolysis in the Non Nok Tha sample was present in only two individuals, one female and one male (Douglas 1996) .

# Summary

While it would be expected that patterns of trauma would be relatively random in a community, this is not entirely the case in prehistoric Thailand. The people of Khok Phanom Di did not show evidence of many fractures. When present they were restricted to the small bones of the hands and feet as well as the clavicle. Spondylolysis was also evident in only a few individuals. The people of Nong Nor showed a wider variety of fractures that appears quite random. However, the evidence of two ulna shaft fractures and a facial fracture of the mandibular body, are, when taken in combination, consistent with direct blows associated with attempting to protect ones face and head with the arm and unsuccessfully defending a blow to the face. Ban Lum Khao also had potential evidence for interpersonal violence with a predominance of forearm fractures and a broken nose in one individual. However, this is not the only possible explanation as falls may also have been common in both of these samples.

The pattern of long bone fractures in the Ban Lum Khao sample showed a pattern that was not necessarily random. Males suffered more fractures than females and the forearm bones were very commonly involved. Spondylolysis was also most common in this sample, again with more males than females affected. This may indicate the presence of a genetic trait present in this community that, in conjunction with back trauma, predisposed them to vertebral fracture.

Little evidence of long bone fracturing was observed in the people of Ban Na Di with one metatarsal and one femoral shaft fractured. Only one individual suffered from spondylolysis.

# 7 Dental Health and Disease

## Introduction

Teeth are usually well preserved from archaeological excavations. Their high component of inorganic material makes them the hardest, and most chemically stable tissue in the body (Roberts and Manchester 1995; Hillson 1996). There is a wide range of useful information that can be gained from analyses of the dentition, including estimations of age at death, dietary inferences, health and disease, and occupational use.

The aim of this study is to obtain a comparative picture of health throughout prehistoric Thailand. Therefore the analysis of the dentition will focus on the presence of a selection of degenerative and infectious dental conditions. The analysis of these conditions will help provide a profile of dental health for each sample. Interpretations concerning how this may have related to both their health and their diet will then be compared among the samples. The apparent relationship between dental health and the diet of a community is a useful tool:

> Given the link between subsistence and dental pathology (Turner, 1979; Lukacs, 1989), it is now possible to use the systematic recording of dental disease in skeletal samples as a base for reconstructing diet and ultimately subsistence patterns. (Littleton and Frohlich 1993: 427-8)

However, the relationship is not as straightforward as it may appear. Interpretations and comparisons among populations can be made difficult by such factors as poor preservation and the age structure of the samples which will be discussed in relation to the Thai samples later. A further problem in using archaeological evidence of subsistence to infer diet is that not all foods are recovered from a prehistoric site. Botanical remains, for example, are seldom fully represented as their high organic component results in a high decomposition rate and therefore their remains are not often recovered.

From the results presented in this chapter, dietary reconstruction will be attempted in conjunction primarily with archaeological evidence, although ethnographic evidence can also be useful. Botanical, faunal, and artefact analyses can indicate both a population's potential diet and the processes that were undertaken to prepare food. Identifying differences in dental pathology prevalences among different samples or between males and females may indicate differences in diet among these groups. This, in turn, may reflect some aspects of their social organisation. For example, differences in diet may be related to divisions of labour as mentioned in Chapter 2. In some populations men, being the hunters have tended to eat more protein, while women gathered and ate more plant material, small animals, and shellfish, as their respective subsistence activities allowed (Larsen 1997).

The level of oral hygiene, i.e. teeth cleaning, is an unknown factor in prehistory. However, it is assumed that oral hygiene practices, if any, were the same over all samples in prehistoric Thailand. It is also assumed that if such practices did occur, it was not undertaken to any significant level. Therefore, the dental problems observed will be more a reflection of the diet (Roberts and Manchester 1995).

There are however, complex interactions between the various types of dental pathologies observable in skeletal remains. Rarely do they occur in isolation from one another as will be discussed later. Lukacs (1989) notes the importance of deciphering these interactions before going on to reconstruct a population's diet. Poor dental health can also have serious consequences for an individual's general health, through such problems as an infection becoming systemic.

The prevalence of advanced attrition, caries, periapical cavities, and antemortem tooth loss will be presented for each prehistoric Thai sample. The analysis of which will help to fulfil the aim of reconstructing and comparing the diet of the different samples. Advanced attrition is a degenerative condition, while caries and periapical cavities provide evidence of infectious conditions, the latter being a secondary response to infection. Antemortem tooth loss may be a consequence of pulp exposure or periodontal disease through caries or advanced attrition. Calculus may also induce antemortem tooth loss through the development of periodontal disease (Lukacs 1989). This highlights the inter-relatedness of such dental pathologies. Each parameter will be discussed more fully in the relevant sections. Calculus and periodontal disease have not been excluded in this analysis for a number of reasons. The preservation of the skeletal material varies considerably among the four samples and many individuals had some postmortem damage that involved the alveolar crest resulting in difficulties in actually assessing the degree of periodontal disease. With respect to calculus observation, some post-excavation cleaning had removed the calculus deposits in some samples. Additionally, methods for quantifying these two parameters, in order to make them comparable among the samples, have not yet been perfected.

There are two common methods of presenting prevalences of dental pathologies, the individual count method and the tooth count method (Lukacs 1989). While the former has advantages in focusing on the individual's dentition, it can yield inadequate sample sizes. The latter method is the method used here as it provides large sample sizes, thereby

Table 7.1: Coding scheme for recording the presence and absence of teeth

| Code | Name | Description |
|---|---|---|
| 0 | postmortem loss | socket remaining |
| 1 | present | in situ or loose |
| 2 | broken | all or part of crown remaining |
| 3 | postmortem loss | root only remaining |
| 4 | antemortem loss | pathological |
| 5 | antemortem loss | ablation |
| 6 | erupting | |
| 7 | unerupted | |
| 8 | agenesis | |
| 9 | unknown | |

increasing the reliability of statistical tests. It also eliminates the problems associated with individuals without the full complement of teeth from postmortem loss.

Although teeth are generally well preserved there are some biases within the dental arcade both in recovery and susceptibility to dental disease. The molars and premolars are multirooted teeth and are therefore more stable in their sockets and more likely to be recovered (Moore and Corbett 1971). Anterior teeth, incisors and canines, only have a single root and may be more easily lost post mortem, increasing their chances of not being recovered during excavation.

Molars have more cusps and fissures in which oral bacteria can harbour than the anterior teeth. This accumulation of bacteria along with their more difficult access for cleaning may make these posterior teeth more vulnerable to the development of caries and other subsequent dental problems (Powell 1985).

Dental pathologies, including tooth wear and dental infections, typically increase with age. An older individual would be expected to have a greater degree of wear and have a higher probability of exposure to infectious oral pathologies (Powell 1988). Therefore comparing two populations with different age structures can lead to errors in interpretations of the cause of dental pathologies. Age structures are thus compared in all the following analyses.

A further limitation in the analysis of dental pathologies in skeletal material is the fact that acute conditions, such as acute infection, will not leave a trace on the tooth or surrounding bone. An acute infection, depending on the host's response, may either kill an individual or be resolved, both occurring quickly.

It is only for convenience that the dentition is considered in a separate section. As cautioned by Houghton (1996), this approach switches the classification of cause to a regional or anatomical one and perpetuates the attitude that the dentition is something rather apart from, and uncoordinated with, the rest of the body. Therefore these dental results are related back to what is seen in the rest of the body in Chapter 8 to integrate it into the picture of community health and disease.

# Dental Census

This section details the number of teeth and tooth positions present in the four skeletal samples. Every tooth position able to be observed was recorded following the coding system presented in Table 7.1. For example, an intact, whole tooth present was recorded as a code 1. However, recognising the difference between an ablated tooth or one lost pathologically, is not straightforward. More reference will be made to this problem in the section analysing antemortem tooth loss.

The dentition from Khok Phanom Di were analysed by Tayles (1992) and that of Nong Nor by V. Hunt (Tayles *et al.* 1998). Methods in recording the teeth present and pathologies present were exactly the same in each sample, despite different researchers. The level of interobserver error between myself and Tayles (1992) is presented in Appendix A. No significant differences were found except in the case of one or two individuals particularly where the skeletal material had degraded in preservation over time through frequent handling.

Subadults were classed as those individuals less than 15 years and their dental census is summarised in Table 7.2. In each sample more than 50% of all subadults had no teeth. The Khok Phanom Di sample had the lowest proportion of subadults with teeth (29.1%), while the subadults from Ban Lum Khao had the highest (47.1%). These results reflect the age structure of the subadults from each sample. For example, the Khok Phanom Di sample had the highest proportion of infants present (Table 3.10 and Figure 3.6). Therefore, given the fact that this age group do not have many teeth fully erupted, this sample from Khok Phanom Di consequently showed the smallest proportion of subadults with teeth.

The deciduous and permanent teeth were considered separately in the following analyses and their census is also found in Table 7.2. No subadult in any of the samples had a complete (n=20) complement of deciduous teeth. However, the mean number of deciduous teeth per individuals with deciduous teeth varied little over the four samples from 7.8 in the Nong Nor sample to 9.1 in the Ban Lum Khao sample.

**Table 7.2:** Summary of the subadult dental census for each of the prehistoric Thai samples

| | Khok Phanom Di | | Nong Nor | | Ban Lum Khao | | Ban Na Di | |
|---|---|---|---|---|---|---|---|---|
| No. subadults | 86 | | 33 | | 51 | | 28 | |
| No. with teeth[1] | 25 | | 15 | | 24 | | 10 | |
| % with teeth | 29.1 | | 45.5 | | 47.1 | | 35.7 | |
| No. with mixed dentition | 5 | | 2 | | 5 | | 2 | |
| | Dec. | Perm. | Dec. | Perm. | Dec. | Perm. | Dec. | Perm. |
| No. with teeth[1] | 20 | 10 | 11 | 6 | 20 | 9 | 8 | 4 |
| No. of teeth | 175 | 162 | 86 | 62 | 182 | 92 | 69 | 58 |
| Mean no. of teeth[2] | 8.8 | 16.2 | 7.8 | 10.3 | 9.1 | 10.2 | 8.6 | 14.5 |
| Range of number of teeth present[2] | 2-17 | 5-27 | 1-18 | 1-21 | 2-19 | 2-27 | 1-13 | 2-26 |

1. deciduous and/or permanent teeth code 1 only; 2. per individual with teeth.
Dec. = deciduous teeth, Perm. = permanent teeth.

**Table 7.3:** The adult dental census for each of the prehistoric Thai samples summarised for whole intact teeth and tooth positions

| *Whole teeth*[3] | Khok Phanom Di | Nong Nor | Ban Lum Khao | Ban Na Di |
|---|---|---|---|---|
| No. adults with age & sex estimates | 68 | 83 | 58 | 36 |
| No. with teeth[1] | 64 | 62 | 43 | 33 |
| % with teeth | 94.1 | 74.7 | 74.1 | 91.7 |
| No. of teeth | 1187 | 1032 | 904 | 516 |
| Mean no. of teeth[2] | 18.5 | 16.6 | 21.0 | 15.6 |
| Range of teeth present[2] | 6-32 | 1-32 | 3-32 | 3-32 |
| *Tooth positions*[3] | | | | |
| No. with tooth positions[4] | 67 | 62 | 43 | 33 |
| % with tooth positions | 98.5 | 74.7 | 74.1 | 91.7 |
| No. of tooth positions | 2046 | 1323 | 1154 | 707 |
| Mean no. of teeth positions[5] | 30.5 | 21.3 | 26.8 | 21.4 |
| Range of tooth positions present[5] | 16-32 | 1-32 | 3-32 | 3-32 |

1. includes only those with one or more whole, complete teeth (code 1 only); 2. per individual with teeth; 3. does not include those adults of unknown sex and age with dentition (n=6 from Nong Nor, n=1 from Ban Na Di); 4. tooth positions includes all codes; 5. per individual with tooth positions.

No subadult would be expected to have a complete set of permanent teeth (n=32) as the third molar generally erupts later than the age of 15 years. The mean number of permanent subadult teeth is quite low in all samples, ranging from 10.2 in Ban Lum Khao subadults to a maximum of 16.2 in those from Khok Phanom Di.

Table 7.3 presents the dental census of all the adults with age at death and sex estimates from each of the prehistoric samples. Both the census of whole teeth (code 1) and of tooth positions (all codes) are provided. Those few adults with teeth unable to be assigned to a sex were excluded as in the following analyses sex differences in dental pathology are investigated. Those unable to be aged are also excluded as the age structures of each sample are an essential component in interpreting age-related dental pathologies. No adults from either the Khok Phanom Di or Ban Lum Khao samples were excluded, six adults were excluded from Nong Nor and one from the Ban Na Di sample.

Only one individual from Khok Phanom Di and three from Ban Na Di had no dental material, therefore these two samples have over 90% of adults able to be included in the dental analysis. Both the Ban Lum Khao and Nong Nor samples had nearly one quarter of adults with no dentition. However, the Ban Lum Khao sample had the highest mean number of teeth per individual with teeth (21.0). Khok Phanom Di had a very high mean number of tooth positions (30.5) able to be recorded, with the Ban Lum Khao sample second highest (26.8). The Nong Nor and Ban Na Di samples had much lower means and this is a reflection of the poorer quality of preservation of material from these two samples.

In the following analyses of dental pathologies the number of teeth observed may vary slightly from the above tables. In some cases some broken (codes 2 or 3) or erupting (codes 6 or 7) teeth were able to be observed increasing the sample size, while some whole teeth (codes 1) were encrusted and unable to be observed and thus reduced the sample size.

The age structure of the individuals with teeth and tooth positions were compared statistically among the four samples for reasons mentioned above. Each sample was divided into two age groups, those aged less than 30 years and those aged equal to or greater than 30 years. This division was used to provide as near equal sample sizes as possible between the

**Table 7.4:** Grading scheme for advanced attrition (adapted from Molnar (1971))

| Wear Category | Incisor and Canine | Premolar | Molar |
|---|---|---|---|
| 1 | unworn | unworn | unworn |
| 2 | wear facets minimal in size | wear facets, no observable dentine | wear facets, no observable dentine |
| 3 | cusp pattern obliterated, small dentine patches may be present | cusp pattern partially or completely obliterated. Small dentine patches. | cusp pattern partially or completely obliterated. Small dentine patches. |
| 4 | Dentine patch (Minimal) | Two or more dentine patches, one of large size | Three or more small dentine patches |
| 5 | Dentine patch (Extensive) | Two or more dentine patches, secondary dentine may be slight | Three or more large dentine patches, secondary dentine, none to slight |
| 6 | Secondary dentine (Moderate to Extensive) | Entire tooth still surrounded by enamel, secondary dentine moderate to heavy | Secondary dentine moderate to extensive, entire tooth completely surrounded by enamel |
| 7 | Crown (enamel) worn away, on at least one side, extensive secondary dentine | Crown (enamel) worn away, on at least one side, extensive secondary dentine | Crown (enamel) worn away, on at least one side, extensive secondary dentine |
| 8 | Roots functioning as occlusal surface | Roots functioning as occlusal surface | Roots functioning as occlusal surface |

two age groups. Restricting the analysis to two adult age groups was necessary as further divisions would have produced very small sample sizes in each division that would not have been able to be statistically compared.

# Advanced Attrition

### Introduction

There are a number of different ways in which teeth can wear. Dental attrition is wear produced by tooth-on-tooth contact and is generally the natural result of chewing (Powell 1985; Hillson 1996). Dental abrasion is wear that is not produced by tooth-on-tooth contact and does not produce clear facets, but is seen as a general loss of surface detail through contact with abrasive particles carried on the cheeks, tongue, and food (Hillson 1996). Erosion is another form of tooth wear and this is where, for example, an acidic polluting environment or high acid-content foods may erode the tooth enamel (Roberts and Manchester 1995). It is not always easy to differentiate between the different types of tooth wear but this section concentrates on occlusal wear which will predominantly be related to dental attrition but some abrasion may also have occurred.

Attrition is an age-related phenomenon and has been used in this study as one of multiple factors in ageing adult individuals (see Chapter 3). The primary concern is the amount of advanced attrition of the occlusal surfaces present in each sample, as it is this level of attrition which has potentially pathological implications. The attrition of teeth is not a pathological condition *per se* but it can predispose an individual to other dental pathologies (Roberts and Manchester 1995). Normally, as teeth wear and the occlusal enamel is worn away, secondary dentine forms at the occlusal surface to ensure the pulp cavity beneath is protected (Powell 1985; Roberts and Manchester 1995). If the formation of secondary dentine is overtaken by the rate of tooth attrition, the pulp cavity becomes open and exposed to colonisation by bacteria from, for example, nearby plaque or carious lesions. Infection and inflammation may result, with the possibility of eventual tooth loss (Powell 1985; Hillson 1996). The inter-relatedness of dental pathologies and its significance will be further outlined in the discussion at the end of this chapter.

A major factor, often forgotten, in the production of occlusal attrition is bruxism. Bruxism is the grinding or tapping of teeth and occurs most frequently whilst asleep (Hillson 1996). However, the level to which this was present in prehistoric populations is unknown and is assumed, in this study, to be the same and minimal over all samples.

Food is one of the other major factors affecting tooth attrition (Roberts and Manchester 1995). The composition of the diet is important in the production of attrition, as is the processing of the food prior to consumption. For example, small particles of stone may become incorporated into the food when a stone mortar is used to grind grain (Roberts and Manchester 1995). In general, the study of attrition severity is used as an indicator of the coarseness of the diet. From this, differences in diet among groups, whether males and females from within a sample or inter-population comparisons, can be inferred. This is usually discussed with reference to archaeological evidence, such as botanical, faunal, and artefact analysis. These aspects may indicate a population's potential diet and the processes undertaken in preparing food. Differences in the severity of wear between males and females may also highlight behavioural variability in tooth use in addition to dietary differences.

Larsen (1997) summarises evidence from a variety of North American skeletal samples that document the change from foragers to late farmers. He states that they "...show appreciable declines in severity of the occlusal surface wear..." (Larsen 1997: 250). This is attributed to the change from hard textured and abrasive foods to a higher proportion of more soft textured grains, which is not only a factor of diet composition changes, but also the way in which the foods were processed. For example, there was a concomitant change with agriculture to the use of wooden grinding implements rather than stone in some areas of North America (Larsen 1997). As mentioned above this can also influence attrition.

There are various grading schemes available for assessing the degree of attrition in teeth (for example Molnar 1971; Scott 1979; Smith 1984). The Scott system (Scott 1979) presents an ordinal dental attrition scoring technique for molar teeth which involves a quadrant system. However, it was not felt necessary to obtain such detailed information for this study. Molnar (1971) provides a less complex scheme with eight grades that was used in this study. This provides enough information for both assisting with estimating age at death and assessing potentially pathological teeth with advanced wear.

## Methods

Attrition was graded on a scale from one to eight based on the scheme by Molnar (1971). Details of the scheme grading the degree of wear are provided in Table 7.4. The diagrams not reproduced here were also very useful. An individual scaling system was given for different tooth types, grouping together the incisors and canines, the premolars, and the molars. The abrasion of the cusps and the appearance of the dentine patches were the primary features used to evaluate the degree of wear (Molnar 1971).

A grade of attrition was recorded for every tooth present in all individuals from each skeletal sample, as described in the age estimate section of Chapter 3. Grades 6-8 were considered to be indicative of advanced attrition where the enamel has been completely abraded from the occlusal surface. The proportion of this degree of wear is what is presented and analysed in the following results. No subadults are included in the results.

## Results

The age structure, divided into those less than 30 years and those equal to or greater than 30 years, is given for each sample in Table 7.5. The age structure of the male and female samples with teeth are also provided as advanced attrition will be analysed by sex as well as by the total sample. There were statistically significant differences in the number of teeth belonging to each age group among the samples as indicated by the $Chi^2$ results (all p-values < 0.01) (Table 7.5). Fisher's Exact p-values highlight exactly which groups were different from each other in Tables 7.6.

In general, the Ban Na Di sample had a different proportion present in each age group compared with the other samples. There were more older adult teeth and less younger adult teeth overall and in both the male and female subsamples. Therefore, if the results of advanced attrition in the Ban Na Di people showed high levels this may have been due solely to the high number of teeth from older adults rather than, for example, any dietary influence.

The Ban Lum Khao age structure showed more teeth from younger adults overall were present, which was a reflection of the high proportion of young females in the whole sample. The male subsample from Ban Lum Khao had more older adults, as did the Nong Nor male subsample. The Khok Phanom Di sample, as well as the total Nong Nor samples and Nong Nor females, all had relatively similar proportions of teeth in each age group. This distribution will be considered in the interpretation of the results.

Table 7.7 and Figure 7.1 present the prevalence of teeth with advanced attrition in each sample and as divided by sex. Table 7.8 provides the Fisher's Exact p-values for comparisons among the samples. Overall the Nong Nor sample showed the highest prevalence of advanced attrition (17.3%) compared with the other samples. This was statistically significant compared with both the Khok Phanom Di and Ban Na Di samples and was very close to being significantly higher than the Ban Lum Khao sample. The lowest levels of advanced attrition were in the teeth of the Khok Phanom Di people (8.0%) and this result was statistically lower than all other samples. In the total prevalences the northeastern samples, Ban Lum Khao and Ban Na Di, were very similar (13.0% and 12.0% respectively).

Both the Nong Nor and Ban Na Di females showed high numbers of teeth with advanced attrition (15.2% and 17.9% respectively). These prevalences were both statistically higher than Khok Phanom Di and Ban Lum Khao females which showed considerably lower levels (4.6% and 6.3% respectively).

The pattern of prevalence in the male samples was somewhat different although the Nong Nor teeth still showed a high level of advanced attrition. The Ban Lum Khao male sample showed the highest level (20.5%) and both these groups, Nong Nor and Ban Lum Khao, had statistically higher levels compared with both Khok Phanom Di and Ban Na Di males.

While the Nong Nor males and females showed consistently high, and similar, levels of advanced attrition, all other groups showed significant sex differences (Table 7.7). Both Khok Phanom Di and Ban Lum Khao males had the higher level compared with their female subsamples, while the reverse was true in the Ban Na Di sample.

**Table 7.5:** Summary of the number of teeth assessed for advanced attrition as divided by the age at death and sex of the individual from which they came

| | Khok Phanom Di | | Nong Nor | | Ban Lum Khao | | Ban Na Di | | Comparison among samples | |
|---|---|---|---|---|---|---|---|---|---|---|
| | N | % | N | % | N | % | N | % | Chi $^2$ | p-value |
| *Total* | | | | | | | | | | |
| <30y | 678 | 47.7 | 432 | 42.6 | 484 | 55.5 | 163 | 32.0 | | |
| ≥30y | 743 | 52.3 | 582 | 57.4 | 388 | 44.5 | 347 | 68.0 | **78.6** | **<0.01** |
| *Female* | | | | | | | | | | |
| <30y | 357 | 47.1 | 219 | 46.2 | 325 | 70.2 | 28 | 20.0 | | |
| ≥30y | 401 | 52.9 | 255 | 53.8 | 138 | 29.8 | 112 | 80.0 | **130.95** | **<0.01** |
| *Male* | | | | | | | | | | |
| <30y | 321 | 48.4 | 213 | 39.4 | 159 | 38.9 | 135 | 36.5 | | |
| ≥30y | 342 | 51.6 | 327 | 60.6 | 250 | 61.1 | 235 | 63.5 | **18.90** | **<0.01** |
| *Male vs Female Fisher's Exact p-values* | | | | | | | | | | |
| | 0.63 | | 0.03 | | **<0.01** | | **<0.01** | | | |

N = number of teeth; **Bold** indicates statistical significance among samples at 1% level

**Table 7.6:** Fisher's Exact p-values for inter-sample comparisons of the age structure of the teeth assessed for advanced attrition

| *Total* | Nong Nor | Ban Lum Khao | Ban Na Di |
|---|---|---|---|
| Khok Phanom Di | 0.01 | **<0.01** | **<0.01** |
| Nong Nor | - | **<0.01** | **<0.01** |
| Ban Lum Khao | - | - | **<0.01** |
| *Females* | | | |
| Khok Phanom Di | 0.77 | **<0.01** | **<0.01** |
| Nong Nor | - | **<0.01** | **<0.01** |
| Ban Lum Khao | - | - | **<0.01** |
| *Males* | | | |
| Khok Phanom Di | **<0.01** | **<0.01** | **<0.01** |
| Nong Nor | - | 0.89 | 0.40 |
| Ban Lum Khao | - | - | 0.51 |

**Bold** indicates statistical significance at 1% level

**Table 7.7:** Summary of the number of teeth with advanced attrition in each sample, for each sex and in total

| | Khok Phanom Di | | Nong Nor | | Ban Lum Khao | | Ban Na Di | | Comparison among samples | |
|---|---|---|---|---|---|---|---|---|---|---|
| | A/N | % | A/N | % | A/N | % | A/N | % | Chi 2 | p-value |
| Total | 113/1421 | 8.0 | 175/1014 | 17.3 | 113/872 | 13.0 | 61/510 | 12.0 | **48.96** | **<0.01** |
| Female | 35/758 | 4.6 | 72/474 | 15.2 | 29/463 | 6.3 | 25/140 | 17.9 | **58.82** | **<0.01** |
| Male | 78/663 | 11.8 | 103/540 | 19.1 | 84/409 | 20.5 | 36/370 | 9.7 | **30.01** | **<0.01** |
| *Male vs Female Fisher's Exact p-value* | | | | | | | | | | |
| FET | **<0.01** | | 0.11 | | **<0.01** | | **<0.01** | | | |

A= affected, N= total number of teeth observed; **Bold** indicates statistical significance among samples at 1% level.

**Table 7.8:** Fisher's Exact p-values for inter-sample comparisons of the prevalence of advanced attrition, for each sex and in total

| *Total* | Nong Nor | Ban Lum Khao | Ban Na Di |
|---|---|---|---|
| Khok Phanom Di | **<0.01** | **<0.01** | **<0.01** |
| Nong Nor | - | 0.01 | **<0.01** |
| Ban Lum Khao | - | - | 0.62 |
| *Females* | | | |
| Khok Phanom Di | **<0.01** | 0.23 | **<0.01** |
| Nong Nor | - | **<0.01** | 0.43 |
| Ban Lum Khao | - | - | **0.01** |
| *Males* | | | |
| Khok Phanom Di | **<0.01** | **<0.01** | 0.35 |
| Nong Nor | - | 0.62 | **<0.01** |
| Ban Lum Khao | - | - | **<0.01** |

**Bold** indicates statistical significance at 1% level

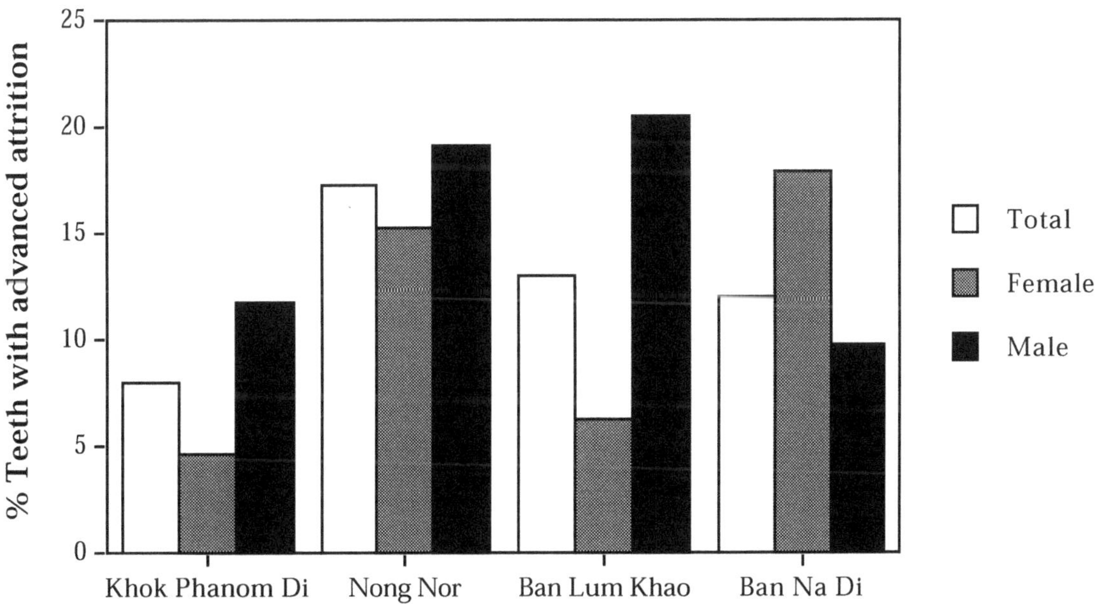


**Figure 7.1:** The percentage of teeth with advanced attrition in each sample, for each sex and in total

# Caries

## Introduction

Dental caries is one of the most prevalent diseases affecting the human body and affects both sexes and all age groups (Aufderheide and Rodriquez-Martin 1998). It is a multibacterial and multifactorial disease of the teeth (Aufderheide and Rodriquez-Martin 1998). Caries are characterised by the demineralisation of the inorganic substances and destruction of the organic substances of the teeth (Aufderheide and Rodriquez-Martin 1998; Soames and Southam 1998). Ultimately, a cavity within the tooth is produced. However, there are many stages in this disease and a carious lesion may present as a slight enamel opacity through to a cavitation that may involve the partial or complete loss of the crown or root (Larsen 1997).

Both dental plaque and fermentable carbohydrates in the mouth are necessary for the development of caries (Hillson 1990). Plaque provides the necessary bacterial colonies and carbohydrates are the bacteria's food source. Bacteria are present in the mouth normally, as they begin to breakdown or digest certain foods. Therefore, they are not always associated with disease. One of the digestion processes bacteria are responsible for is the conversion of carbohydrates (starches and sugars) into more simple sugars. During this process, bacteria release acid which can, when the pH is reduced low enough (approximately 5.5 (Soames and Southam 1998)), begin to destroy the enamel, dentine, or cementum of the teeth.

The *Streptococci mutans* group and lactobacilli are the bacteria most frequently found associated with carious lesions (Hillson 1996; Soames and Southam 1998) but a number of other streptococci and actinomyces species may also be present (Hillson 1996). *S. mutans* and lactobacilli can withstand stronger acidic conditions than other bacteria and produce acid rapidly (Hillson 1996).

Not all teeth are equally susceptible to developing caries as physical characteristics of the tooth influence the development of caries. The posterior teeth with deep occlusal pits and fissures are excellent sites for trapping food and harbouring bacteria, making these teeth more vulnerable than the incisors. The enamel structure and composition are also important (Soames and Southam 1998). For example, a tooth with a defect in the enamel, such as that which occurs with enamel hypoplasia, may provide an easier route for enamel destruction. The carious lesion may form in a complete band with the enamel hypoplasia and is called circular enamel caries (Hillson 1990).

Caries can also develop differently at different sites on the tooth and the recording of the location of the caries can be useful (Hillson 1996). Occlusal caries are those which develop in the fissure systems of molars and premolars and are common sites for caries to develop in modern populations. Interproximal caries may develop deep cavities and their position makes them hard to observe without radiographs. Smooth surface caries are not very common (Hillson 1996). Root caries are associated with root exposure through periodontitis and they frequently develop secondary to chronic periodontal disease (Aufderheide and Rodriquez-Martin 1998). Genetic predisposition plus environmental conditions associated with diet will dictate the common type of caries within a population.

Factors extrinsic to the tooth are other considerations in the development of a carious lesion and these include the saliva and diet of the individual as well as their level of immunity (Soames and Southam 1998). A key factor is the length of time the teeth are exposed to acid during the acid phase which occurs after eating. The longer the acid phase the more destruction possible. The length of the acid phase is influenced by a number of factors which includes how frequently food is consumed, and the composition and flow of saliva (Larsen 1997).

There are two types of caries, chronic and arrested. Chronic caries develop slowly but are active, unlike the arrested caries which may remain stable for months or years (Hillson 1996). Because of the slow progress in the arrested type of caries, there is time for defence mechanisms to come into action to protect the pulp cavity (Soames and Southam 1998). A carious lesion may alternate between both these types during its lifetime. There is an additional third rare type of caries, known as rampant caries which progress rapidly, particularly in children (Hillson 1996).

Caries and diet

"The clearest single factor in caries epidemiology is...sugar " (Hillson 1996: 282). It is widely accepted that a direct relationship exists between dental caries and fermentable carbohydrates in the diet but not all carbohydrates have the same cariogenicity (Soames and Southam 1998). For example, sucrose is significantly more cariogenic than other sugars, although all common sugars (glucose, maltose, fructose and sucrose) are able to produce a relatively rapid pH decline in comparison with starch (Hillson 1996; Soames and Southam 1998).

Starch by itself does not have the same high cariogenic properties as sugars. Starch is composed of larger molecules that must be broken down before being able to be incorporated into plaque (Hillson 1996). The combination of starch and sugar in food is highly cariogenic as this combines a marked pH depression with a longer duration (Hillson 1996).

The presence of fluoride ions may provide some defence against the development of caries. Fluoride available during the development of the dentition becomes incorporated into the enamel matrix and decreases the solubility of dental tissues (Hillson 1996). It can also provide protection in its environmental form after the teeth have formed as it can inhibit bacterial metabolism and enhance remineralisation (Hillson 1996; Soames and Southam 1998).

*1) Estimated number of teeth lost due to caries.*
[Number of teeth lost antemortem] x [proportion of teeth with pulp exposure due to caries]

*2) Total estimated number of teeth with caries.*
[estimated number of teeth lost due to caries] + [number of carious teeth observed]

*3) Total number of original teeth.*
[number of teeth observed ] + [number of teeth lost antemortem]

*4) Corrected caries rate.*
[total estimated number of teeth with caries] / [total number of original teeth]

**Figure 7.2:** Procedure for calculating the corrected caries rate (adapted from Lukacs 1995)

**Table 7.9:** The proportion of deciduous and permanent teeth with caries in subadults

| Site | Deciduous Teeth | | Permanent Teeth | |
|---|---|---|---|---|
| | A/N | % | A/N | % |
| Khok Phanom Di | 12/179 | 6.7 | 10/158 | 6.3 |
| Nong Nor | 0/86 | 0 | 1/62 | 1.6 |
| Ban Lum Khao | 4/182 | 2.2 | 1/92 | 1.1 |
| Ban Na Di | 12/69 | 17.4 | 0/58 | 0 |

A= affected, N= total number of teeth observed

### Methods

Every observable tooth was assessed for the presence or absence of carious lesions. The location of a lesion was recorded but has not been used in the following analysis. This method does not record multiple lesions on one tooth. More importance is placed on the proportion of teeth with carious lesions and the differences and similarities in prevalence among the samples.

The calculation developed by Lukacs (1995) for presenting the prevalence of caries among a sample is used here. This is called the corrected caries rate and it incorporates an estimate of the number of teeth lost antemortem from caries to enable a more accurate prevalence of caries to be obtained. The formula for calculating the caries correction factor (Lukacs 1995) is presented in Figure 7.2.

To calculate the corrected caries rate following Lukacs' (1995) method, the proportion of teeth with pulp exposure due to caries is required (Figure 7.2 (1)). The pulp of a tooth can be exposed through either caries or advanced attrition. Carious lesions that had specifically exposed the pulp were not recorded however, carious lesions involving the tooth crown to the extent that it obscured the site of the original caries were recorded as massive caries and are used in the corrected caries rate calculation. These are very likely to have exposed the pulp cavity. This number is divided by the total number of teeth with pulp exposure (number of teeth with advanced attrition plus the number of teeth with massive caries). Other parts of the calculation are more straightforward and can be followed in Figure 7.2.

The observed caries rate was also calculated. This is simply the proportion of the number of carious teeth over the total number of teeth observed. This was calculated in order to make some comparisons with other studies which had used this method of presenting caries prevalence.

### Results

Subadults

Caries prevalences in the subadult samples are presented as percentages by tooth (observed caries rate) (Table 7.9) and not as corrected caries rates as there is no antemortem tooth loss in these subadults. Sample sizes are small in most groups, particularly Ban Na Di and Nong Nor. Therefore statistical analysis is not feasible and interpretations must be made with caution. The mean age at death among the samples ranges from 3-4 years for the deciduous samples and 9-11 years for the permanent dentition.

The percentage of carious deciduous teeth is highest in the Ban Na Di sample (17.4%) but the 12 caries are distributed among only two subadults. Khok Phanom Di subadults have the second highest number of teeth with caries in the deciduous teeth (6.7%) and the highest, and similar, number in the subadult permanent teeth (6.3%). Both the Nong Nor and Ban Lum Khao subadults have minimal or no caries.

Adults

The age structure, subdivided by those teeth belonging to individuals less than 30 years and greater than or equal to 30 years, is provided for those able to be assessed for caries in Table 7.10. There were statistically significant differences in the age structures among the samples as the $Chi^2$ results indicated. Fisher's Exact p-values in Table 7.11 highlighted exactly which groups were significantly different from each other.

The Ban Na Di sample showed a significantly higher proportion of older adults and this was particularly evident in the female subsample. This was in direct contrast to the Ban Lum Khao females which showed much higher numbers of teeth from young adults. Both the age

**Table 7.10:** Summary of the number of teeth observed for caries as divided by the age at death and sex of the individual from which they came

| | Khok Phanom Di | | Nong Nor | | Ban Lum Khao | | Ban Na Di | | Comparison among samples | |
|---|---|---|---|---|---|---|---|---|---|---|
| | N | % | N | % | N | % | N | % | Chi 2 | p-value |
| *Total* | | | | | | | | | | |
| <30y | 663 | 51.8 | 432 | 42.5 | 499 | 56.4 | 163 | 31.7 | | |
| ≥30y | 618 | 48.2 | 585 | 57.5 | 386 | 43.6 | 352 | 68.3 | **99.5** | **<0.01** |
| *Female* | | | | | | | | | | |
| <30y | 348 | 53.0 | 220 | 46.6 | 339 | 71.1 | 28 | 19.0 | | |
| ≥30y | 308 | 47.0 | 252 | 53.4 | 138 | 28.9 | 119 | 81.0 | **138.3** | **<0.01** |
| *Male* | | | | | | | | | | |
| <30y | 315 | 50.4 | 212 | 38.9 | 160 | 39.2 | 135 | 36.7 | | |
| ≥30y | 310 | 49.6 | 333 | 61.1 | 248 | 60.8 | 233 | 63.3 | **25.73** | **<0.01** |
| *Male vs Female Fisher's Exact p-values* | | | | | | | | | | |
| | 0.37 | | 0.02 | | **<0.01** | | **<0.01** | | | |

N = number of teeth; **Bold** indicates statistical significance at 1% level.

**Table 7.11:** Fisher's Exact p-values for inter-sample comparisons of the age structure of the teeth assessed for caries

| *Total* | Nong Nor | Ban Lum Khao | Ban Na Di |
|---|---|---|---|
| Khok Phanom Di | **<0.01** | 0.04 | **<0.01** |
| Nong Nor | - | **<0.01** | **<0.01** |
| Ban Lum Khao | - | - | **<0.01** |
| *Females* | | | |
| Khok Phanom Di | 0.03 | **<0.01** | **<0.01** |
| Nong Nor | - | **<0.01** | **<0.01** |
| Ban Lum Khao | - | - | **<0.01** |
| *Males* | | | |
| Khok Phanom Di | **<0.01** | **<0.01** | **<0.01** |
| Nong Nor | - | 0.95 | 0.53 |
| Ban Lum Khao | - | - | 0.51 |

**Bold** indicates statistical significance at 1% level

**Table 7.12:** Summary of the corrected caries rates in each sample, for each sex and in total

| | Khok Phanom Di | | Nong Nor | | Ban Lum Khao | | Ban Na Di | | Comparison among samples | |
|---|---|---|---|---|---|---|---|---|---|---|
| | A/N | % | A/N | % | A/N | % | A/N | % | Chi $^2$ | p-value |
| Total | 194/1465 | 13.2 | 70/1075 | 6.5 | 49/944 | 5.2 | 29/553 | 5.2 | **68.86** | **<0.01** |
| Female | 169/799 | 21.2 | 48/512 | 9.4 | 43/508 | 8.5 | 19/178 | 10.7 | **57.70** | **<0.01** |
| Male | 49/666 | 7.4 | 23/563 | 4.1 | 9/436 | 2.1 | 12/375 | 3.2 | **20.11** | **<0.01** |
| *Male vs Female Fisher's Exact p-value* | | | | | | | | | | |
| | **<0.01** | | **<0.01** | | **<0.01** | | **<0.01** | | | |

A= affected, N= total number of teeth observed; **Bold** indicates statistical significance among samples at 1% level.

**Table 7.13:** Fisher's Exact p-values for inter-sample comparisons of the corrected caries rates, for each sex and in total

| *Total* | Nong Nor | Ban Lum Khao | Ban Na Di |
|---|---|---|---|
| Khok Phanom Di | **<0.01** | **<0.01** | **<0.01** |
| Nong Nor | - | 0.22 | 0.33 |
| Ban Lum Khao | - | - | >0.99 |
| *Females* | | | |
| Khok Phanom Di | **<0.01** | **<0.01** | **<0.01** |
| Nong Nor | - | 0.66 | 0.66 |
| Ban Lum Khao | - | - | 0.37 |
| *Males* | | | |
| Khok Phanom Di | 0.01 | **<0.01** | **<0.01** |
| Nong Nor | - | 0.10 | 0.60 |
| Ban Lum Khao | - | - | 0.38 |

**Bold** indicates statistical significance at 1% level

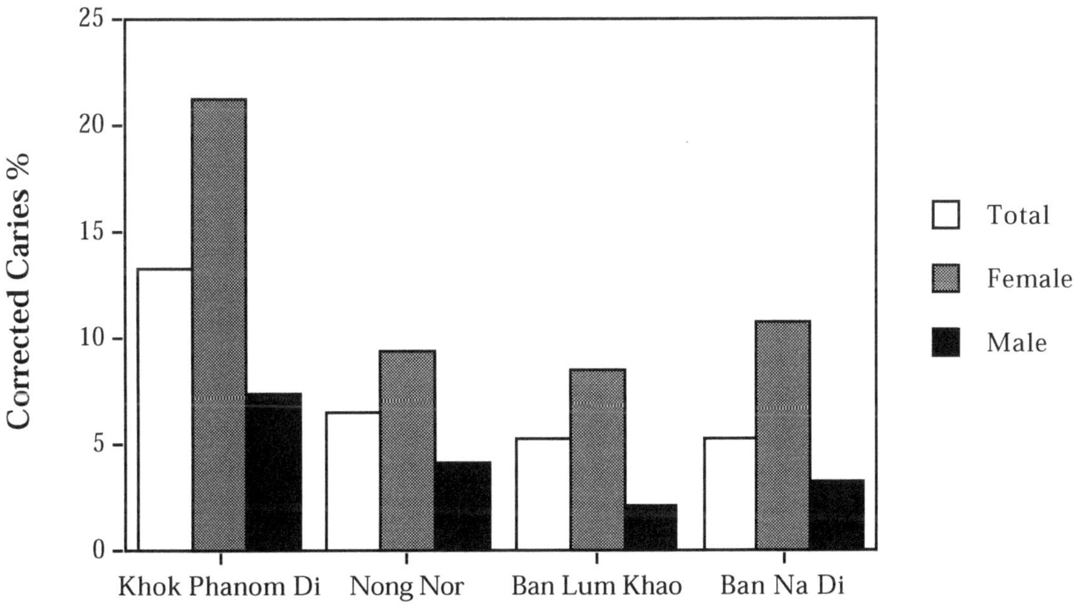


**Figure 7.3:** The corrected caries rate of each sample, for each sex and total

structures of Ban Lum Khao and Ban Na Di females were statistically different from each other and both the other samples (Table 7.11).

Overall in the total samples, Khok Phanom Di and Ban Lum Khao were quite similar with near equal ratios of young and old teeth. However, young adult females from Ban Lum Khao with teeth far exceeded the number of older females as mentioned, and this pattern was the reverse in the Ban Lum Khao males.

Nong Nor adults overall had slightly less young adult teeth than older adult and this pattern was the same in female subsample. The males also had more teeth from older individuals but the difference was more marked than in the females. The males from the Ban Lum Khao and Ban Na Di samples also showed a marked higher number of teeth from older individuals. No statistical differences were present between these three male groups (Table 7.11).

The observed caries rates per tooth are provided in Appendix E to enable comparisons with those studies that have not used Lukacs calculation for correcting caries prevalences. Corrected caries rates are presented in Table 7.12 and Figure 7.3 for the total sample and as divided by sex. Fisher's Exact p-values, for comparisons among the samples, are presented in Table 7.13.

Over the total adult samples and the female subsamples, Khok Phanom Di adults showed the highest rate (13.2% and 21.2% respectively). This is statistically higher compared with all other samples (Table 7.13). The other three samples all had quite similar caries rates, differing by a maximum of only 1.3% in the total sample and 2.2% in the female subsamples.

Considering the male subsamples alone, differences among the samples were not so striking. Khok Phanom Di males still showed the highest corrected caries rate (7.4%) but this was only statistically higher than the northeastern samples (Table 7.13). The Nong Nor males had a rate 3.3% lower than that of the Khok Phanom Di male teeth and this was very close to being a statistically significant difference with a p-value of 0.01.

Sex differences were apparent in all samples as Fisher's Exact p-values in Table 7.12 indicate. In all cases females had the higher corrected caries rate.

# Secondary Dental Infections and Inflammation

### Introduction

The observation of periapical bony cavities are frequently included in analyses of the dental health of skeletal samples as they represent an inflammatory reaction to infection of the periapical tissues. However, the terms periapical abscess or abscess cavity are used to denote all periapical cavities in the majority of the dental anthropology literature. This terminology is incorrect as it implies all periapical cavities contained abscesses. In fact,

> Abscess formation is only one of a number of possible inflammatory responses to infection of the dentition and its supporting structures and it is in fact most unlikely that periapical cavities are formed by abscesses. (Dias and Tayles 1997: 548)

More commonly a periapical cavity has contained a relatively benign granuloma or a periodontal cyst. Recent work by Dias and Tayles (1997) has highlighted how these different pathologies can be differentially diagnosed in skeletal material and the following is a summary of that work.

Infection of periapical tissues begins with oral bacteria entering an exposed pulp cavity and migrating down into the tooth roots and eventually the periapical region. Dental pulp may become exposed through such problems as advanced attrition and caries. The inflammatory response of the periapical tissues including the surrounding alveolar bone may be acute or chronic. A resolved acute infection will not produce bony changes. However chronic inflammation will and this is what is observed in skeletal material.

A periapical granuloma is the most common response to chronic inflammation and is characterised in dry bone by the formation of a small, smooth walled cavity. Given time, a granuloma may develop into an apical periodontal cyst. This can be differentiated from a granuloma only by its larger size. However, a cyst, unlike the granuloma, will not resolve if the source of infection is removed.

Although a less common process, a granuloma *may* progress into an acute or chronic abscess. Again, only the chronic form will be visible in bone. A chronic abscess can be differentiated from other conditions by the presence of a bony channel from the cavity to the skin or oral mucosa through which pus is discharged. However, an acute periapical abscess may, if the infection is virulent enough, lead to osteomyelitis (inflammation of the bone marrow and adjacent bone). Acute osteomyelitis can be fatal if septicaemia develops. Chronic osteomyelitis can create a bony periapical cavity but the presence of other bony reactions typical of osteomyelitis will differentiate it from other reactions.

Typically, only the acute and unidentifiable conditions discussed produce any severe clinical symptoms such as extreme pain and possible systemic effects such as fever. Periapical granulomas and periodontal cysts are associated with only mild, if any, pain and no systemic effects (Dias and Tayles 1997).

Populations with high rates of caries and/or advanced attrition may be more vulnerable to developing periapical infections, particularly if the immune response is lowered by, for example, an inadequate diet. Most often cavities contain benign lesions that will not be a significant health problem for the individual. A contained infection is in fact suggestive of a healthy immune system able to respond and suppress the effects of such an infection. However, such an infection represents the *potential* to lead to more severe and life threatening types of systemic infection such as septicaemia and osteomyelitis (Dias and Tayles 1997; Larsen 1997).

## Methods

Periapical cavities were recorded in all possible dental material through observations of the alveolar bone surrounding each tooth position. No attempt, however, at this stage has been made to differentially diagnose the cavities, but the majority are assumed to be granulomas or periodontal cysts.

## Results

The age structure for each sample is provided in Table 7.14. This data also applies for the analysis of antemortem tooth loss in the following section. There were statistically significant differences in the number of tooth positions belonging to each age group among the samples as the $Chi^2$ results in Table 7.14 indicates (p-values < 0.01). Fisher's Exact p-values (Tables 7.15) further indicate exactly which groups are different from each other.

In the total samples, both the Ban Na Di and Nong Nor samples were similar in that they had significantly lower proportions of younger adult tooth positions and higher proportions in the older adults. Khok Phanom Di and Ban Lum Khao had relatively equal sample proportions among the age groups although they were statistically different from each other and the other two samples.

The female age structures were slightly different and Ban Lum Khao had higher proportions of tooth positions in young females while all the other samples had higher numbers in the older age range. This was particularly marked in the Ban Na Di females where more than 80% of the sample was from older adults.

The tooth positions in the male groups consistently showed higher numbers in the older age group compared with the younger group. However, the Khok Phanom Di sample had near equal proportions which was statistically different from the other samples.

Table 7.16 and Figure 7.4 present the prevalences of teeth with periapical cavities in each sample and divided by sex. Table 7.17 provides the Fisher's Exact p-values for comparisons among the samples. Over all samples there were generally only low prevalences of periapical cavities, ranging from 0.5% to 7.1%. Khok Phanom Di people showed the highest percentage in the total samples (6.0%) and this was statistically higher than all others. This was the same in the female subsamples. However, the male prevalences of periapical cavities in the Khok Phanom Di sample (4.7%) was statistically higher than both Ban Lum Khao and Nong Nor but not higher than Ban Na Di (2.8%).

The Nong Nor people showed consistently low prevalences of periapical cavities in both males (1.2%) and females (0.5%). These levels were not much lower than those of the Ban Lum Khao and Ban Na Di samples.

No sample showed significant sex differences in the prevalence of periapical cavities (Table 7.16).

**Table 7.14:** Summary of the number of tooth positions observed for periapical cavities as divided by the age at death and sex of the individual from which they came

| | Khok Phanom Di | | Nong Nor | | Ban Lum Khao | | Ban Na Di | | | |
|---|---|---|---|---|---|---|---|---|---|---|
| | N | % | N | % | N | % | N | % | Chi 2 | p-value |
| *Total* | | | | | | | | | | |
| <30y | 929 | 45.4 | 508 | 38.4 | 593 | 51.4 | 233 | 33.0 | | |
| ≥30y | 1117 | 54.6 | 815 | 61.6 | 561 | 48.6 | 474 | 67.0 | **78.2** | **<0.01** |
| *Female* | | | | | | | | | | |
| <30y | 468 | 42.5 | 261 | 41.4 | 407 | 67.1 | 44 | 18.3 | | |
| ≥30y | 634 | 57.5 | 369 | 58.6 | 200 | 32.9 | 197 | 81.7 | **193.9** | **<0.01** |
| *Male* | | | | | | | | | | |
| <30y | 461 | 48.8 | 247 | 35.6 | 186 | 34.0 | 189 | 40.6 | | |
| ≥30y | 483 | 51.2 | 446 | 64.4 | 631 | 66.0 | 277 | 59.4 | **43.31** | **<0.01** |
| *Male vs Female Fisher's Exact p-value* | | | | | | | | | | |
| | **<0.01** | | 0.03 | | **<0.01** | | **<0.01** | | | |

N = number of teeth; **Bold** indicates statistical significance at 1% level
N.B. This is the same for antemortem tooth loss.

**Table 7.15:** Fisher's Exact p-values for inter-sample comparisons of the age structure of the tooth positions assessed for periapical cavities

| *Total* | Nong Nor | Ban Lum Khao | Ban Na Di |
|---|---|---|---|
| Khok Phanom Di | **<0.01** | **<0.01** | **<0.01** |
| Nong Nor | - | **<0.01** | 0.02 |
| Ban Lum Khao | - | - | **<0.01** |
| *Females* | | | |
| Khok Phanom Di | 0.68 | **<0.01** | **<0.01** |
| Nong Nor | - | **<0.01** | **<0.01** |
| Ban Lum Khao | - | - | **<0.01** |
| *Males* | | | |
| Khok Phanom Di | **<0.01** | **<0.01** | **<0.01** |
| Nong Nor | - | 0.59 | 0.10 |
| Ban Lum Khao | - | - | 0.04 |

**Bold** indicates statistical significance at 1% level

**Table 7.16:** Summary of the number of tooth positions with periapical cavities in each sample, for each sex and in total

| | Khok Phanom Di | | Nong Nor | | Ban Lum Khao | | Ban Na Di | | Comparison among samples | |
|---|---|---|---|---|---|---|---|---|---|---|
| | A/N | % | A/N | % | A/N | % | A/N | % | Chi 2 | p-value |
| Total | 122/2046 | 6.0 | 11/1323 | 0.8 | 15/1154 | 1.3 | 15/707 | 2.1 | **92.71** | **<0.01** |
| Female | 78/1102 | 7.1 | 3/630 | 0.5 | 5/607 | 0.8 | 2/241 | 0.8 | **78.66** | **<0.01** |
| Male | 44/944 | 4.7 | 8/693 | 1.2 | 10/547 | 1.8 | 13/466 | 2.8 | **20.58** | **<0.01** |
| *Male vs Female Fisher's Exact p-value* | | | | | | | | | | |
| | 0.02 | | 0.23 | | 0.19 | | 0.10 | | | |

A= affected, N= total number of teeth observed; **Bold** indicates statistical significance among samples at 1% level.

**Table 7.17:** Fisher's Exact p-values for inter-sample comparisons of the prevalence of periapical cavities, for each sex and in total

| *Total* | Nong Nor | Ban Lum Khao | Ban Na Di |
|---|---|---|---|
| Khok Phanom Di | **<0.01** | **<0.01** | **<0.01** |
| Nong Nor | | 0.32 | 0.02 |
| Ban Lum Khao | | | 0.19 |
| *Females* | | | |
| Khok Phanom Di | **<0.01** | **<0.01** | **<0.01** |
| Nong Nor | | 0.50 | 0.62 |
| Ban Lum Khao | | | >0.99 |
| *Males* | | | |
| Khok Phanom Di | **<0.01** | **<0.01** | 0.11 |
| Nong Nor | | 0.35 | 0.05 |
| Ban Lum Khao | | | 0.40 |

**Bold** indicates statistical significance at 1% level

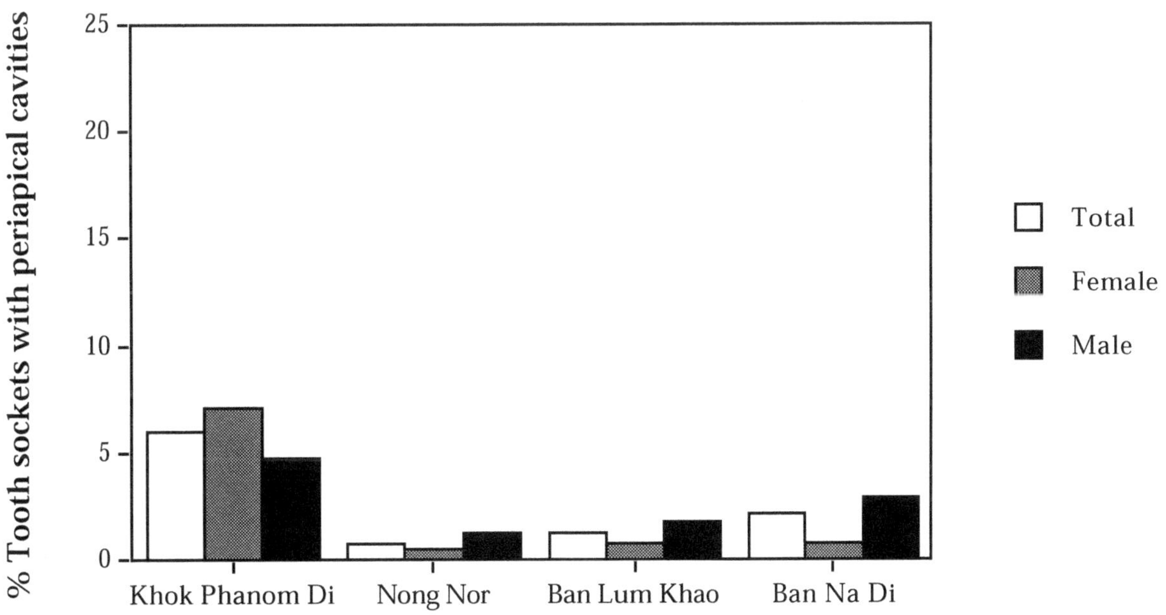


**Figure 7.4:** The percentage of tooth positions with periapical cavities in each sample, for each sex and in total

# Antemortem Tooth Loss

## Introduction

Antemortem tooth loss is the premature loss of a tooth. Teeth can be lost through a number of pathological problems which includes caries and advanced attrition. As previously mentioned, these conditions either expose the dental pulp and/or provide the oral bacteria a route in which to enter the periapical tissues. The subsequent infection may eventually result in the loss of the tooth. The cause is not always going to be obvious as the evidence is often lost when the tooth is lost.

There is also a non-pathological type of antemortem tooth loss, ablation. This represents intentional tooth removal which is a custom in some populations. The difference between a tooth lost pathologically or intentionally is not always obvious from dry bone evidence. The key, however, is in identifying repeated and often symmetrical patterns of tooth loss among individuals. Evidence of ablation in the people of Khok Phanom Di has previously been reported (Tayles 1996b). The people of Khok Phanom Di appeared to favour the removal of the incisors, although the canines were sometimes also involved. In the following analyses the cases of ablation in each individual were identified (code 5, see Table 7.1) and excluded. The possibility of ablation in the other samples was investigated. No cases could be confirmed as there was no obvious pattern of symmetry that is typically seen with ablation. The antemortem tooth loss prevalences presented here are assumed, after some consideration, to be pathological.

## Methods

Every available tooth position was assessed for pathological antemortem tooth loss. The number of teeth lost divided by the total number of tooth positions observed gives the percentage of teeth affected. Adult males and females are analysed both separately and as a group. Comparisons were made among the samples and between sexes within each sample as with other dental pathologies.

## Results

The age structure for the total teeth assessed for antemortem tooth loss is the same as that provided for periapical cavities (Tables 7.14-15). This was discussed in the results section of the assessment of periapical cavities.

The number of teeth lost antemortem is provided for each sample in Table 7.18 and Figure 7.5. Table 7.19 details the Fisher's Exact p-values for comparisons among the samples. In the total adult comparisons, the Khok Phanom Di samples had the highest prevalence of antemortem tooth loss (9.0%), and this was statistically higher than all other samples. The other three samples showed very similar prevalences of antemortem tooth loss compared to each other, varying by no more than 1%.

In the female subsamples, Khok Phanom Di females still showed the highest prevalence (13.0%), Ban Na Di females also had a relatively high prevalence (12.9%), which was almost identical to the Khok Phanom Di females. Both Nong Nor and Ban Lum Khao females had similarly low levels of antemortem tooth loss which were both statistically significantly lower compared with Khok Phanom Di and Ban Na Di females.

In comparison, male levels of antemortem tooth loss were much lower than the female levels, and Ban Na Di males showed the lowest (1.5%) prevalence. This was statistically lower than both Khok Phanom Di and Ban Lum Khao, but only 1.1% lower than that in the Nong Nor sample.

Within each sample, except Ban Lum Khao, sex differences were significant (Table 7.18), with each group showing more antemortem tooth loss in females than in males.

**Table 7.18:** Summary of the number of tooth positions with antemortem tooth loss in each sample, for each sex and in total

| | Khok Phanom Di | | Nong Nor | | Ban Lum Khao | | Ban Na Di | | Comparison among samples | |
|---|---|---|---|---|---|---|---|---|---|---|
| | A/N | % | A/N | % | A/N | % | A/N | % | Chi 2 | p-value |
| Total | 184/2046 | 9.0 | 58/1323 | 4.4 | 59/1154 | 5.1 | 38/707 | 5.4 | **35.89** | **<0.01** |
| Female | 143/1102 | 13.0 | 40/630 | 6.3 | 31/607 | 5.1 | 31/241 | 12.9 | **39.58** | **<0.01** |
| Male | 41/944 | 4.3 | 18/693 | 2.6 | 28/547 | 5.1 | 7/466 | 1.5 | **13.22** | **<0.01** |
| *Male vs Female Fisher's Exact p-value* | | | | | | | | | | |
| | **<0.01** | | **<0.01** | | >0.99 | | **<0.01** | | | |

A= affected, N= total number of teeth observed; **Bold** indicates statistical significance among samples at 1% level.

**Table 7.19:** Fisher's Exact p-values for inter-sample comparisons of the prevalence of antemortem tooth loss, for each sex and in total

| *Total* | Nong Nor | Ban Lum Khao | Ban Na Di |
|---|---|---|---|
| Khok Phanom Di | **<0.01** | **<0.01** | **<0.01** |
| Nong Nor | - | 0.40 | 0.32 |
| Ban Lum Khao | - | - | 0.83 |
| *Females* | | | |
| Khok Phanom Di | **<0.01** | **<0.01** | >0.99 |
| Nong Nor | - | 0.39 | **<0.01** |
| Ban Lum Khao | - | - | **<0.01** |
| *Males* | | | |
| Khok Phanom Di | 0.08 | 0.52 | **<0.01** |
| Nong Nor | - | 0.02 | 0.22 |
| Ban Lum Khao | - | - | **<0.01** |

**Bold** indicates statistical significance at 1% level

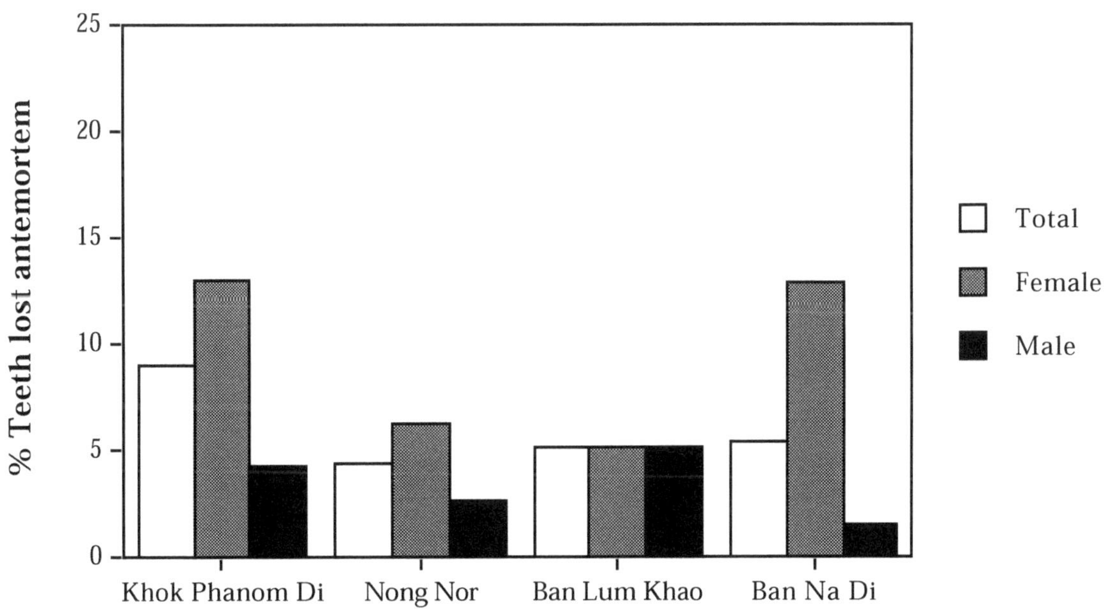


**Figure 7.5:** The percentage of tooth positions with antemortem tooth loss in each sample, for each sex and the total sample

**Table 7.20:** Dental pathology profiles for the prehistoric Thai samples

| Sample | Advanced Attrition | Caries | Periapical Cavities | Antemortem Tooth Loss |
|---|---|---|---|---|
| Khok Phanom Di | Low | **High** | **High** | **High** |
| Nong Nor | **High** | Low | Low | Low |
| Ban Lum Khao | Moderate | Low | Low | Low |
| Ban Na Di | Moderate | Low | Low | Low |

# Results Summary

**Advanced attrition**

The prevalence of advanced attrition in both the Ban Na Di and Ban Lum Khao samples was most likely age related. The sample from Ban Lum Khao was comprised of mostly young adult females and older adult males. This explains the low level of advanced attrition in the females and the high levels in the males in this sample (and the subsequent significant difference between males and females).

The Ban Na Di sample, in general, had more older adults, particularly females, and this would explain the high levels of advanced attrition in the females and lower levels in the males in this sample (and the subsequent significant difference between males and females). The total samples from both Ban Na Di and Ban Lum Khao, however, do provide a general indication of the level of advanced attrition in these samples. Both Ban Lum Khao and Ban Na Di had very similar moderate levels of advanced attrition overall.

The results of advanced attrition in the Khok Phanom Di and Nong Nor samples, and particularly the differences between them, did not appear to be age related. Nong Nor males and females both showed the highest proportion of teeth with advanced attrition. The Khok Phanom Di total adult sample showed considerably lower levels compared to the other samples, particularly in the females. This sample showed a significant difference between males and females with females having just under half the percentage of teeth with advanced attrition of males.

**Caries**

The amount of carious teeth in the subadults was minimal in all samples. The high proportion of caries in the Ban Na Di deciduous teeth was probably a reflection of the small sample size. Khok Phanom Di subadults had the highest proportion of teeth with caries in the subadult permanent teeth.

In the adult samples, the Khok Phanom Di people showed a significantly higher prevalence of caries in the total sample, while the other three samples showed similar lower levels. This was the same pattern in the female and male subsamples although the differences were not as significant in the males.

All samples showed significant sex differences with the females more frequently affected than the males. The higher prevalence of caries in the Ban Na Di females compared with the males may be a reflection of the age structure which showed Ban Na Di had more older females. The results of the other three samples do not appear to be age related.

**Secondary Infection and Inflammation**

Overall the prevalence of periapical cavities was low compared with the other pathologies. The Khok Phanom Di adults showed a considerably higher proportion in both males and females in comparison with the other samples. The other three samples all had very minimal numbers of periapical cavities. There were no significant sex differences within any of the samples. The prevalence of periapical cavities does not appear to have been influenced by the age structure in any of the samples.

**Antemortem tooth loss**

The Khok Phanom Di people showed the highest percentage of teeth affected by antemortem tooth loss. All other samples had quite similar levels of teeth lost antemortem. The Ban Na Di female level was elevated due to the high proportion of older females. However, the differences between the sexes in the Nong Nor and Khok Phanom Di samples were unrelated to age and both showed more female teeth lost antemortem than male teeth. The males and females from Ban Lum Khao showed exactly the same percentage of teeth lost antemortem.

The dental pathology profiles of each sample are summarised in Table 7.20. However, the terms 'high', 'moderate', and 'low', were used here as relative terms comparing the four samples only. They do not relate to the definition of such terms in other studies (for example Lukacs 1989; Littleton and Frohlich 1993).

# Discussion

In attempting to reconstruct and compare the health and diet of prehistoric populations, the dentition can provide significant information. Presenting data for each skeletal sample in the form of a "Dental Pathology Profile" as suggested by Lukacs (1989: 261) provides an indication, easily comparable, of each sample's dental health. This simply involves assigning relative terms of low, moderate, or high for each dental pathology of each sample relative to the other samples. This information is presented in Table 7.20 for the samples analysed in this thesis.

The following discussion will focus on how the diet of each population may have differed to influence the dental pathology profiles of the samples. This must first include a discussion of how the age structures of each sample have influenced the results, before the influences of diet can be considered.

The prevalence of advanced attrition, caries, and antemortem tooth loss in the people of Ban Na Di may be a reflection of the age structure of this sample. In general, there is a higher proportion of older adults, particularly in the female subsample, than in any of the other three samples. This explains the high level of advanced attrition, caries, and antemortem tooth loss in females as these pathologies are all positively correlated with age. Therefore, in the following comparisons, the results from the Ban Na Di sample, in particular the data from the male and female subsamples, will not be discussed with reference to environmental or cultural factors. However, the results of the total adult sample from Ban Na Di will be used as a general indicator of the prevalence of these dental pathologies.

The problem with the Ban Lum Khao dental sample is that it was comprised of a significantly higher proportion of young adult females (15-29 years) than any other sample and a comparatively high proportion of older males. Therefore the low level of advanced attrition in the females and high in the males can be explained by this age structure. No other dental pathology prevalence from Ban Lum Khao appears to have been affected by the age structure of the sample. Neither of the Khok Phanom Di and Nong Nor results appear to have been influenced by their respective age structures.

There are three dental pathology profiles amongst the four samples (Table 7.20). The first is from the people of Khok Phanom Di. They have a high level of caries which may be due to a complex variety of factors discussed below. The people of Khok Phanom Di had low levels of advanced attrition and this implies the majority of the diet was relatively soft in texture. The high level of periapical cavities are most likely to have been caused by caries, given the high prevalence of this and the low prevalence of advanced attrition as an alternative route of infection. The high level of antemortem tooth loss is also probably a reflection of both the high caries percentage and the subsequent infection of the periapical tissues that has advanced severely enough to eventually cause premature tooth loss. However, periodontal disease cannot be ruled out as a contributing factor to tooth loss.

The people of Nong Nor also have a distinctive dental pathology profile. This sample stands out as different from the others primarily because of a high prevalence of advanced attrition. This is most likely indicative of an abrasive diet although genetic factors cannot be eliminated. The low caries prevalence suggests their diet did not have a high proportion of cariogenic foods. An alternative possibility is that the advanced wear removed the potential sites of caries development. This interaction of dental pathologies will be discussed later in this chapter. There are few periapical cavities and few teeth lost prematurely in the Nong Nor people, indicating few individuals suffered from infection advanced enough to cause a bony response and subsequent tooth loss. However, in the few individuals that did suffer from periapical pathologies, it was most likely caused by pulp exposure through advanced attrition.

Both samples from the northeast, Ban Lum Khao and Ban Na Di, have strikingly similar dental pathology profiles over their total samples, defining the third main pattern. Low caries rates suggest that, like the Nong Nor people, cariogenic foods did not make up a large component of their diet. The diet was, however, moderately coarse in texture, less so than Nong Nor, but more so than Khok Phanom Di samples. The prevalence of both periapical cavities and antemortem tooth loss was low and it is not obvious whether these were wear or caries induced; both probably contributed.

There are three general points that have arisen from the dental pathology profiles of these four samples. The first point highlights the general decline through time in the prevalence of all dental pathologies which is equivalent to an improvement in general dental health with time. The people of Khok Phanom Di show the highest prevalences of all dental pathologies analysed except for advanced attrition, in comparison with the other three samples. The high level of advanced attrition in the people of Nong Nor in comparison with the other three samples is the second point. The third general point is the consistent similarity in the total prevalences of all dental pathologies of the people of Ban Lum Khao and Ban Na Di in the northeast.

The first question raised, therefore, is why is there an improvement in dental health through time? In order to address this question, the differences among the samples that can influence dental health need to be considered. Diet is one of the single most important influences on the dentition. Others include age and this has been discussed already.

From archaeological evidence, and now dental evidence, it can be inferred that the diet of the people of Khok Phanom Di was quite different from that of the other samples. The site of prehistoric Khok Phanom Di was adjacent to coastal, estuarine, and mangrove environments, while all the other sites were inland. The Khok Phanom Di people ate more fish and shellfish and less terrestrial animal meat. The people of Nong Nor did have some access to the coastal environment but were not as close as those at Khok Phanom Di. Additionally, the local environment of Khok Phanom Di was not particularly favourable for growing large quantities of rice at the time of occupation, while at the later sites, rice agriculture was well-established, especially by early Iron Age at Ban Na Di.

Superficially, it appears there are differences in the diet that correspond to differences in each sample's dental pathology profile. However, in considering the relationship in more detail and linking, more precisely, actual food types to the dental health of these groups, it is not immediately obvious what the answer is. This is particularly evident when the evidence for caries is considered. General trends (although not universal) in other regions of the world, document an increase in individuals with carious lesions for populations undergoing a transition to the intensive agriculture of some carbohydrate staples, for example, maize in North America (Larsen 1997). Caries research in prehistoric populations has been given considerable emphasis in the analysis of changing diets with agricultural intensification. This has occurred because of the apparent relationship between carbohydrate consumption and the development of caries. The development of agriculture led to the increased consumption of carbohydrates such as maize, wheat, and other grains, as well as roots and tubers such as taro and yams. Thus, the increase in caries seen in many parts of the world has been explained in this manner as it is assumed all carbohydrate staples are cariogenic.

Therefore, it could be hypothesised that the Khok Phanom Di people, being the least reliant on rice agriculture ,would exhibit the lowest caries rates and an increase in caries prevalences would be observed though the Bronze and Iron Age as reliance on rice agriculture increased. However, in direct contrast to many other world trends (Leigh 1925; Turner 1979; Larsen 1982; Lukacs 1992) the intensification of agriculture in prehistoric Thailand appears to have coincided with a *decrease* in the prevalence of caries, and a general improvement in dental health. This, in fact, makes sense when the actual cariogenicity of rice is compared with other staple carbohydrates such as maize and wheat.

Sreebny (1983) showed an increase in rice consumption had no affect on the number of carious lesions. Sreebny's study also indicated that an increase in the consumption of other carbohydrates, maize and wheat, have a negative and positive association with caries, respectively. Obviously not all carbohydrate cereals have the same cariogenic nature. However, the results of this study have been compounded by the influence of sugar consumption that differed among the groups studied. Those from wealthier countries tend to have higher sugar consumption levels. The data suggest that at best the variation demonstrated by these cereals accounts for about 20% of caries compared with sugar which accounts for approximately 50% of caries prevalence (Sreebny 1983).

Turner (1979) collated mean caries rates from around the world and concluded there was a general increase in caries from hunter-gatherer societies to agricultural societies. However, there a number of factors not adequately accounted for in making this rather simple association between caries and agriculture. Most importantly, the effects of age are not taken into consideration despite changes in the age structures of these samples that almost certainly would have occurred with the adoption of agriculture. Longevity, fertility, and population size were all likely to have increased during this time, thus affecting the demographic profiles of the samples. Despite this and other problems with Turner's (1979) study, many subsequent studies have accepted the association between caries and agriculture and gone as far as using it as evidence of an agricultural lifestyle (for example Roosevelt 1984).

A review of 75 archaeological dental samples from eastern North America (Larsen *et al.* 1991) showed all but three samples with a maize component in their diet had a caries rate of greater than 7%, while the foraging populations all had less than 7% of teeth affected with caries. In general, the increased reliance on maize does appear to have led to an increased number of caries. This is probably due to the high sucrose content of maize. Sucrose is a simple sugar and therefore more easily and quickly metabolised by oral bacteria (Larsen 1997). The relationship between maize and caries has been adopted by many researchers as applying to all types of carbohydrate agriculture without considering that different grains or tubers may not be as highly cariogenic.

Tayles *et al.* (2000) have recently investigated the clinical and experimental literature to determine whether or not rice actually is cariogenic. This arose after the research presented in this thesis showed a decrease in caries with rice agriculture and an additional Thai prehistoric sample, from the later Iron Age (300BC - 300AD), Noen U-Loke, continued the downward trend.

Rice eaten in prehistoric Southeast Asia was most likely to have been in the unpolished state which would have provided a certain degree of coarseness in the diet (Tayles *et al.* 2000). A coarse diet can assist with the mechanical cleaning of teeth by removing food particles that may otherwise have remained to be metabolised by oral bacteria. As discussed above, a coarse diet may also create higher levels of advanced wear, thus removing the potential sites for the development of caries.

The cariogenicity of rice is difficult to assess in humans because of the non-reversibility of caries. However, the extent of the acid drop after consumption of some carbohydrates has been studied (Lingström *et al.* 1993). Parboiled rice induced one of the smallest pH reductions in comparison to other sugars and starches such as sucrose, bread and potatoes (Lingström *et al.* 1993). This would imply rice had a relatively low cariogenecity.

There are a number of other factors to consider in the development of caries, many of which cannot be assessed in skeletal populations. For example, the people from Khok Phanom Di may have had a greater genetic predisposition to developing caries. Genetics can influence the presence or absence of a person's, or population's, particular oral bacterial flora, with some bacteria able to produce acid more rapidly than others (Hillson 1996). Salivary flow can also influence the development of caries and this, too, can be related to a person's genetic makeup. Salivary flow influences the clearance rate, the speed at which oral pH rises again and food is removed from around the teeth (Thylstrup and Fejerskov 1994). The faster the clearance rate, the less time the teeth are exposed to the acid phase. Some people may also have a greater immunity against the various bacteria.

Additionally, caries development may be influenced by the frequency of eating, and the level of dental hygiene, both further unknown factors in prehistoric populations. One other factor that may have influenced the relatively high level of caries in the people of Khok Phanom Di is their high level of enamel hypoplasia (Tayles 1999). Enamel hypoplasia in the form of pits and grooves around the teeth provide extra potential sites for the harbouring of bacteria and food particles (Powell 1985). Enamel hypoplasia also represents defective enamel which may be more easily destroyed.

Environmental fluoride can provide some degree of protection against caries development as mentioned earlier (Hillson 1996; Soames and Southam 1998). A survey of dental caries and dental fluorosis in Thailand in the 1960's found "...caries experience was low at all ages up to 45 years...Concurrent with the low caries experience...dental fluorosis was common throughout Thailand" (Leatherwood *et al.* 1965: 1797). Dental fluorosis was assessed in all regions of Thailand but an analysis of the prevalence in the southeast ('eastern' in the study) compared with the northeast ('northeast' and 'northern') showed very little difference, with 57.5% in the northeast compared with 60.2% in the southeast (Leatherwood 1966). However, when they analysed the fluoride content in the drinking water only small amounts were found, in contrast to high urinary excretion. They concluded fluorides were derived from foods (Leatherwood *et al.* 1965). It was suggested in a later study that the high consumption of sea salt and fish, may have been a high source of fluoride in the diet (Hadjimarkos 1966). However, an analysis of the fluoride content of sea salt in Ceylon claimed sea salt would be a negligible contributor to fluoride in the diet (Senewiratne *et al.* 1974). This study also analysed the fluoride content of rice, but this to was found to contain only low concentrations (Senewiratne *et al.* 1974). An analysis of the fluoride content of dental tissues from each of the prehistoric Thai samples may provide further information. However, over time fluoride can accumulate in bone and teeth, leaching in from the surrounding soil (Littleton 1999).

Over all it is apparent there is not always a straightforward relationship between the percentage of caries and diet. However, there must have been something of high cariogenicity in the Khok Phanom Di diet that was subsequently not available or less relied on in later times as the Nong Nor, Ban Lum Khao, and Ban Na Di samples all show lower caries rates. High numbers of carious lesions have been observed in hunter-gatherer populations, particularly those that rely on starchy plant material (high in carbohydrate) with a sticky texture. Many fruits and some nuts come under this category and these were not necessarily grown agriculturally in prehistory (Larsen 1995). Bananas, palm sugar, yams, and taro are all possible candidates for additional carbohydrates in the diet (Tayles *et al.* 2000) that may have contributed to the caries in all samples, but particularly so in the Khok Phanom Di people. These foods may have been less relied on in the later sites of Ban Lum Khao, Nong Nor, and Ban Na Di as they had a more reliable food resource in their rice agriculture. In modern rural Thailand, Leatherwood *et al.* (1965) found a difference in caries rates between 'northern' Thailand (including northeastern regions) and 'southern' Thailand (including the southeastern region) with higher DMF (Decayed, Missing and Filled permanent teeth) teeth in the latter in all age groups. The chewing of betel nut is relatively common in Thailand today and this has been shown to be cariostatic in a small study in Central Thailand (Kamalanathan *et al.* 1960a cited in Tayles 1999) and in Papua New Guinea (Schamschula *et al.* 1978 cited in Tayles 1999). While this evidence is not conclusive by any means it is possible the people from prehistoric Thailand were betel nut chewers, thus conveying a cariostatic component in the diet. However, no skeletal sample showed any evidence of the characteristic staining that accompanies this habit.

The composition of the food and its method of preparation has influenced the increasing prevalence of caries with agriculture. This includes the fact that much of the plant material cultivated is boiled into a soft gruel which not only promotes the growth of bacterial colonies but does not provide any mechanical cleaning by either the action of mastication (Larsen 1995) and/or the grit in the food.
An abrasive diet, established by the level of advanced wear in prehistory, can act to cleanse teeth mechanically:

> Diets that regularly contain rough-textured foods and/or abrasive particles (eg., grit in stone-ground flour) are less cariogenic than are those more refined in texture, because of the differential stimulus to natural oral cleansing. (Powell 1985: 314)

Thus emphasising the importance of considering the range of dental pathologies as they interplay.

High levels of advanced attrition have been commonly seen in populations with a reliance on a marine diet and also includes low caries and a lack of antemortem tooth loss (Littleton and Frohlich 1993; Sealy *et al.* 1992). In other regions of the world, such as the Arabian Gulf, prehistorically a marine diet produced high levels of advanced attrition thought to have been the result of the higher amount of grit and sand in the food, and low caries prevalences through the lesser amount of carbohydrates in the diet (Littleton and Frohlich 1993). Sealy *et al.* (1992) also accounted for the higher level of wear as a factor for reduced caries and additionally suggested seafood may have cariostatic effects although this has not been experimentally proven. This 'marine' dental pathology profile is, however, in direct contrast to that of the Khok Phanom Di people who were marine dependent to some extent. The marine dental profile would seem to fit better with the Nong Nor sample.

The high levels of severe attrition in the Nong Nor people reflect a coarseness of diet that is not present in the other samples. Perhaps the people of Nong Nor had better access to coastal resources than is thought from archaeological evidence, or their foods were relatively unprocessed and tough, such as some plant foods can be. Additionally, caries may have existed at lower prevalences in the people of Nong Nor as the sites for harbouring bacteria were removed through advanced attrition. The high attrition at Nong Nor did not have subsequent effects on tooth infection and loss in this population.

The dentition can show high levels of caries and high levels of wear together. While wear removes the occlusal sites in which caries often develop, the wear enlarges the interproximal spaces. This promotes the trapping of food particles between the teeth which can lead to caries developing in this region (Powell 1985). Further investigation into the predominant location of caries within each Thai sample may provide further information in this regard.

The reduction in the prevalence of periapical cavities and antemortem tooth loss through time indicates a reduced susceptibility to infection. However, given these are secondary developments to caries, advanced attrition, or periodontal disease, it would be expected the reduction in the secondary pathologies would be related to an equivalent reduction in caries, advanced attrition and periodontal disease. While caries prevalence reduces over time, advanced attrition does not follow the same pattern, although it does decrease from the Bronze Age to the Iron Age.

**Table 7.21:** Summary of dental pathology prevalences (percentage by tooth) from a selection of prehistoric skeletal samples from Southeast Asia

| Sample | Advanced Attrition % | Caries* % | Periapical Cavities % | Antemortem Tooth Loss % |
|---|---|---|---|---|
| Khok Phanom Di[1] | 8.0 | 10.9 (13.2) | 6.0 | 9.0 |
| Ban Chiang[2] | 15.4 | 7.1 | 6.1 | 6.7 |
| Nong Nor[3] | 17.3 | 6.5 (6.5) | 0.8 | 4.4 |
| Ban Lum Khao[3] | 13.0 | 4.5 (5.2) | 1.3 | 5.1 |
| Non Nok Tha[2] | 6.3 | 2.8 | 1.9 | 7.4 |
| Ban Na Di[3] | 12.0 | 4.7 (5.2) | 2.1 | 5.4 |
| Noen U-Loke[4] | 14.6 | 3.0 (4.7) | 1.8 | 8.4 |

1. (Tayles 1992); 2. (Douglas 1996); 3. This study; 4. (Nelsen 1999; Tayles *et al.* 2000). *uncorrected caries rates, with corrected rates provided in brackets when available

The similarity between the people of Ban Lum Khao and Ban Na Di in their dental pathology profile could indicate a number of similarities existed between these samples. First, the diet may well have been comparable, and given the similarity in the natural environment of these two sites, this makes sense. However, while the actual composition of the diet may have been similar, it is thought that the proportions of certain foods may have been different. The people of Ban Na Di, compared with the Ban Lum Khao people, are believed to have had a greater knowledge and reliance on rice agriculture. Therefore they were consuming a significant proportion of soft, starchy foods. However, given the low cariogenicity of rice, this diet may have only resulted in low levels of advanced attrition, which is evident in these people, although not at a significantly lower prevalence than the people of Ban Lum Khao. Other factors that may have been influencing the similarity in these northeastern groups dental pathology profile is genetics. It cannot be ruled out that these two samples are from the same gene pool and therefore had the same predisposing factors to dental health.

Sex differences were apparent in some or all of the samples in advanced attrition, caries, and antemortem tooth loss. No sex differences were apparent in the periapical cavity analysis. Some differences in prevalences have been explained by the age structure of each sex within each sample. This includes the sex differences in advanced attrition in the people of Ban Lum Khao and Ban Na Di, and the sex difference in caries and antemortem tooth loss in the Ban Na Di people alone.

All samples showed females with significantly higher caries prevalences than males. Given this consistent trend, Ban Na Di's age structure may only partly explain the difference in this sample. Sex differences may be either physiological or behavioural (Walker 1986 cited in Larsen *et al.* 1991). Physiological factors include the earlier eruption of teeth in females, thereby exposing them to cariogenic factors for a longer period of time. However, the time difference is rarely more than a year and would therefore unlikely contribute significantly (Larsen *et al.* 1991). Behavioural differences include the sexual division of labour that is evident in many populations. Ethnographic evidence from the southeastern Atlantic coast of the United States indicates females tended the crops as well as gathering wild plants and preparing the food. Men on the other hand were predominantly hunters (Larsen *et al.* 1991). This division of labour may have influenced the type of food and the frequency of food eaten. Females may have consumed more carbohydrates and less protein and also ate more frequently increasing the length of time the teeth are exposed to the acid phase. The longer the acid phase the more destruction possible. The possibility of a sexual division of labour in prehistoric Thailand will be further discussed in Chapter 8 in conjunction with other information such as joint disease and ethnographic evidence.

The Khok Phanom Di people were the only sample which showed a significantly different prevalence of advanced attrition between the sexes (unrelated to age structure), with the males having the higher rate. This fits with the sex differences in caries rates, as an individual is more likely to have high attrition with low caries (males) or low attrition and high caries (females) given the interactions discussed previously.

Both the Khok Phanom Di and Nong Nor samples show significantly higher rates of antemortem tooth loss in the female subsamples than the males. This mirrors the caries sex differences and may indicate the females from both samples were more likely to lose their teeth prematurely, with caries as the initiating factor.

A brief comparison of dental health with other prehistoric Thai sites is presented in Table 7.21. The sites are listed in chronological order according to their initial settlement dates. However, Ban Chiang, in particular, has a long occupation and spans the early agricultural to early Iron Age periods. There may be an element of interobserver error among researchers in these results.

Trends through time are particularly apparent in the caries and periapical cavities prevalences, both showing a general decline through time. Levels of advanced attrition are quite varied over all samples, showing no association with time. The prevalence of antemortem tooth loss does not show any consistent pattern with time. It is highest in the people of Khok Phanom Di with 9.0% and lowest during the Bronze Age, particularly in the Nong Nor sample with 4.4%. However, later in the Iron Age, in the sample from Noen U-Loke, the prevalence has increased to 8.4%. A more detailed analysis of changes through time, within each sample, where possible, may further explain these patterns.

## Summary

Dental pathology profiles of each sample indicate three different patterns of dental health. The samples from Khok Phanom Di and Nong Nor stand alone, while the Ban Lum Khao and Ban Na Di profiles are very similar. There are distinct trends with time, with a general improvement in dental health through time. The people of Khok Phanom Di had the highest prevalences of most of the pathologies considered with the exception of advanced attrition. The Nong Nor sample stood out with the highest prevalence of advanced attrition, but otherwise these people had similar dental health to the northeastern samples which showed generally low prevalences of all dental pathologies.

In a general way this information can indicate the type of diet consumed by these people. The dental pathology profile of the people of Khok Phanom Di indicates a possible higher proportion of cariogenic and soft foods compared with the other samples. Nong Nor people with high advanced attrition prevalences were possibly consuming a coarser diet compared to others. The Ban Lum Khao and Ban Na Di samples dental pathology profile suggested few cariogenic foods and a moderate amount of coarse foods were consumed.

There were some differences between the sexes within each sample, although all those within the Ban Na Di sample are a result of their age structure, as is the presence of higher advanced attrition in males from Ban Lum Khao compared with females. Sex differences in caries were evident in all samples with females consistently having the higher prevalence. In the Khok Phanom Di sample the lower prevalence of caries in males may be related to their higher level of advanced attrition than females. Both Nong Nor and Khok Phanom Di showed females lost their teeth prematurely more often than males. These sex differences may indicate a difference in diet between the sexes particularly in the Khok Phanom Di and Nong Nor samples, but also with respect to cariogenic foods in Ban Lum Khao and possibly Ban Na Di.

# 8 Discussion and Conclusions

The main purpose of this research was to provide a synthesis of the variation in health among four prehistoric Thai skeletal samples. A number of hypotheses were developed concerning how each sample's health status would relate to the others. The variations in health were hypothesised to be due to the differences in the natural and cultural environments of the sites from which the samples came. Factors associated with these divisions can directly and indirectly affect human health as previously discussed in Chapters 1 and 2. The available diet and the pathogen load both stem from the natural environment. The cultural environment influences health through social complexity, technological ability, subsistence mode, and the level of interaction with other communities.

This chapter directly addresses the hypotheses presented in Chapter 1 in light of the skeletal evidence presented in Chapters 3 to 7. As each of the skeletal indicators of health and disease were addressed, the salient points regarding the main differences and similarities among the four samples were discussed in detail. It is now possible to integrate this skeletal data with the archaeological evidence from Chapter 2, using a biocultural approach. Additional information is introduced from ethnographic studies on village life in rural Thailand to attempt to 'fill in the gaps' where no skeletal or archaeological evidence can provide information. This ethnographic evidence is included with caution, however, as it is acknowledged that not all information will be directly relevant to prehistory.

Genetic heterogeneity among the samples is also a further potential source of variation in individual and population health that is not addressed by either the natural or cultural environmental variables. Gross morphological observations of skeletal material cannot convincingly address the question of genetics and obtaining genetic material from these bones has proved difficult (Pierson 1998). While it is an underlying assumption that the four skeletal samples were genetically homogeneous, both among the samples and within each sample, it has been acknowledged where genetic heterogeneity may be a possible explanation for health differences. By careful analysis it may be possible to discover individuals with an underlying 'differential frailty' in their response to the risks of daily life. This was carried out in the assessment of growth and growth disturbances in Chapter 4. However, whether this reflects inherent genetic differences or 'acquired' frailty, such as through nutritional deprivation, cannot be established.

Many of the factors affecting health are interrelated. For example, an individual of lower status in a hierarchical community may be more likely to fall ill because of inadequate nutrition. They may also be less likely to fully recover and therefore be chronically exposed to the same conditions. Explanations can become complicated with an array of possible interpretations because of these interacting factors, not all able to be identified by the biological anthropologist. However, some of the most important influences on health are nutrition, exposure to infection, and the cultural environment.

The discussion begins with a summary of the skeletal evidence for subadult and adult health, disease, and mortality. This includes a combination of non-specific (such as enamel hypoplasia) and specific (such as joint disease) indicators of morbidity. As mentioned in Chapter 1, there are a significant number of diseases that cannot be diagnosed in the skeleton alone.

> ...in all probability, acute, often lethal infections were probably the most frequent in antiquity...yet these usually leave few recognizable skeletal changes. Those infections to which the body has developed at least sufficient immunity to prolong its coexistence as a chronic infection are the most apt to generate obvious skeletal lesions, though quantitatively these chronic infections probably had lesser demographic impact... (Aufderheide and Rodriquez-Martin 1998: 117-118)

Therefore, a survey of the clinical literature focusing on particular health risks in modern developing rural communities in tropical countries is presented. Through this information it may also become apparent that there are particular environmental differences in the presence of certain pathogens. This may help with explaining the patterns of health in the northeast and southeast of prehistoric Thailand. The importance of these influences on infants and children is particularly emphasised.

> Monitoring the health of infants and children can provide the prehistorian with a rich variety of information about the health of a community. As this segment of the population is very sensitive to environmentally and culturally produced insults, changes in morbidity in this segment of the population should provide one of the first signs of changes in environment and culture. (Goodman and Armelagos 1989: 239)

The main influences on child health are nutrition and pathogen load. The latter includes the type and quantity of bacteria, parasites, and viruses in the local environment that children may have been exposed to. This will be dependent mostly on their natural environment, but also the population size, density, and level of hygiene. One of the most important factors determining a person's ability to cope with such an illness or an injury is their diet, or more specifically the nutritional adequacy of the diet. There is some archaeological and skeletal evidence for the potential prehistoric diet of the four samples, as presented in Chapter

2, and this will be summarised here. The complete range of edible foods in prehistoric times is probably impossible to determine, because only those with some inorganic component will be preserved. Once again, ethnographic evidence may be useful in indicating potential food sources not identified archaeologically.

After this discussion of the pathogen load and nutrition in prehistoric and modern Thailand, the hypotheses based on the influence the natural environment on health are directly addressed. There were three hypotheses suggested, the first two based on the theory that a similar natural environment will coincide with a similar health profile. Therefore, the Khok Phanom Di and Nong Nor samples from the southeast, would have had a similar level of health, as would the Ban Lum Khao and Ban Na Di samples, both from the northeast. The third hypothesis was concerned with comparing the health of the two regions, and given the dissimilarities in the natural environment, it was hypothesised that the northeast samples would have a different level of health from the southeast samples.

Finally, the cultural environment in prehistoric Thailand is reviewed. The social structure of each prehistoric sample differed in some ways from the others, and this may also have had a significant impact on their health. The social structure and complexity along with the technological level of a population can have an impact on adult health in terms of workload and the types of activities undertaken by each group. Socio-economic factors may also influence diet and care of the sick. Population size and density may affect hygiene which can affect the pathogen load in some communities. Therefore, as hypothesised, there may be a trend for an improvement in general health through time. The Khok Phanom Di sample would be the least healthy and the Ban Na Di sample the healthiest. The health of the Bronze Age samples, Nong Nor and Ban Lum Khao, would fit somewhere in between.

# Skeletal Evidence Summary

### Childhood health, morbidity, and mortality: age at death and growth

Subadult mortality was highest in the Khok Phanom Di sample. The Ban Lum Khao sample had the second highest subadult mortality, with Ban Na Di third. In comparison, the Nong Nor sample had a very low subadult mortality that is most likely to have been related to preservation factors.

Attempting to explain the differential subadult mortality in these samples is not straightforward. Through the evidence of growth disruptions it may be possible to draw inferences concerning the nature of childhood health problems. Were children more likely to be chronically ill or suffer acutely and fatally? The accumulation of chronic illness or malnutrition may also lead to premature death. By analysing the enamel hypoplasia data in subadults and adults inferences concerning childhood morbidity can be made. Combined with evidence for adult stature, their ability to overcome problems can also be compared. However, in this approach only a few variables are relied on as skeletal evidence is limited. Suggestions as to what the health problems may have been are discussed later when the possible sources of infection and nutritional deficiencies are considered.

Childhood at Khok Phanom Di appeared to be a more vulnerable period than at the other sites. The skeletal evidence suggests this skeletal sample was comprised of more frail and medium frail children, as the combination of mortality, enamel hypoplasia, Harris lines, and stature estimates suggested. Health problems in childhood at Nong Nor were also common but the majority of children coped well with such episodes, suggesting the presence of a high proportion of strong and medium frail children. Few died as subadults at Nong Nor and those that did had a moderate amount of enamel hypoplasia. Adults at Nong Nor attained taller statures than both the Khok Phanom Di and Ban Lum Khao samples. Ban Lum Khao children did not fair as well as those from Nong Nor. They had a relatively high proportion dying as subadults. Many others suffered chronically into adulthood, as the shorter statures suggest. This sample probably had a mix of more frail and medium frail children. Many Ban Na Di subadults also suffered chronically. This sample had the same prevalence of enamel hypoplasia in the adults as at Ban Lum Khao. However, the taller mean statures of the Ban Na Di adults suggests they coped better with these problems, possibly because of a higher proportion of stronger children.

### Adult health, morbidity and mortality

Age at death

All samples showed a generally similar pattern of age distribution among adults. All samples had few or no adults aged between 15-19 years, and the majority in all samples died between the ages of 20-39 years. Few individuals survived beyond 40 years in each of the samples. While the pattern was similar, there were some differences in the actual percentages among the samples. Of particular note was the high proportion of young women, aged between 20-29 years, in the Ban Lum Khao sample. Additionally, there was a general increase through time in the percentage of people surviving beyond 40 years, with the Ban Na Di sample comprised of nearly twice as many adults of this age than Khok Phanom Di.

Near equal sex ratios were calculated for each sample; all except Ban Na Di had slightly more females than males. Parturition, based on pelvic observations, was almost universal in both the Khok Phanom Di and Ban Na Di samples. This was not the case for the Ban Lum Khao sample.

Bone Mass

The adult samples from Khok Phanom Di, Ban Lum Khao, and Ban Na Di all maintained a good bone mass as evidenced by their cortical indices. No data were available from the Nong Nor sample. Neither the male or female means were significantly different among the samples, nor were there any sex differences within each sample. However, a wide range of bone mass within the Khok Phanom Di males and females and the Ban Lum Khao males was apparent. The larger sample size may explain the range in the

Khok Phanom Di males and females and the Ban Lum Khao males was apparent. The larger sample size may explain the range in the Khok Phanom Di sample. Males from Ban Lum Khao showed a wide range of bone mass indices compared to females. This may indicate an inequality within this sex, perhaps in terms of nutrition, that has led to these differences.

### Joint Disease

Two main points arose from the analysis of osteoarthritis. Firstly, the pattern of joints most frequently affected by osteoarthritis differed among the samples. Khok Phanom Di adults had high prevalences of osteoarthritis in the shoulders, knees, and joints of the hands. The Ban Lum Khao males showed high prevalences of osteoarthritis in the joints of the feet and the elbows. Ban Lum Khao females showed very little evidence of degeneration. Ban Na Di adults showed a high prevalence in the wrist joints and females alone showed a relatively high prevalence in the joints of the feet. No evidence was able to be recorded in the Nong Nor sample.

The second main point is that overall there appears to be a decrease in the presence of osteoarthritis through time. In general, the people of Khok Phanom Di had higher prevalences of osteoarthritis than either the Ban Lum Khao or Ban Na Di samples. Both of these trends may be linked to changes in activities among these people.

The prevalence of vertebral osteophytosis showed the same pattern of distribution among the regions of the vertebral column in each sample. All samples showed an increase in vertebral osteophytosis from the cervical region to the lumbar region. However, the actual percentages were different among the samples. In general, both Khok Phanom Di and Ban Na Di had the higher prevalences.

Confounding problems in the interpretation of such patterns of osteoarthritis and vertebral osteophytosis included the age and sex structures of some of the samples. Additionally, while some explanations have been suggested concerning the types of activities influencing these patterns, it is emphasised that these are purely suggestions. There are possibly many other alternative explanations that include genetic factors.

### Trauma

The main point to arise from the analysis of trauma was that the pattern of bones affected differed among the samples and that not all patterns appeared random. The people of Khok Phanom Di showed evidence of only a few fractures. The people of Nong Nor showed a wider variety of fractures. The evidence of some fractures in this sample may be indicative of interpersonal violence. The Ban Lum Khao sample also had potential evidence for interpersonal violence. However, this is not the only explanation, falls may also have been common in both of these samples. The pattern of long bone fractures in the Ban Lum Khao sample showed a pattern that was not necessarily random. Males suffered more fractures than females and the forearm bones were very commonly involved. Spondylolysis was also most common in this sample, again with more males than females affected. Little evidence of long bone fracturing was observed in the people of Ban Na Di.

### Dental Health

There are three general points that arose in the analysis of the dentition. The first point is the general improvement in dental health with advancing time. The people of Khok Phanom Di had the poorest dental health with high prevalences of caries, periapical cavities, and antemortem tooth loss. By the early Iron Age, in the Ban Na Di sample, caries had decreased, as had the evidence for secondary infection and tooth loss.

The second main point concerns the prevalence of advanced attrition in the sample from Nong Nor. This is considerably higher than any of the other samples. Thirdly, it was evident that the Ban Lum Khao and Ban Na Di samples, both from the northeast, have very similar dental health profiles with low to moderate prevalences of all dental pathologies.

### Anaemia

The skeletal evidence for anaemia was not presented in this thesis. This subject was given considerable attention in previous work on the Khok Phanom Di people by Tayles (1996a) . This evidence does have significant implications for the health of this sample and thus it is summarised here.

In a previous detailed investigation of the health of the people of Khok Phanom Di (Tayles 1996a, 1999) a number of individuals were identified to have suffered from both mild and severe anaemia. For example, cribra orbitalia was observed in 10 children, all between the ages of 1-14 years, and three adults all less than 30 years at death, two males and one female. In combination with other skeletal evidence it was concluded that it was most probably of genetic origin. One (or more) of the haemoglobinopathies, possibly thalassaemia, was suggested as the cause of the majority of the severe and mild cases of anaemia evident in these skeletal remains. Mild cases of anaemia were also likely to have resulted from malaria (Tayles 1999) .

Evidence of skeletal changes associated with anaemia in the other samples was minimal to nonexistent. Therefore a detailed discussion, given the preservation problems, of this type of specific pathology would have been difficult to accurately investigate. The evidence that was observed is briefly described here. Cribra orbitalia was present in three adults in the Nong Nor sample, a 30-39 year old male, a 30-39 year old female, and one adult of unknown sex. Had the skeletal material been preserved better further evidence may have been identified. Adults with evidence of anaemia were not necessarily anaemic at the time of death, but this indicates they suffered from anaemia during childhood. In the Ban Lum Khao sample only one individual, an infant, was observed with cribra orbitalia. However, the layer of concretions over many boneshave inhibited observations. No evidence of cribra orbitalia was found in the Ban Na Di skeletal material.

Given the lack of pathognomonic features of any of the haemoglobinopathies or thalassaemia syndromes in the Nong Nor, Ban Lum Khao, and Ban Na Di samples, the few cases of cribra orbitalia indicating anaemia were probably more likely to be acquired anaemia associated with cases of such pathogens as hookworm infections or iron deficiency in the diet.

**Table 8.1:** Examples of some common pathogens found in tropical countries and their effect on human, particularly child, health*

| Class of pathogen | Pathogen | Effect on humans |
|---|---|---|
| Bacteria | *Shigella* | acute diarrhoea |
| | *Salmonella* | acute diarrhoea |
| | *Escherichia coli* | acute diarrhoea |
| | *Streptococcus pneumoniae* | respiratory infection |
| Parasites | | |
| Protozoa | *Giardia lamblia* | acute diarrhoea |
| | *Plasmodium falciparum* | malaria: anaemia, fatigue, acute diarrhoea |
| | *Entamoeba histolytica* | amoebiasis: diarrhoea or dysentery |
| Helminths | Hookworm | iron deficiency, malnutrition, reduced growth, B12 and folic acid deficiencies |
| | *Ascaris lumbricoides* | intestinal pain, protein-energy malnutrition |
| Nematode | Filariases (lymphatic) | fever, groin pain, oedematous swelling of legs |
| Trematodes | Liver flukes (eg. *Opisthorhis viverrini*) | pain, loss of appetite, bile duct cancer |
| | Intestinal flukes (eg. fasciolopsiasis) | vitamin B12 deficiency, small intestinal ulcers, malnutrition, diarrhoea, fever, anaemia |
| Viruses | Rotavirus | acute diarrhoea |

*References: Jelliffe 1970; Ortiz 1970; Sornmani *et al.* 1984; Harinasuta 1987; Chongsuvivatwong *et al.* 1996; Gilles 1996; Hart 1996; Haswell-Elkins and Elkins 1996; Nontasut *et al.* 1996; Suwanto 1997; Aufderheide and Rodriquez-Martin 1998.

# The Natural Environment

### Pathogen load

A tropical environment, such as that which exists in Thailand, harbours a wide variety of pathogens capable of affecting the morbidity and mortality of a population. This is the case in modern times (for example Jelliffe 1970) and would have been similar in prehistoric times. However, extrapolating the modern evidence of pathogens in tropical countries to prehistoric times is not always possible. This is because the interactions between a pathogen, its host, and the environment are changing all the time.

The relationship between these three are dependent on a variety of factors. The pathogen itself, being any microorganism that can cause tissue damage or disease, has a certain virulence which is the degree to which it can be pathogenic (Ellner and Neu 1992). A pathogen's virulence is influenced by a number of intrinsic factors. For example, it must be present in sufficient numbers and capable of surviving transmission from host to host. The use of an insect vector is a useful way for a pathogen to avoid the external environment. Otherwise, a pathogen may have to be resistant to heat, UV light, and drying, for example, in order to survive. Once in the host a pathogen requires the ability to adhere to the host and protect itself against the host's defence system (Evans 1998).

The occurrence of an infection depends on the host being exposed to the pathogen and the susceptibility of the host. The former is modified by behavioural factors such as occupation, familial exposure, hygienic habits, and socioeconomic level. The latter includes the age of the individual at infection, the dosage and duration of the pathogen, the presence of coexisting diseases, especially those that are chronic, the genetic makeup and its influence on the immune response, the immune state of the individual at the time of infection and the nutritional status (Evans 1998).

The environment provides the setting in which the pathogen and host interact (Evans 1998). The environment contains both the physical and biological mechanisms for the spread of the pathogen. The latter includes water, food, and air. The former includes animal, bird, or insect vectors. Vectors usually have strict requirements for their survival and variations in such factors as humidity, temperature, and vegetation can all affect the maintenance of the species in adequate numbers, and thus the transmission of the pathogen. For example, many pathogens are seasonal. Warm weather and tropical climates are favourable for many insects and for people to be outside, thereby increasing the incidence of mosquito-borne diseases, such as malaria. Colder weather brings individuals inside and closer together which tends to promote infections spread through the respiratory system (Evans 1998).

There are some pathogens, such as those requiring large population sizes and those introduced from other countries, that would not have been present in small, relatively isolated prehistoric Thai communities. Some pathogens may have been more predominant in the northeast or southeast regions of Thailand depending on the type of environment each pathogen or its vector requires to survive. The relative effects of infectious diseases are dependent, to a large extent, on animal reservoirs and insect vectors which can imply a relationship to particular geographical regions and seasonal variations (Jelliffe 1970).

**Table 8.2:** Prevalence of acute bacteria, dysentery and malaria in 1980, Thailand

| Region | Population size | Acute Diarrhoea | | Malaria | |
|---|---|---|---|---|---|
| | N | N | % | N | % |
| Thailand | 47173000 | 224477 | 0.48 | 123495 | 0.26 |
| Southeastern[1] | 14801000 | 76487 | 0.52 | 68679 | 0.46 |
| Near Khok Phanom Di and Nong Nor[2] | 1606000 | 11433 | 0.71 | 42793 | 2.66 |
| Northeastern | 16670000 | 73331 | 0.44 | 13963 | 0.08 |
| Near Ban Lum Khao[3] | 4227000 | 25974 | 0.61 | 4230 | 0.03 |
| Near Ban Na Di[4] | 4448000 | 14388 | 0.32 | 1246 | 0.03 |

1. Includes Bangkok; 2. Chon Buri, Rayong, Chanthaburi, Trat; 3. Nakhon Ratchasima, Buriram, Surin; 4. Khon Kaen, Sakon Nakhon, Nong Khao, Udon Thani. Ref: Ministry of Public Health, Thaildand 1980.

Table 8.1 presents some details of pathogens common in tropical countries and likely to have been present in prehistory. These are outlined in order to obtain an idea about what was causing the differential morbidity and mortality among the prehistoric Thai samples.

Firstly, the explanation for differential morbidity and mortality among the subadults will be considered. There are three main contributors to the premature mortality of infants in tropical countries. They are diarrhoeal disease, malnutrition, and respiratory tract infections (Jelliffe 1970). In a Thai village in the 1970s, children had an average of five episodes of respiratory infection and two episodes of diarrhoea in one year that required drugs for treatment (Chandrapanond *et al.* 1973). As mentioned above, very few of these problems will leave diagnostic skeletal lesions on a child. Therefore, it can only be suggested what was causing the mortality and morbidity in prehistoric Thailand. By comparing the prevalence rates in different regions within modern Thailand, pathogens specific to each area may account for some variation in child mortality and morbidity in prehistory.

The main causes of death during natal and neonatal periods in tropical regions are obstetrical trauma, neonatal infection from unclean obstetrical practices, and premature birth (Jelliffe 1970). Many of the causes of infant mortality can be traced back to maternal ill-health during pregnancy. For example, foetal malnutrition can be caused by food taboos for pregnant women. This has been observed in Burmese women whose diets are restricted in the hope of producing a smaller baby and therefore an easier birth (Jelliffe and Bennet 1962 cited in Jelliffe 1970).

### Infectious diarrhoea

Diarrhoeal disease is probably the leading cause of mortality during the first year of life in tropical countries (Jelliffe 1970). Acute diarrhoea is one of the major public health problems in Thailand where it ranks first for morbidity in children under five years and second for mortality (Suwatano 1997) (the first being acute respiratory tract infections (Hart 1996)). The mortality rate for diarrhoeal disease ranges from 0.05-0.07% in Thailand (Suwatano 1997). The annual incidence of diarrhoea in developing countries varies from 2 episodes to 7.9 episodes per year (Chandrapanond *et al.* 1973; Hart 1996). Nutritional sequelae and dehydration are common problems with diarrhoea in tropical countries (Hart 1996).

In general, paediatric diarrhoea is most often due to viral enteropathogens such as rotavirus (Hart 1996). Other common causative agents of gastroenteritis are bacteria and parasites, but a psychogenic origin is also possible (Ortiz 1970). Bacteria of the *Shigella* and *Salmonella* groups and enteropathogenic *Escherichia coli* are common agents, as well as parasites such as *Giardia lamblia* (Ortiz 1970; Suwatano 1997). Falciparum malaria may also cause acute diarrhoea in young children (Ortiz 1970). Clinical manifestations depend on the organism, host, and the environment as discussed earlier. For example, an immune response compromised due to malnutrition affects the frequency and severity of the disease (Hart 1996).

The causative agents of acute diarrhoea vary according to geographic region (Suwatano 1997) given their different requirements for survival. A 1980 epidemiological report by the Ministry of Public Health in Thailand presented the number of cases reported for common public health problems in different regions within Thailand. Data only was provided and no interpretations for the differences and similarities in the incidence of diseases among regions were offered. A selection of health problems from this report that are relevant to prehistory and to the areas studied in this thesis are presented in Table 8.2. The statistics for the occurrence of acute diarrhoea are 0.52% for the southeastern area population and 0.44% for the northeastern population. It is apparent the rate of acute diarrhoea is only slightly less for the northeast than the southeast. However, when the regions are divided further into the areas closest to the prehistoric sites, the difference between the two northeastern sites is significant. Twice as many cases (0.61% of the population) of acute diarrhoea were reported from the area surrounding Ban Lum Khao compared with Ban Na Di (0.32% of the population). Extrapolating this back to prehistory is not directly possible as other factors such as population size and environmental variations need to be considered. However, it does provide a potential variation between the two samples.

In a study in the Phetchaburi province of southern Thailand, a survey of children less than five years old with acute diarrhoea and admitted to hospital, was undertaken. It did not include children with other chronic or systemic underlying disease or other associated infections. Rotavirus was the single highest cause of acute diarrhoea, with 17.2% patients affected, but *mixed* infections were the most common cause, with 31.3% of patients affected by more than one pathogen. The most common associated problems were fever (92.2% affected) and dehydration (84.4%)

(Suwatano 1997). These types of secondary effects indicate how cumulative problems can have significant health consequences.

'Weanling diarrhoea' is a syndrome characterised by alimentary infection, poor nutrition, and an indigestible, abrasive diet during late infancy and early childhood (Jelliffe 1970). This correlates to the period when the child is gradually weaned from a complete diet of breast milk and begins eating solid food. The alimentary infection is often caused by micro-organisms that are actually non-pathogenic in later life (Jelliffe 1970), so adult carriers and food can pass on such pathogens to young children who are not yet capable of an adequate immune response (Ortiz 1970).

Total or even partial undernutrition can provoke and sustain diarrhoea. An infant adjusting to a different diet is commonly affected with some undernourishment. In skeletal samples a period of reduced growth around the age of two to three years may indicate weanling problems such as those mentioned above. The growth rate data available from the prehistoric Thai samples did not identify a 'weaning age' in any of the samples, however, the sample sizes were quite small. It is likely there were problems associated with weaning in prehistory and this is further discussed with reference to the weanling diet later.

### Respiratory infections

Respiratory infections are associated with high infant mortality. Particular risk factors are certain living conditions, where ill-ventilated, smoky housing and cross infection from overcrowding are the main problems (Jelliffe 1970). Overcrowding was probably not a significant problem among the autonomous communities of prehistoric Thailand. From evidence for housing based on archaeological excavation at Ban Kao, a sedentary 'Neolithic' site in central Thailand, Sørenson and Hatting (1967) concluded that "...the people probably lived in small groups, possibly in houses built on piles" (cited in Shoocongdej 1996: 209). This is the same as modern rural housing where many houses are on piles and the cooking is carried out below the house, which may be walled off or open air. Further archaeological evidence of the association of hearths with postholes, indicating structures, may provide further information regarding housing in prehistory in Thailand.

Pneumonia is an acute infection of the lower respiratory tract and the very young and old are particularly vulnerable. However, the newborn infant is partially protected for the first few months by maternal antibodies (Aufderheide and Rodriquez-Martin 1998). Many infectious agents can cause pneumonia, including bacteria, such as *Streptococcus pneumoniae*, and viruses. However, the latter require reasonably large populations to survive (5000-45000 people) (Aufderheide and Rodriquez-Martin 1998) that may not have been present in prehistoric Thailand. Before treatment was available for bacterial pneumonia the death rate was between 23.9-32.8%. Crowding, malnutrition, physically fatiguing occupations, trauma, age extremes, and dust inhalation are all further risk factors of pneumonia infection (Aufderheide and Rodriquez-Martin 1998).

In tropical areas, respiratory diseases are associated with the dry, hot season. Respiratory infections can be transmitted from person to person and the dryness of the atmosphere allows the desiccation of droplets of infection. The lack of humidity allows the pathogens to be transmitted in dust form. "Because of this factor of absolute humidity in these types of infection there is a very marked geographical and seasonal distribution of the diseases" (Barton 1977: 91). The prevalence of such directly transmitted diseases, including respiratory infections, along with intestinal helminthics, and dysentery, are affected by the general sanitary state of the environment, although low economic status and public ignorance are of greater significance in modern tropical countries (Barton 1977).

### Parasites

There are a number of parasitic infections that are particularly virulent in tropical countries, including protozoa associated with malaria, helminth infections such as hookworm, and also the trematodes (flukes). Each are discussed in turn.

### Parasitic protozoa and malaria

The protozoa *Plasmodium sp.,* carried by mosquitoes of the *Anopheles* genus, cause malaria. Malaria is restricted to subtropical and tropical areas and is characteristically seasonal. It still results in around 3000 people dying per day (mostly African children) today (Aufderheide and Rodriquez-Martin 1998). Malaria is dependent on many factors, including the existence of the right environment for the vector, *Anopheles sp.* mosquitoes. Each species has a preference for certain water conditions for breeding, activity temperature, rainfall, host blood type, and forest level (canopy/ground). Host factors also influence malaria. These include genetic factors such as sickle cell anaemia, behavioural factors such as occupational exposure, site of residence, and also acquired immune resistance (Aufderheide and Rodriquez-Martin 1998).

There are four main species of Plasmodium associated with malaria. *P. vivax*, *P. falciparum*, *P. malaria*, and *P. ovale.* The parasite can consume 25-75% of the red blood cells haemoglobin content and this level is directly related to the severity of the effects the parasites have on the host. Fever, chills, headache, and muscular pain are some common symptoms. Falciparum malaria is commonly fatal. Before the age of about five years, nonimmune children are most often affected acutely. After this age, if the child survives, they are more likely to have developed sufficient immunity not to suffer acutely if reinfection occurs. However, transient insults from other pathogens or episodes of malnutrition, can produce transient partial immuno-suppression and the disease becomes clinical again (Aufderheide and Rodriquez-Martin 1998).

*P. falciparum* is by far the most common source of malarial infection in Thailand, with nearly three quarters of cases attributed to this parasite (Harinasuta 1987). Falciparum malaria generally affects those aged between six months and three years in hyperendemic regions and is commonly associated with severe anaemia (Jelliffe 1970). Diarrhoea can also be a problem, as can the development of septicaemia of skin infections from insect bites (Jelliffe 1970). In epidemic regions, where malaria is seasonal, infants are more likely to die during the mosquito season (Jelliffe 1970).

An adult with chronic malaria is not a healthy individual. They have constant parsitaemia and iron deficiency anaemia, leading to acute recurrences of the infection, and reduced energy, productivity, and fertility. These consequences will have an effect on the whole population when people are able to contribute equally to daily workloads.

The occurrence of malaria shows regional differences within Thailand as was apparent in an analysis of the number of cases occurring in 1980 (Table 8.2). There was a considerably higher prevalence in the southeastern region compared to the northeast. The coastal areas within southeastern Thailand and near the sites of Khok Phanom Di and Nong Nor were higher still with 2.66% of the population affected in 1980 (Ministry of Public Health, Thailand). Both regions near the two northeastern sites have the same, very low, occurrence of malaria, at 0.03% of the population. This epidemiological data only presents the number of cases that were reported in 1980. Therefore, this does not include the number of individuals already suffering from chronic or recurrent malaria. These prevalences cannot be related directly to prehistory as they represent post-spraying programmes in attempts to remove the vector from the environment. The northeastern area has also be extensively deforested since prehistory. However, this data does portray potentially important regional differences relevant to this study.

There are no diagnostic skeletal changes with malaria. Instead, pathological consequences associated with malarial infection combined with environmental evidence, may lead to a possible diagnosis of the existence of malaria in a skeletal sample. For example, anaemia is one such consequence that can be identified skeletally. However, skeletal markers of anaemia, such as cribra orbitalia, are of a general nature and may be present with any type of anaemia, including iron deficiency anaemia, the haemolytic anaemias, or genetic anaemia.

There are other protozoan infections commonly seen in tropical countries today. This includes amoebiasis, which is a further contributor to acute or chronic diarrhoea or dysentery.

### Parasitic helminth infections

The helminth parasites include hookworm, a soil-transmitted parasite. A helminth infection can have substantial affects on the health of the host, including iron deficiency anaemia, as the adult worm derives its nutrition by sucking blood from a mucosal capillary in the intestine (Aufderheide and Rodriquez-Martin 1998). This infection is particularly bad in children where the normal haematopoietic marrow is barely able to meet the rapidly increasing oxygen demands of growth, let alone a pathological condition in addition (Aufderheide and Rodriquez-Martin 1998).

Children with heavy parasite loads of hookworm are more likely to develop not only iron deficiency anaemia, but also malnutrition. This may induce a period of reduced skeletal growth (Stephenson 1985 cited in Nontasut *et al.* 1996). Vitamin B12 and folic acid deficiencies have also been associated with hookworm infection (Nontasut *et al.* 1996). There appears to be no preponderance of sex or age in this infection. As hookworm lives in the soil, individuals who wear shoes have lower infection rates. Contamination of the soil is particularly high in villages where latrine use is low (Chongsuvivatwong *et al.* 1996) and would also be influenced by population density. In a northeastern village in the 1970s, 36.8% of preschool children and 61% of school children were infected with parasites. Hookworm infection ranked highest. Both growth and the degree of malnutrition were affected by parasitic infestation in these children (Chandrapanond *et al.* 1973).

Another soil-transmitted helminth is *Ascaris lumbricoides*, commonly known as roundworm. Infection with this helminth causes ascariasis, which occurs all over the world but is highly prevalent in China and Southeast Asia (Gilles 1996). Children over two years of age are more commonly affected by this infection. Some immunity to heavy infections can develop, therefore, adult reinfections will generally be lighter (Gilles 1996). Larvae migrate via the lungs to the small intestine and can cause ascaris pneumonia. The adult worm in the small intestine must be present in high numbers to cause symptoms. Such symptoms include intestinal obstruction which can be fatal. The most common complaint in mild cases is intestinal pain (Gilles 1996). Ascariasis can also contribute to protein-energy malnutrition and vitamin A deficiency (Gilles 1996).

The incidence of hookworm and other soil-transmitted helminths, such as ascaris, has a varied distribution among the regions of Thailand. The incidence of hookworm in the northeast (19.5%) is almost twice that of the central (which includes the southeast) region (11.2%) (Harinasuta and Vajrasthira 1964). For the larvae to survive in the soil, the soil must be damp. Although the people of the northeast lived in a drier region they lived adjacent to rivers prone to flooding in the wet season, rather than the salt water in coastal Khok Phanom Di which may have inhibited the survival of hookworm.

The incidence of ascaris is completely opposite to that of hookworm. It has an incidence three and a half times higher in the central region (26.4%) compared to the northeast (7.5%) (Harinasuta and Vajrasthira 1964). Ascaris favours regions of moderate temperature and high humidity. Its incidence is low in the northeast because the ova are dispersed during the rainy season when flooding occurs and the summer is too hot and dry and the winter too cold for the development of the ova (Harinasuta and Vajrasthira 1964).

### Parasitic trematode infections

Liver and intestinal flukes (trematodes) are also present in tropical countries, with the former particularly prevalent in oriental countries (Aufderheide and Rodriquez-Martin 1998). Opisthorchiasis is caused by infection with human liver flukes, *Opisthorchis felineus*, *O. viverrini* and *Clonorchis sinesis* (Haswell-Elkins and Elkins 1996). The latter two species are endemic in Thailand and are associated with the leading cause of death in modern northeast Thailand, bile duct cancer (Haswell-Elkins and Elkins 1996). Opisthorchiasis infection arises from eating raw fish which is a popular dish in northeast Thailand (Sornmani *et al.* 1984). *O. viverrini* is estimated to infect 35% of people in northeast Thailand (Haswell-Elkins and Elkins 1996). Chronically infected people have few specific signs or symptoms, but generally pain in the region of the liver, diarrhoea, loss of appetite, and indigestion are common (Sornmani *et al.* 1984). Severe cases can lead to weight loss and weakness and are associated with a 15-fold risk of bile duct cancer (Haswell-Elkins and Elkins 1996). Whether the latter was a problem in prehistory is another unanswered question. Echinostomiasis, caused by an intestinal flukes, is often associated with opisthorchiasis as they are both associated with the consumption of raw fish and molluscs (Haswell-Elkins and Elkins 1996).

Fasciolopsiasis is caused by the giant intestinal fluke, *Fasciolopsis buski* (Haswell-Elkins and Elkins 1996). Infection is largely confined to Asian countries, and within Thailand infection is largely confined to low lying areas in the central regions with heavy rainfall and extensive flooding, which leads to faecal contamination of water (Haswell-Elkins and Elkins 1996). Prevalences are also high in areas with cultivation or year-round availability of aquatic vegetation (Haswell-Elkins and Elkins 1996). Community based prevalences are around 20%, with a typical peak occurring in children over five years (Haswell-Elkins and Elkins 1996). The parasites attach to the jujunal or duodenal wall and can result in ulcers. Vitamin B12 serum levels are lowered and large infections are associated with general malabsorption that can lead to malnutrition. Clinical signs include diarrhoea, hunger pains, poor appetite, fever, and abdominal pain. Late, severe cases present with such symptoms as anaemia, anorexia, and general weakness (Haswell-Elkins and Elkins 1996).

Schistosomiasis is a parasitic infection caused by blood flukes, *Schistosoma sp.* However, only two species have been identified in the vicinity of Thailand. *S. mekongi*, is found only in the valley of the Mekong River (Davis 1996) and a *S. japonicum*-like parasite is restricted to the south of Thailand (Harinasuta 1983). Therefore these parasites were probably not a problem to the prehistoric Thai people studied in this thesis.

Filariasis is the result of infection with one of the lymphatic filariases, a parasitic nematode. Symptoms include fever, groin pain, and oedematous swelling of legs (elephantiasis) (Harinasuta 1986 cited in Tayles 1999). However, it appears this nematode group are restricted to the south and west of Thailand (Harinasuta 1986 cited in Tayles 1999).

The mortality of older children can be the result of cumulative disease episodes during infancy. They then succumb later to malnutrition during the prolonged weaning period and to infection, especially diarrhoeal diseases, intestinal parasites, malignant malaria (Jelliffe 1970).

"In the majority of cases, it is incorrect and impossible to assign a single cause to a child death in a tropical country" (Jelliffe 1970: 19). This statement, although from a modern clinical point of view, is even more relevant in prehistory. This emphasises the frequent multiple causes, and the synergistic interaction of nutrition and infection, in the outcome of an individual's and community's portrayal of health.

## Nutrition

> The profound importance of good nutrition to normal growth, development, and resistance to disease cannot be overemphasized in discussions of critical determinants of human health. (Powell 1988: 58)

The previous discussion has outlined the potential pathogens that may have affected a prehistoric population in Thailand. The effects of these various pathogens often leave no diagnostic trace on the skeleton. Non-specific skeletal changes, such as growth disruptions, may be apparent, but these can also be linked with inadequate nutrition. As the discussion turns to the role of nutrition on health and disease in prehistory, it is again apparent that little direct or specific evidence is available. However, there are some avenues that can be pursued. The following discussion aims to outline the prehistoric diet and its possible differential effects on the health of these four prehistoric Thai communities. It begins by describing the direct evidence from archaeological faunal and botanical analyses. However, not all foods eaten by these people will have been preserved. In fact it was only through meticulous screening and flotation during the excavation and subsequent publication of the research, that the palaeobotanical information is available for Khok Phanom Di. Therefore, the discussion considers the evidence for diet and its nutritional quality through observations of the modern Thai diet, particularly that in the rural, more isolated communities.

### Prehistoric diet and nutrition

The main points concerning the comparative analysis of the dental health of the four prehistoric Thai samples have already been summarised earlier in this chapter and discussed in detail in Chapter 7. This has suggested differences and similarities in the diet existed among the samples. The archaeological evidence for the diets from each of the prehistoric samples was described in Chapter 2 and is briefly revisited here.

The people of Khok Phanom Di represent an early agricultural group. From the botanical remains it is known that rice was cultivated by these people. However rice agriculture was probably not the predominant subsistence activity and possibly did not occur throughout the occupation of the site. It would appear greater reliance was placed on the exploitation of coastal resources, gathering, for example, fish, shellfish, from the surrounding marine, estuarine, and mangrove environments. The majority of biological remains were those favouring the aquatic environment (Higham and Bannanurag 1991b). It was concluded, from archaeobotanical and ethnographic evidence, that the mangrove environment in which the Khok Phanom Di site was sited did not provide abundant plant remains (Thompson 1996). However, the adjacent freshwater ecosystems may have provided many wild vegetables that grow along streams or in freshwater swamps (Thompson 1996). Terrestrial animal meat was not commonly eaten, only a few bones of pig, chicken, deer, cattle, water buffalo, macaque, and domesticated dog were found (Higham and Bannanurag 1991b).

The floodplain environment surrounding the site of Nong Nor would have been very favourable for the cultivation of rice. The position of the site relative to the coast would also have enabled the collection of associated marine resources. However, the large river and floodplain were closer and would have provided most of the aquatic foods such as fish and shellfish. Faunal remains identified were pig, cattle, and chicken, as well as domestic dog (Higham 1998).

Wet rice agriculture would have been possible for the prehistoric people of the Mun River Valley (Welch 1985). In addition, there was indirect evidence, in the form of rice temper used in pottery, that suggests it was a subsistence activity undertaken by the people of Ban Lum Khao. Mammalian species present were water buffalo, deer, domesticated dog, cattle, and pigs, and also freshwater fish and shellfish. It has been suggested, based on the large size of the bones and shells, that the area had previously been unexploited (O'Reilly 1999) and therefore it is likely such resources were in abundance.

Numerous biological remains were identified from the excavation of Ban Na Di, particularly the domesticated pig and dog, cattle and deer. There were also many fish species, some shellfish, and smaller animals such as hares and rats. Rice agriculture would have been the predominant subsistence activity, more so than at any of the other sites (Higham and Kijngam 1984c). The surrounding environment, like that of Ban Lum Khao, appeared to have provided a good, wide ranging diet. However, seasonal effects may have been felt more harshly in the northeast compared to the southeast.

## Modern diet and nutrition

While the archaeological evidence concerning the prehistoric diets of the four samples studied have provided an idea regarding the basic components of their diets, they cannot establish the actual proportions of each food consumed. They also fail to identify those remains with no organic component that have not been preserved. Therefore, ethnographic evidence may provide further information regarding not only potential sources of other foods consumed by a rice agricultural community, but also the reliance placed on certain foods, and the overall nutritional value of a typical rice diet.

From nutritional surveys in rural and isolated villages in Thailand, information regarding the dietary habits of Thai people reliant on rice agriculture has been obtained. One large study of a village northeast of Bangkok, Bang Chan has been particularly fruitful (Hauck *et al.* 1959a, 1959b; Hauck and Sudsaneh 1959; Hanks 1972) but also some in the northeast (Chandrapanond *et al.* 1972, 1973) and northwest (Kunstadter 1985). The main components of the modern Thai diet will be presented with a discussion of their nutritional value and importance in the diet as has been determined from such studies of rural agricultural Thai villages.

### *Rice*

"In Thai, the general term meaning "to eat" is translated literally "to eat rice," and all food items are divided into two classes: rice and "with rice" " (Hauck *et al.* 1959a: 1145). Traditionally, rice is considered the only requirement that the human body needs and 'with rice' dishes are only added to make rice more palatable so more is eaten (Suvarnakich 1950). However, rice is not a complete food and the rice diet is generally considered to be not varied enough to adequately provide all essential nutrients.

Rice provides over 80% of the caloric intake in a rice diet (Hauck *et al.* 1959a), therefore rice provides the majority of the energy requirements. More than half of the protein in a typical 'rice diet' is also derived from rice (Hauck *et al.* 1959a). While the actual protein content in rice is low compared with other cereals, such as wheat and whole maize, it is of a higher biological value (F.A.O. 1954). This has been correlated with the relative infrequency of protein malnutrition in rice eating countries (F.A.O. 1954). However, rice does not provide lysine, one of the eight essential amino acids that cannot be synthesised by the body (Wilson 1985).

Rice also does not provide much vitamin A, C, D, and riboflavin, it is also low in fat and poor in iron and calcium (F.A.O. 1954). The washing of rice before cooking also results in the loss of other nutrients, particularly thiamine (F.A.O. 1954). The processing of rice between harvesting and consumption can also result in nutrient losses. Milled and polished rice results in significant losses of thiamine. However, the more traditional, home-pounded rice, involving a foot or hand operated mortar and pestle, is associated with the least nutrient losses. Rice is also coarser in its unpolished form and this was the type most likely eaten in prehistoric times.

There are two main varieties of rice eaten in Thailand, glutinous or sticky rice and non-glutinous, ordinary rice. Glutinous rice is particularly favoured in the northeast and northern regions (Suvarnakich 1950). Whether this preference existed in prehistoric times is unknown.

### *Fish*

Fish provides the most animal protein in a typical Thai diet. This is because of its abundance and availability almost anywhere (Suvarnakich 1950). In total, fish and other foods of animal origin provide the remaining 20% of the total protein of a rice diet (Hauck *et al.* 1959a). Nutritionally, fish is adequate as the principal source of dietary protein (Geiger and Borgstrom 1962). Fish is a particularly good source of lysine (Geiger and Borgstrom 1962) and when rice and fish are eaten together, as they usually are, all amino acids are made available at the same time. This enables the synthesis of protein to occur within the body (Wilson 1985).

Fish are prepared in many different ways for consumption, including roasted, boiled, dried, salted, and fermented (Hauck *et al.* 1959a). Small fish are commonly eaten whole, and therefore, with consumption of the bones, are a good source of calcium (Suvarnakich 1950).

However, those communities that rely on the sea for their supply of fish may be vulnerable to periods of deficiency during the monsoon when seas are rough. Therefore, in better times, excess fish are commonly dried and stored for times of shortages (Wilson 1985). In a study of a coastal fishing village in Malaysia (Wilson 1985), children were found to have a significantly greater muscle mass than a comparative inland group where fish was eaten infrequently. Blood tests suggested this was a direct reflection of the coastal village's better protein intake (Wilson 1985).

*Other animal products*

Apart from fish, animal meat is generally consumed in very small quantities in traditional rural Thailand. Animal products such as eggs, and also pork and birds are the most frequently eaten. Chickens live underneath the house and provide eggs, however chicken meat was not commonly eaten in the Bang Chan village (Hauck *et al.* 1959a) although it was more commonly eaten in the northwest (Kunstadter 1985). Also field rats, water buffalo, eels, frogs, cobra and other snakes were eaten. However, little beef was eaten as the cattle are preserved for farm work (Suvarnakich 1950). Meat stews were generally reserved for ceremonial occasions or on days of heavy work in a rural village in northwestern Thailand (Kunstadter 1985).

*Vegetables and Fruits*

Hundreds of varieties of fruits and vegetables are in abundance all year round throughout Thailand and these are consumed in a quantity second only to rice (Suvarnakich 1950). Fruits and vegetables such as bananas, coconuts and coconut milk, palm sugar, swamp cabbage, pineapple, mango, tamarind papaya, and lotus seeds were all commonly consumed in central Bang Chan (Hauck *et al.* 1959a). These components of the diet provided the main source of calcium, and other minerals, as well as vitamins A, C and some of the B-complex vitamins (Suvarnakich 1950). Coconut also provides a small amount of fat (Suvarnakich 1950).

Vegetables such as swamp cabbage and *phak kached*, a green water plant, were cultivated or gathered wild from creek-side winter gardens or along the edge of the canals (Hauck *et al.* 1959a; Kunstadter 1985). These were used to produce a vegetable stew (maybe with a small quantity of dried or fermented fish added for protein) and provided the most common dish to be eaten with rice in northwestern rural Thailand (Kunstadter 1985). Uncooked leafy vegetables or herbs, such as onions, mint, coriander, raw bamboo shoots, were also part of the diet. Maize, taro, yams, and manioc, were also seasonal supplements (Kunstadter 1985).

*Infant and weanling diet*

Breast milk was the predominant milk for infant feeding in an isolated rural Thai village (Suttapreyasri *et al.* 1975). However, while this is usually considered to provide adequate nutrition for an infant, there were some foods considered taboo for the pregnant and lactating mother. "Nutritious foods, i.e. egg and meat, vegetables and fruits, rice and flour are rather the major food taboos of the rural women during pregnant, lactating and puerperal periods..." (Suttapreyasri *et al.* 1975: 358). However, no more specific information, outlining exactly what particularly types of meat or vegetables were prohibited, was provided. Supplementary foods were given to the infant from birth and consisted of rice and banana (Suttapreyasri *et al.* 1975). This is a potential source of pathogens. Milk never enters the adult diet as this is regarded as food for nursing babies only (Suvarnakich 1950).

Calorie deficiency in the weanling diet was suggested to be causing growth retardation between 12-18 month old children in rural villages in northern Thailand in the 1970s (Gershoff *et al.* 1977 cited in Bailey *et al.* 1984). "Paradoxically, such calorie insufficiency occurs in an environment with abundant supplies of rice" (Chandrapanond *et al.* 1972 cited in Bailey *et al.* 1984: 555). It is possible that the childhood diet was not adequate in the right proportions. Children need a high energy, protein, and lipid diet but their small stomachs cannot handle a large volume (Bogin 1999). Rice is a bulky food and probably needs to be consumed in reasonable quantities to obtain the required energy.

Nutritional problems with the modern Thai diet

In spite of the apparent abundance of food, malnutrition does exist in modern Thailand and is associated with high morbidity and mortality in infants and young children (Chandrapanond *et al.* 1973). Suvarnakich (1950: 289), as mentioned above, believes "this is due to the general belief handed down from ancient days...that rice is the only essential food that the human body needs".

Nutritional problems evident in modern Southeast Asia that are of particular importance for infant health, although some also affect adult health, include protein-energy malnutrition, and iodine, vitamin A, and iron deficiency diseases (Heywood and Marks 1993). Protein-energy malnutrition in children 0-4 years of age was assessed by UNICEF to exist in 26% of children in Thailand (Heywood and Marks 1993). These nutritional deficiencies are not normally a significant problem within the first six months of life as breastfeeding tends to provide nutritional requirements and breastfed tropical babies usually thrive during this time (Jelliffe 1970).

Ariboflavinosis (riboflavin deficiency), iron deficiency anaemia, and moderate degrees of protein-calorie malnutrition, were observed in a northeastern Thai village (Chandrapanond *et al.* 1973). Using nutritional anthropometric measurements, Chandrapanond *et al.* (1973) observed 76% of 215 children (aged 1-6 years) were malnourished based on general standards. Heywood and Marks (1993: 133) state that "there is now a strong consensus that the growth potential of children throughout the world is similar...". Loka *et al.* (1994) attempted to investigate whether Southeast Asian children were smaller in stature because of their genetic potential or poor nutrition. They compared the growth of Southeast Asia children (Vietnam, Cambodia, and Laos) born in Australia with those born overseas. Those born overseas were 1% shorter and 10% lighter in weight. They concluded that environmental factors accounted for the smallness of Southeast Asian children, preventing them from reaching their full genetic potential that they had been able to be reached in those born in Australia (Loka *et al.* 1994).

Hauck *et al.* (1959b) observed height and weight values for age and sex for children of Bang Chan below published standards for both urban Thai children and American children. Possible signs of nutritional deficiency were also observed "...although little or no malnutrition of the sort that causes marked apathy, and no clearly defined deficiency disease was found among them" (Hauck *et al.* 1959b: 14).

In a 1934 health survey (cited in Suvarnakich 1950), no clear cut malnutrition was observed in rural areas in Thailand. Underweight and anaemic individuals were observed, but their condition could almost always be attributed to parasites and tropical diseases, such as malaria and dysentery (Suvarnakich 1950).

> Approximately 20-35% of 100 school children were found to be deficient in vitamin B2 and B6 in the northeast of Thailand: major factors involved were suggested to be the insufficient intake of nutrients and a high prevalence of parasitic infections. (Schreurs 1976 cited in Nontasut *et al.* 1996: 47)

Anaemia, as previously discussed, can be caused by malnutrition and infections such as hookworm and malaria. But severe anaemia can result when these conditions create cumulative stress on an individual. Having one problem often makes a person more likely to develop another problem that can exacerbate the condition (Jelliffe 1970): the nutrition-infection synergism. Iron deficiency anaemia has an effect on maternal mortality and morbidity through low haemoglobin levels and increase the risk of haemorrhage at the time of delivery (Heywood and Marks 1993).

Iodine deficiency is related to a wide spectrum of disorders which include goitre at all ages, endemic cretinism, increased rates of stillbirths, and neonatal and infant mortality. In Southeast Asia it is estimated that 280 million people live in iodine deficient areas; 100 million people are estimated to have goitre. (Heywood and Marks 1993). In areas of mild endemicity it manifests selectively in groups exposed to physiological stresses, such as school-going children around the age of puberty, and pregnant and lactating women; female-to-male ratios of prevalence are high (Ramalingaswami 1973). The 'goitre belt' of endemicity includes the northern and northeastern provinces of Thailand, and, here, several surveys carried out since 1955 showed prevalences of 15% to 72% (Ramalingaswami 1973). It is estimated that nearly 3.5 million people are at risk in the goitre regions of Thailand (Ramalingaswami 1973). When the goitre rate reaches above 50% in a population, developmental disturbances such as cretinism and deaf mutism begin to appear and this has been seen in the endemic regions in Thailand (Ramalingaswami 1973). Environmental iodine deficiency is the primary cause in the north and northeast (Ramalingaswami 1973).

Vitamin A deficiency is a significant public health problem in all countries of Southeast Asia. With respect to morbidity, there is clear evidence that vitamin A affects the course and outcomes of infectious disease in children (Heywood and Marks 1993). There are a number of other specific nutritional deficiencies present in Thailand, however not all of them would have been a problem in prehistoric Thailand. For example, beriberi, caused by low thiamine in the diet, only became a significant problem when the modern processes of milling and polishing of rice began. These processes resulted in the significant loss of this nutrient (Suvarnakich 1950).

Overall it is apparent that the evidence for malnutrition in Thailand is not consistent. This is possibly related to regional variations and research methodologies. There is also the confounding and synergistic factor of the difference in the prevalence of infectious diseases affecting nutritional status. Additionally, different infections affect nutritional status in distinct ways:

> The relationship between the nutritional status of children and morbidity due to infectious disease varies according to which of the most important infectious diseases of young children - acute lower respiratory tract infections, diarrhoea and malaria - are being considered. Low nutritional status is associated with increased risk and severity of acute respiratory infections and with increased severity of diarrhoea. However, there is no evidence that malnutrition increases the risk of malaria. (Heywood and Marks 1993: 134-135)

### Hypotheses: comparisons based on the natural environment

Southeast Thailand health: 'Khok Phanom Di = Nong Nor'

It was hypothesised that there would be some similarities in the health of the two southeastern samples, Khok Phanom Di and Nong Nor, on the basis of some similarities in their natural environment.

Both samples had evidence of childhood growth disturbances but it appears these problems had different outcomes in each sample. The Khok Phanom Di sample appeared to be comprised of more frail children than Nong Nor. This has resulted in an increased susceptibility to childhood illness or malnutrition and an increased likelihood of suffering acutely and fatally at Khok Phanom Di. This is probably related to either heterogeneity in frailty within and between samples, which may have been the result of nutritional inadequacies, or disease, or genetic heterogeneity, all reducing the effectiveness of the immune response.

Modern evidence, previously discussed, indicates malaria is particularly prevalent in this region of Thailand. Prehistoric skeletal evidence indicates the possibility of thalassaemia in the people of Khok Phanom Di. The presence of thalassaemia indicates the population has had a long exposure to malaria, and the genetic disorder has developed in order to reduce the effects of malarial infection. However, those with thalassaemia or those with malaria had anaemia to cope with. This has had a detrimental effect on a number of children, while the thalassaemia protects some from severe malarial infection, the anaemia has made them more susceptible to other infections and malnutrition. This is nutrition-infection synergism at work.

If the Nong Nor collection had been better preserved, a more accurate indication of how they coped with the malarial parasite may have been obtained. The samples could then have been compared with respect to their exposure to the malarial parasite. While there is very little direct evidence for anaemia in the people of Nong Nor (three individuals had cribra orbitalia), the evidence for childhood growth suggests that it was somewhat better than for the people of Khok Phanom Di. For example, children who survived childhood and its associated risks at Nong Nor, appeared to be able to compensate for childhood growth disruptions as the people in this sample attained a taller mean adult stature than those from Khok Phanom Di. It is possible this was based on the reliability of food resources, perhaps associated with the better conditions for rice agriculture at Nong Nor. The site of Nong Nor was situated slightly further inland compared to Khok Phanom Di and the freshwater floodplains were perhaps more conducive to sustainable rice agriculture. The possibility that these two samples were not genetically similar can also provide an explanation for these differences.

A higher proportion of older adults were present in the Nong Nor sample suggesting a longer adult longevity compared with the Khok Phanom Di people. This correlates with the evidence for better health in the Nong Nor subadults.

Dental health was poorer in the Khok Phanom Di people. They had more caries, periapical cavities, and antemortem tooth loss than the Nong Nor people. However, the Nong Nor people had a higher prevalence of advanced attrition. These results could indicate some dietary differences existed between the two sites, with the people of Khok Phanom Di consuming a more cariogenic and softer diet, while those at Nong Nor had a coarse diet with less cariogenic components. As was discussed in Chapter 7, this is likely to be partly related to the quantity of rice consumed. Bananas, palm sugar, yams, and taro are all possible candidates for additional possibly cariogenic carbohydrates in the Khok Phanom Di diet (Tayles *et al.* n.d.). These foods may have been less relied on at Nong Nor, as they had a staple food resource in their rice agriculture. Unmilled rice, being coarse in texture and of low cariogenicity, could explain the dental pathology profile of the Nong Nor sample.

The skeletal samples from Khok Phanom Di and Nong Nor are separated by nearly 800-900 years. Considering this difference it is difficult to separate the effects of cultural changes with time from the effects of the natural environment on the health of these two communities. A further confounding problem is the presence of sampling bias within the Nong Nor sample, related to the skeletal material's poor preservation. A number of the health parameters were not able to be recorded for the Nong Nor sample, including joint disease, bone mass, and Harris lines. This has made it very difficult to compare these two samples with any degree of confidence. However, from the available evidence, it appears this hypothesis has been disproved.

### Northeast Thailand health: 'Ban Lum Khao = Ban Na Di'

The skeletal samples of Ban Lum Khao and Ban Na Di are separated by only a few hundred years, and in fact they may have overlapped in time. On the basis of their natural environments, similarities are evident and, as hypothesised, this may be related to their generally consistent similarities in health. Both samples appeared to have similarly abundant food resources. Both samples undertook rice agriculture, but perhaps more predominantly at Ban Na Di. This would have provided both samples with a stable staple carbohydrate.

The sample from Ban Lum Khao had a higher subadult mortality level than Ban Na Di. This, and the prevalence of enamel hypoplasia and stature estimates, suggests the Ban Lum Khao sample was comprised of more frail and medium frail children, while the Ban Na Di sample had more medium frail and strong children. Both samples had evidence of health problems throughout childhood, but they reacted differently to them, either because of nutritional or genetic differences.

The Ban Lum Khao people attained a shorter mean adult stature than those from Ban Na Di. This may be related to better nutrition at the right time to compensate for reduced growth during periods of illness during childhood at Ban Na Di. Given the archaeological evidence for faunal and botanical remains, it would appear unlikely that malnutrition was a significant problem at either site. According to archaeological evidence, Ban Lum Khao was newly settled in a previously unexploited area and Ban Na Di was very familiar with rice agriculture and the care of domesticated stock.

Nutritional inadequacies may have arisen through infection. Children, and adults, in the northeast in modern times have high prevalences of hookworm, opisthorchiasis, respiratory infections, and acute diarrhoea. Nutritional sequelae can arise from most of these conditions, which in turn can have detrimental affects on the effectiveness of an individual's immune response; the nutrition-infection synergistic relationship establishes itself.

Twice as many cases of acute diarrhoea were reported from the area surrounding Ban Lum Khao compared with Ban Na Di in a 1980 survey by the Public Health Ministry in Thailand (Table 8.2). No suggestions for the differences among areas were provided. Given the greater subadult mortality and shorter statures in the Ban Lum Khao sample, the difference in the prevalence of acute diarrhoea, as an indicator of pathogen load, is one possible explanation for the higher morbidity and mortality in the Ban Lum Khao people. The frequency of infectious episodes in the children of Ban Lum Khao may have been greater resulting in chronic nutritional problems. These in turn would have compromised their immune response to what may otherwise have been non-threatening infections. An immunocompromised individual with an infection is more likely to develop symptoms and disease as the infection dosage does not have to be as high.

The young women from prehistoric Ban Lum Khao were also dying in higher proportions than at Ban Na Di. This too may be related to the possible higher infectious pathogen load and subsequent nutritional sequelae, increasing the likelihood of infections arising from childbirth and their inability to overcome them, for example. Poor maternal health can lead to poor infant health, which may have resulted in the presence of more frail children at Ban Lum Khao.

The northeast of Thailand is a drier area than other parts of Thailand, and there is geoarchaeological evidence to suggest climatic conditions were even more arid prior to the Iron age (Habberfield-Short 1999). These dry conditions may have exacerbated the spread of respiratory infections among the prehistoric people of Ban Lum Khao by allowing pathogens to be transmitted with dust. Subadults may have been especially affected given they had the second highest subadult mortality percentage of all samples. There is evidence that "...a wet cycle persisted from the early to late Iron Age before a return to seasonally arid conditions characterising the present" (Saturagsa 1987; Chong 1988; Nutalaya *et al.* 1989; Udomchoke 1989; all cited in

Habberfield-Short 1999: 218). This correlates with part of the time of occupation of prehistoric Ban Na Di. This may partly explain the lower subadult mortality at Ban Na Di, given the lower amount of respiratory infections occurring within this relatively wet climate.

As mentioned earlier subadult morbidity and mortality in a tropical country is more likely the result of a combination of factors having a cumulative and negative affect on a child. In the prehistoric northeast of Thailand it is possible that there was a higher pathogen load in the vicinity of Ban Lum Khao. Additionally, the people of Ban Na Di, with their intensive rice agriculture and knowledge of animal husbandry, may have had a more reliable and controllable diet, therefore providing children with the nutritional requirements for coping with such a pathogen load.

Joint disease in the northeast shows some similarities and some differences. Over all joints, the Ban Lum Khao and Ban Na Di samples had a similar proportion of individuals affected by osteoarthritis. The sex differences apparent are most likely related to the age structures of these two samples and cannot be commented on further. The pattern of joints affected by osteoarthritis showed some differences and some similarities as discussed in detail in Chapter 5. Ban Na Di people showed the higher prevalence of vertebral osteophytosis. This may indicate these people were undertaking heavier physical labour with biomechanical stress to the back.

Fracture patterns in these two northeastern samples showed no similarities. Ban Na Di people had very few, while the Ban Lum Khao people had a relatively high risk of trauma. The pattern of bones affected in the Ban Lum Khao people did not appear totally random and may, in part, have been related to interpersonal violence, a culturally related phenomenon discussed later. Alternatively, the natural environment within the vicinity of Ban Lum Khao may have had more rougher terrain.

Dental health is very similar between these two samples, further supporting the similarity in diet and therefore environment. The presence of caries, advanced attrition, periapical cavities, and antemortem tooth loss were all very similar between the two samples. There were some sex differences evident, but these are not considered further as they may have been related to the age and sex structure differences between the samples. The dental pathology profile from these two samples indicates few cariogenic foods were consumed and that the diet was moderately coarse in texture. Unpolished, unmilled rice, as was likely to be the form eaten in prehistory, can be relatively coarse. This may have worn the teeth to this moderate level and, as a consequence, removed the sites for caries to develop, thus explaining the low caries rates.

The hypothesis suggesting that the health of the northeastern samples would be similar is confirmed by some of these results and not by others. When differences have arisen, it is the Ban Lum Khao sample that showed more detrimental health problems. However, these two samples are more similar to each other than the samples from the southeast are to each other.

Regional health: 'Southeast ≠ Northeast Thailand'

This hypothesis suggested that the health of the samples from the northeast and southeast would be quite different and the differences could be related to the differential effects of the natural environment. From modern clinical studies of health in the two regions, it is apparent that the southeast and northeast have different pathogens and different pathogen loads. The effect they are able to have on a community depends largely on the community's diet and their degree of frailty.

The southeast has a higher prevalence of malaria than the northeast in modern times. This is likely to have been the case in prehistory as supported by the skeletal evidence from Khok Phanom Di. If the Nong Nor collection had been better preserved an indication of whether they also had been exposed to the malarial parasite may have been obtained.

Apart from malaria, one of the leading causes of anaemia in Thailand is hookworm infection. This is more common in the northeast than the southeast according to health reports for modern villages. The skeletal evidence from the northeast does not show high prevalences of anaemia, but childhood morbidity could be explained by the presence of such infections as hookworm and opisthorchiasis, which are also common in modern Thailand. These can both be associated with nutritional sequelae that can set up chronic malnutrition and an inefficient immune system. Childhood morbidity and mortality was also a common occurrence in the southeast and in this region it may have been related to ascaris, and possibly fasciolopsiasis, as well as malaria.

The northeast of Thailand is a much drier area than the southeast, and, as mentioned above, climatic conditions may have been even more arid prior to the Iron age (Habberfield-Short 1999). These conditions may have exacerbated the spread of respiratory infections among the prehistoric people of the northeast. The effects of the monsoonal dry season may have been felt more significantly in northeast. Fresh water and aquatic biological resources may have been depleted, unless preserved and stored during the wet season. The northeast has a good source of salt that may have been used for preservation and therefore reduced the effects of the dry season. A community with a reliable source of fish has been seen to be better off nutritionally as was evident in a Malaysian fishing village (Wilson 1985). This would have been easily met for Khok Phanom Di and possibly Nong Nor but dependent on the season and knowledge of preservation in the northeast.

A lack of environmental iodine has led to endemic goitre in the northeast. This may have been a further contributor to neonatal and infant mortality in the prehistory of this region, that was not present in the southeast.

The effects of the diet on the dentition of these people was different between the two regions. However, they were also different between the southeastern samples. In general, the Khok Phanom Di sample stood out with the poorest dental health. The other three samples were all much similar with the exception of the high advanced attrition in the Nong Nor sample. As mentioned in Chapter 7, Leatherwood *et al.* (1965) found a difference in caries rates between modern rural 'northern' Thailand (including northeastern regions) and 'southern' Thailand (including the southeastern region) with higher DMF (Decayed, Missing and Filled permanent teeth) teeth in the latter. This pattern reflects that seen in prehistoric Khok Phanom Di, the sample with poor dental health. Overall, there are number of differences in the dental health between Khok Phanom Di and the other three samples, which may be more a reflection of cultural advances, such as subsistence mode, than related to the natural environment.

Osteoarthritis appeared to have affected those in the southeast more frequently than in the northeast. However, the absence of osteoarthritis data from Nong Nor had resulted in the inability to determine if this difference was related to the natural environment and subsistence activities, or cultural differences related to subsistence activities. The pattern of joints affected is different between the two regions and there are also differences within the northeast as discussed earlier. This implies activities were different and they may be related to the activities required for a coastal settlement to acquire food, such as in boats, or it may relate to cultural differences discussed below. Genetic differences among samples is also a possibility. Vertebral osteophytosis is generally highest in the Ban Na Di and Khok Phanom Di samples, therefore showing no association with the natural environment.

Trauma patterns among the samples also show no association with the natural environment. Each sample has its own distinctive pattern that, as discussed later, may be more related to the cultural environment.

If the comparisons of these two areas are restricted to those from the same time period, the potentially confounding effects of the cultural environment may be reduced. The Nong Nor and Ban Lum Khao samples were both from the Bronze Age. Unfortunately, any comparison with the Nong Nor skeletal material is difficult given its preservation. There are some differences, however, that were identified. While the lower subadult mortality in the Nong Nor sample may be explained by a sampling bias, the difference in the prevalence of enamel hypoplasia indicates childhood health differences. Subadults appeared to have died from acute problems at Nong Nor, while the higher prevalence of enamel hypoplasia at Ban Lum Khao suggests more chronic problems. If the stature means are compared, the Ban Lum Khao people, perhaps as a result of chronic suffering, did not attain as tall statures as the Nong Nor people. In this respect, it could be stated that the Nong Nor people who survived childhood were healthier than those from Ban Lum Khao. Paradoxically, this could also be interpreted as an *adaptation* by the people of Ban Lum Khao to cope with certain environmental problems.

This 'small but healthy' concept was proposed by Seckler (1982 cited in Bogin 1999). It theorised that small size was a homeostatic adjustment without any detrimental effects by those living with malnutrition. This is based on the idea that small bodies require less food for survival. However, it fails to acknowledge that these people were malnourished and thus relatively *unhealthy* to begin with. It also implies that the causative factors for adaptation to small size, that is chronically poor diets and frequent infections, are desirable, as an 'adaptation' is usually considered a positive response (Goodman 1994). But this is not a 'cost-free' adaptation. For example, stunted girls who grow up small have an increased risk of giving birth to growth-retarded infants, who, in turn, have an increased risk of dying prematurely (Goodman 1994). Additionally, people who were undernourished as children can have an impaired work capacity throughout their life (Bogin 1999).

While small stature is probably an *adjustment* to poor nutrition it has been shown it is unlikely to be a *genetic* adaptation. Bogin's (1999) work with the Maya of Guatemala, showed that they have been suffering from malnutrition, heavy workloads, and disease for many generations and they are also relatively short statured. However, Mayan immigrants growing up in the United States have achieved much higher statures than Mayan children growing up in Guatemala. As mentioned earlier, this has also been shown in a comparison of Southeast Asian children growing up in Australia and those born in Southeast Asia (Loka *et al.* 1994). The immigrants were taller for their age indicating their genetic potential had not been reached in their own country. Therefore small size is not necessarily a genetic adaptation, but a phenotypic adaptation as a response to environmental conditions.

Comparing the health of the southeast and northeast samples in Thai prehistory has not been straightforward. This hypothesis was dependent on the previous two hypotheses being true, but, as has been discussed, there are very few similarities in the health of the southeastern samples. This makes it difficult to compare the southeast with the northeast when both of the southeastern samples have their own specific health profiles. Time differences are also present that make it difficult to differentiate between the influence of the cultural and natural environment on health. Therefore, it cannot be stated from this evidence alone whether this hypothesis was correct.

# Cultural environment

> When one has said all there is to say about animal reservoirs, the characteristics of protozoal and parasitic infections, the direct and indirect effects of climate, it is still true to say that human disease is a function of human behaviour, particularly of community or crowd behaviour (Smith, 1941), and no survey of health...would be complete without reference to the way in which culture can affect disease. A community's cultural and behaviour patterns can encourage or minimize the transmission of disease indirectly, through the type of house that they build and live in, the distribution of the family in the various rooms throughout the house, the clothes that they wear or the lack of them, the diet, and customs related to feeding. (Barton 1977: 93)

The cultural influence on disease and nutrition is a further important factor to consider when comparing populations and is the second major variable on which a hypothesis was developed in Chapter 1. As has been explained in Chapter 2 the cultural environment has changed through time in prehistoric Thailand. Cultural changes, including the development of rice agriculture, the domestication of animals, the development of bronze and iron metallurgy, as well as changes in social complexity, can all potentially influence a community's health. It was hypothesised that there would be a general improvement in health through time. The sample from Khok Phanom Di would be the least healthy and Ban Na Di the healthiest, with the samples from Nong Nor and Ban Lum Khao providing the intermediate step.

The development of the domestication of animals and rice agriculture have effects on the reliability of food and the prevalence of zoonoses in human groups. However, the development of agriculture may not necessarily be associated with a more nutritionally adequate diet. A hunter-gatherer diet may be of a better nutritional quality as it is more varied, although energy expenditure may be higher in obtaining this diet. The knowledge of metallurgy can lead to the production of more efficient tools for clearing and ploughing land in the intensification of agriculture. This may increase the quantity of food available and consequently support an increasing population size. The population size and density and the type of housing, can affect the prevalence of certain respiratory infections and many other diseases, as mentioned earlier in this chapter. The workload of a community will also be affected. The presence of a hierarchy within a community can have an affect on the organisation, division, and type of labour undertaken. This may have differentially affected community health through different physical workload requirements by subgroups. The presence of significant sex differences in the prevalence of joint disease may indicate this. The differences may became more marked through time and imply an increasing social complexity. A more socially complex community may also limit access to food and other resources to some subgroups. This would see greater intrasite variability in health. Sex differences in dental health, growth, and growth disturbances, may be suggestive of such subdivisions. Additionally, inequalities within sexes may be indicative of divisions within males or females. This may be indicated by a wide range of stature, for example, within one sex.

These are some of the main points that were introduced in Chapters 1 and 2 that reflect the influence of time on community health in prehistory. Whether this was the case in prehistoric Thailand will be addressed in light of the skeletal evidence presented in Chapters 3 to 7. First, the available archaeological evidence concerning the cultural environment of each Thai sample will be summarised. This will then be tied in with the skeletal evidence in order to directly address the hypothesis above.

### Archaeological evidence

Cultural aspects of the four prehistoric Thai samples have changed through time. The changes have been concerned with subsistence mode, metallurgy, and social structure. The main points are summarised below from Chapter 2. There is considerably more archaeological evidence available for the site of Khok Phanom Di. This has made the discussion of this skeletal sample easier.

The people of prehistoric Khok Phanom Di were not familiar with metallurgy, but were highly skilled potters, with males and females possibly contributing to different aspects of this process. This would imply some sexual divisions of labour were part of this community. The social structure was not entirely egalitarian, with some men and women achieving a higher status as has been inferred from grave good analyses, although this is somewhat restricted to late in the sequence of burials. The development of rice agriculture was at its earliest stages and subsistence activities more commonly involved fishing, hunting, and gathering based around a marine and coastal environment. These people were not completely isolated as the presence of exotic stone artefacts implies their participation in an exchange network.

The people of Nong Nor were part of a community that, like Khok Phanom Di, were not particularly socially complex. Grave goods analysed identified no particularly wealthy individuals or group. Pottery manufacturing was a common local activity. Bronze artefacts were also in abundance. Rice agriculture would also have taken up more of these people's time. Other subsistence activities would have included exploiting the coastal and freshwater environments accessible from the site, as well as the raising of domesticated stock. The wide range of exotic artefacts indicate that Nong Nor's exchange network extended further than that of Khok Phanom Di.

The social structure of Ban Lum Khao was also reasonably egalitarian as the mortuary ritual generally implies. Similar activities as occurred at Nong Nor would have been undertaken by the people of Ban Lum Khao. Rice cultivation, tending domesticates, fishing, gathering plants, and some hunting would have all been part of daily life in the Bronze Age. Technological knowledge included pottery manufacturing and bronze working. An exchange system was also in place.

Ban Na Di would have been a community busy with rice agriculture and care of domesticated stock, as well as hunting, fishing, and gathering. The production of bronze artefacts was undertaken. A wide ranging exchange network, more so than at Ban Lum Khao, was in place. Familiarity with iron forging also began in the later stages of occupation of Ban Na Di. Their social structure also appears to have been developing in complexity, more so than in the other samples. Grave good analysis differentiated between the two areas excavated at Ban Na Di, particularly on the presence of exotic goods. This may mean the presence of biases, possibly in terms of diet and labour, were part of life at Ban Na Di.

### Hypothesis: comparisons based on the cultural environment

Health: 'Khok Phanom Di < Nong Nor/Ban Lum Khao < Ban Na Di'

Addressing both health trends and health variability through time formed the basis of the cultural hypothesis. It was suggested that health would improve through time, but that there would also be a concomitant increase in the variability of health, although the latter has been made difficult by the age and sex structures of the northeastern samples, which have reduced the reliability of assessing sex differences through time. This hypothesis was partially dependent on the hypothesis that Nong Nor and Ban Lum Khao had a similar health profile, given their similar position in time. However, this was not proved entirely correct.

Childhood was a vulnerable period at all sites as suggested by the evidence for subadult morbidity and mortality. However, there are some health trends through time that may be related to an improvement in living conditions. The earliest sample, Khok Phanom Di, had the highest subadult mortality, the highest enamel hypoplasia in adults, and the shortest statures. Those that survived childhood were compromised in their growth. The proportion of subadults dying in each sample does not decrease consistently through time. However, if the Nong Nor subadult mortality is accepted as the result of sampling bias and it is removed from the equation, the trend for decreasing subadult mortality is evident, with Khok Phanom Di having the highest, Ban Lum Khao intermediate, and Ban Na Di with the lowest subadult mortality.

There is also an improvement in how well each sample coped with childhood. Ban Lum Khao, Nong Nor, and Ban Na Di all have taller statures than the Khok Phanom Di sample. The latter two samples were also taller than the Ban Lum Khao sample. It would therefore appear that Nong Nor and Ban Na Di were the samples that coped the best with growth disturbances during childhood. This may be related to improved nutrition and a more reliable diet. The Ban Lum Khao sample was comprised of more frail and medium frail children that may have been related to the possibility that they were some of the first settlers to the Mun River area. Time is required for communities to become familiar with their local environment in order to exploit it maximally. No pre-agricultural or early agricultural communities have been found in this region. The people of Ban Lum Khao were better off than those from Khok Phanom Di who had to cope with the malarial parasite. In contrast, both the Nong Nor and Ban Na Di samples were people from a community that had some prior knowledge of their respective environments. The occupation phase, preceding the cemetery phase, at around 2500 BC at Nong Nor, and the fact that Ban Na Di was near the earlier settled site of Ban Chiang implies that these people had previous knowledge of their environments.

The stature estimates, as an indication of nutritional quality, appear to indicate biases were present in males in the Bronze Age (the stature variation in the females from Khok Phanom Di was due to the presence of one very short woman). No such bias was evident in the more socially complex sample of Ban Na Di. This goes against the hypothesis for increased variability of health with time within each sample. It is possible that although social divisions existed in Ban Na Di, because of their potentially more reliable food sources (considering their level of rice agriculture and animal domestication), no subgroups were *nutritionally* disadvantaged. Perhaps the lower status group was biased in other ways, such as their workload. Alternatively, the sample size may not have been large enough to differentiate between such groups. While there is no archaeological evidence to infer the Bronze Age groups were as socially complex as Ban Na Di, skeletal evidence may indicate there was. It is possible the Bronze Age culture did not confer a person's status through mortuary ritual.

A further indication of nutritional quality is in the analysis of bone mass. Each sample had a similar, and adequate, bone mass index (excluding the Nong Nor sample which was unable to be assessed). There was no obvious trend with time. Khok Phanom Di males and females and Ban Lum Khao males all had quite a range of variation in bone mass. However, the former may be explained simply by the larger sample size. The latter may add evidence to the existence of inequalities within the male group from Ban Lum Khao, as was also indicated by the stature estimates.

A further trend, possibly indicating an improvement in health with time, is evident in the adult age at death distribution. There was a consistent increase with time in the number of adults surviving to older age (40+ years). The proportion of adults aged 40-49+ years in the Ban Na Di sample was nearly twice as high as the earliest sample, Khok Phanom Di.

As mentioned above, evidence of different patterns of joint disease may highlight divisions of labour within a sample. Changes in the prevalence over time may indicate changes in the amount of physical work and its efficiency. As summarised earlier, it is evident that the pattern of joints affected among the samples were different. This would appear to represent contrasting patterns of joint susceptibility, possibly as a result of different patterns of usage. It has already been tentatively postulated as to the varying possibilities of these activities in Chapter 7 and, given that these are purely suggestions, they will not be discussed further. However, it was also evident that the sample from Khok Phanom Di generally displayed a higher prevalence of osteoarthritis, in both males and females, than the Ban Lum Khao and Ban Na Di samples. The people of Khok Phanom Di, therefore, appeared to have been exposed to a higher level of biomechanical stress. Over cultural changes with time a reduction in osteoarthritis occurred. Genetic differences may also explain differences in osteoarthritis among the samples. Unfortunately, the analysis of the sex differences within the northeastern samples was confounded by the age and sex structure of these samples. A difference among the sexes within these samples may have been able to expand the picture of social structure and associated divisions of labour.

The evidence for trauma among the samples has provided interesting information regarding patterns of risk over time. The Bronze Age stood out as being a particular period of risk for trauma, both in the northeast and southeast. There is no definitive archaeological evidence for generalised warfare during this time in Thailand, however this does not rule out the possibility of interpersonal violence which may have occurred in some individuals during the Bronze Age. Alternatively, these samples may have been undertaking more activities associated with such risks, such as the clearance of land for rice agriculture and hunting.

There is an improvement in dental health with time that may have been associated with a change in diet among the samples. In general, the Khok Phanom Di sample had the poorest dental health, while the other samples were all quite similar to each other. The exception was the high prevalence of advanced attrition in the Nong Nor samples. The three later samples all probably had more rice in their diet than Khok Phanom Di. Rice in the unmilled form has a coarse texture. This abrasiveness may have resulted in sufficient attrition to remove potential sites for carious lesions to develop, therefore resulting in a dental pathology profile with low caries rates and moderate to high advanced attrition. The people of Nong Nor may have been consuming some additional abrasive food or have had a genetic difference that resulted in accelerated tooth wear compared with the northeastern samples. As the later samples had a more reliable food resource with rice agriculture, the reliance on other previously used foods that may have been more cariogenic, may have been reduced in comparison with their consumption at Khok Phanom Di.

Overall, it is evident that the cultural hypothesis has not been proved entirely correct. However, a slight adjustment to it, where the health of Ban Lum Khao is between that of Khok Phanom Di and Nong Nor, rather than equal to that of Nong Nor, would make it more accurate: 'Khok Phanom Di < Ban Lum Khao < Nong Nor < Ban Na Di'.

# Conclusions

The initial impetus for this research came from a recognition that it was now possible to begin forming a large scale synthesis of health and disease in prehistoric Thailand. A number of single population studies had paved the way for this and the acquisition of new skeletal samples made it feasible. Four skeletal samples from southeastern and northeastern regions within Thailand covered the period of late prehistory, from 2500BC to 400BC.

Health, morbidity, and mortality have been investigated through the use of a selection of nonspecific and specific parameters that have enabled a *representation* of the health and disease of the four samples to be obtained. Mortality, growth, joint disease, trauma, and dental health have all been extensively compared among the samples. Using a biocultural approach this information has been integrated with archaeological and ethnographic evidence to determine the predominant influences on health in each natural and cultural environment.

The variation in the natural and cultural environments of the samples provided a basis on which to discuss changes in health over time. A number of hypotheses were developed in order to systematically address these variations. With respect to the natural environment, it was hypothesised that those skeletal samples from similar natural environments would have a similar level of health and those samples from different natural environments would differ in their health status. Additionally, it was hypothesised that as the cultural environment changed through time, so to would health.

The natural environment has had a significant influence on the health status of the people of prehistoric Thailand, but not always in the way hypothesised. There were few similarities evident in the health status between the southeastern samples. The Khok Phanom Di sample consistently showed a poorer level of health. They had significant childhood illnesses and fatality, shorter adult stature and poorer dental health. The comparison between these southeastern samples was hindered by the poor preservation of Nong Nor's skeletal material and cultural differences between the samples.

In contrast the northeastern skeletal samples, Ban Lum Khao and Ban Na Di, did have a similar health status in some respects. Both samples had similar dental pathology profiles and levels of osteoarthritis. However, the period of childhood appears to have been more detrimental to children at Ban Lum Khao. They had a higher mortality and growth disturbances have led to shorter adult statures being attained. Subtle environmental differences, such as the suggested climate changes over the period of occupation of these sites, may have influenced the apparent health differences. The drier climate of the Bronze Age may have been more detrimental, with water and food shortages a significant problem in the dry season. The wet cycle during the Iron Age may have thus been an advantage for the Ban Na Di people.

Between the two regions studied in Thailand, the southeast and northeast, no consistent trend was evident for either area to be associated with better conditions for health. Both areas had detrimental and beneficial influences. For example, the southeast was a malarious environment but dietary resources were thought to be in abundance. The northeast, although without the high risk of malaria, was prone to more contrasting seasons with the monsoonal climate that may have affected the source of food. The existence of marked differences between the samples within each region has led to difficulties in addressing this last natural environment hypothesis.

The cultural environment changed significantly through the time period studied. The transition to, and particularly the intensification of, rice agriculture was one of the influential changes. It was hypothesised that an improvement in health would be evident with advancing time and the skeletal evidence suggests this was correct. The people of the early agricultural site of Khok Phanom Di were the least healthy. Although they had only incipient rice agriculture, evidence suggests their diet was generally nutritionally adequate. In this case it may have been the presence of the malarial parasite and subsequent morbidity that has influenced their poor health status. An improvement in health is evident in the Bronze Age, although more so in the Nong Nor sample than in the Ban Lum Khao sample, for reasons discussed below. The people of the late Bronze to early Iron Age at Ban Na Di, continued the trend for improving health.

Khok Phanom Di and Ban Lum Khao had a poorer health status than Nong Nor and Ban Na Di, and this may be linked with the possibility that they were settling new environments. The samples with the relatively better health, Nong Nor and Ban Na Di, were the people who were more likely to have had prior knowledge of their natural environments. This appears to have enabled them to exploit it more efficiently. The people of Khok Phanom Di and Ban Lum Khao required time to adjust to their new environments in order to obtain the necessary resources from it and adapt to various aspects, such as the pathogen load.

Overall, the health status gradually improved from the early agricultural times at Khok Phanom Di to the Bronze Age at Ban Lum Khao. The people of Nong Nor were healthier again, as were the people of the early Iron Age at Ban Na Di. This trend for an improvement in health through time, is consistent with the cultural advances that occurred in later periods. The establishment of hierarchical and non-autonomous communities ruled by a centralised political regime occurred with the establishment of the Angkorian civilisation in the first millennium AD in Southeast Asia. In order for this to have occurred it was necessary for their ancestors to have been in a position to cope with such changes, assuming that these communities underwent a linear development. Communities would have had to be healthy, fertile, and had high energy levels. With such communities becoming more complex, people would have to have the energy to support other members undertaking specialised tasks, unrelated to daily subsistence activities necessary for survival. It has been previously suggested that this state was reached during the late Iron Age (Higham 1996). From this *selective* view of prehistoric health in Southeast Asia it would appear, from the available skeletal evidence, that the people were becoming ready for the transformation to a state by this stage suggested. However, in the larger scheme of the entire region of Southeast Asia, this research has been selective, only representing a small number of individuals from Thailand. Thus, it has only just begun to be able to address this issue and more work selecting samples from other areas around Southeast Asia, including Vietnam and Cambodia, would be required to increase the sample size and obtain an even larger, and more representative picture of health through time in prehistory in Southeast Asia.

Other influences on the health of a community includes their genotype. As yet, this aspect has not been able to be investigated among these four Thai samples. It is, however, an alternative explanation for the differences in health among the samples. This may be the case among all samples, but may be particularly relevant to the Khok Phanom Di sample. These people consistently had the poorest health, although despite this, they did in fact survive and reproduce. It cannot be easily determined if Khok Phanom Di was settled by a genetically different group from the other samples. However, it cannot be ruled out as a possibility.

Further limitations in this research included the small number of samples, as mentioned above. The differential preservation was also limiting in the comparability of information among samples. The lack of information with respect to some aspects from the Nong Nor skeletal material has lead to some unanswered questions. This is particularly noticeable regarding the influence of their potentially malarious environment. An additional limitation is related to the small interobserver error in the analysis of joint disease between the Khok Phanom Di sample and the others. This may be overcome in the future by a reanalysis of a larger portion of the Khok Phanom Di sample. With respect to the osteological paradox, this research has countered this as much as possible by the interpretation of multiple variables in order to obtain the most appropriate interpretation.

With three different time periods, two different environments, and some confounding sampling biases to take into account, the interpretation of this information has not been a straightforward task. Each skeletal sample, representing a community, has interacted with their natural and cultural environment in an individual way, as subtle and some not so subtle differences in these variables existed among all samples.

Previous work on prehistoric Thai skeletal remains have been mentioned throughout and have provided a comparison with the present study. However, comparisons have been complicated, especially with respect to trends in health with time as the dating of the samples from Ban Chiang and Non Nok Tha are controversial. Douglas (1996) concluded that her first hypothesis, that the early inhabitants of northeast Thailand were generally healthy, was accepted and this would generally concur with the health status of Ban Na Di which is in the immediate area of Ban Chiang. The second hypothesis by Douglas (1996) that there was a decline in health with time, is less supported by the skeletal evidence, particularly in the Non Nok Tha sample (Douglas 1996). A more extensive integration of the data from Non Nok Tha, Ban Chiang, and Noen U-Loke with the four samples analysed here will further extend the knowledge of health in prehistoric Thailand.

The comparison of the trend in health status with the development and intensification of agriculture in other regions of the world has not been the focus of this thesis. This has been avoided for a number of reasons but primarily because the nature of the resource base varies over the different regions of the world. For example, as has been shown by this research, a rice diet can have quite different effects on health, particularly dental health, compared with other carbohydrate staples, such as maize. It has been generally, although not consistently, hypothesised that the adoption of agriculture leads to a decline in health as the result of increasing population size and density, less varied diet and other factors. However, while this may be the case in some areas of America for example (for example Cohen and Armelagos 1984b; Larsen 1997), this has been shown not to be the case in the southeast and northeast of prehistoric Thailand.

This research has fulfilled its initial aim of providing an extensive synthesis of the variation in health in selected regions of late prehistoric Thailand. Many aspects of health, morbidity, and mortality have been detailed and interpretations offered for the variations and similarities in each sample's health profile. This research has provided a significant step in advancing the knowledge of health in prehistoric Thailand.

# Appendix A

## Khok Phanom Di test for interobserver error

The Khok Phanom Di skeletal sample was first examined by Dr Nancy Tayles and an extensive database was established for the sample. As well as estimating age at death of all individuals and sex estimation for adults, it provided information on indicators of health and disease and physical appearance. In order to compare this data with the other skeletal samples, it needed to be determined that there was little interobserver error in the collection of data between the Khok Phanom Di sample and the other three skeletal samples, from which I collected the data. The following described how this test of interobserver error was carried out and the results.

Random number tables were used to select ten numbers between one and 154. These numbers established the ten burials from Khok Phanom Di that would be re-examined. A protocol had been established in the data collection from Nong Nor, Ban Lum Khao, and Ban Na Di, and this was used to analyse the ten Khok Phanom Di skeletons. The results are presented below, comparing my results with the results of Dr Tayles.

## Age and sex estimation

Table A.1 provides a comparison of the results of the age at death estimation of ten individuals as well as sex estimation in five adults, between myself and Tayles' analysis. Very few differences are apparent in the two analyses. All five subadults were aged the same and all adults were within the range estimated by myself, except for one. Burial 64 was estimated to be between 22-25 years on the basis of epiphyses by myself, however, Tayles estimated this adult to be slightly younger at 21 years. This is not a significant difference as in all the analyses in this present study in which ages are used pool together all young adults aged between 15-29 or 20-29 years. I could not attempt to estimate an age at death in single years as this would require seriation of the entire sample. Sex estimation is consistent, with no discrepancies between the two analyses in any adult individual

## Growth and disturbances of growth

### Subadult diaphyseal lengths

Measurements of the diaphyseal lengths of a selection of subadults are presented in Table A.2, comparing my measurements with those of Tayles (1992). It is evident that there are no significant differences between these measurements. The largest difference totalled 4 mm in the left femur of Burial 12.1. In this case the proximal epiphysis was concreted to the proximal end of the bone and has therefore affected obtaining the maximum length of the diaphysis. In two cases, that of the right radius of Burial 51 and the left ulna of Burial 75, I decided a measurement was possible whereas Tayles (1992) did not. The reverseoccurred in the measurement of the right radius of Burial 12.1.

A paired t-test of the measurements which were measured by both researchers indicates there were no significant differences between the two sets of measurements (t-value = 1.54, p-value = 0.15).

**Table A.1:** Khok Phanom Di age at death and sex estimation results from the author and Tayles (1992)

| Burial | This study | | Tayles 1992 | |
|---|---|---|---|---|
| | Age (years) | Sex | Age (years) | Sex |
| 12.1 | 12 | - | 12 | - |
| 38 | 30-34 | male | 33 | male |
| 39 | 25-29 | female | 25 | female |
| 51 | 0 | - | 0 | - |
| 64 | 22-25 | female | 21 | female |
| 72 | 28-35 | male | 32 | male |
| 75 | 0 | - | 0 | - |
| 93 | 40-44 | male | 42 | male |
| 96 | 4 | - | 4 | - |
| 98 | 0 | - | 0 | - |

**Table A.2:** Khok Phanom Di diaphyseal lengths (mm) of subadults compared

| Burial | Humerus | | Radius | | Ulna | | Femur | | Tibia | |
|---|---|---|---|---|---|---|---|---|---|---|
| | Left | Right | Left | Right | Left | Right | Left | Right | Left | Right |
| This study | | | | | | | | | | |
| 51 | | | | 49.0 | | | | | | |
| 75 | 63.0 | 64.0 | 54.0 | 55.0 | 60.0 | 62.0 | 75.0 | 74.0 | 66.0 | |
| 98 | 66.0 | | | | | | | 77.0 | | |
| 96 | | | | | | | | | | 150 |
| 12.1 | | | | - | | | 265 | | | |
| Tayles 1992 | | | | | | | | | | |
| 51 | | | | - | | | | | | |
| 75 | 64.0 | 64.0 | 54.0 | 54.0 | | 60.0 | 75.0 | 74.0 | 66.0 | |
| 98 | 66.0 | | | | | | | 77.0 | | |
| 96 | | | | | | | | | | 149 |
| 12.1 | | | | 143 | | | 261 | | | |

**Table A.3:** Khok Phanom Di long bone lengths (mm) and stature estimates compared

| Burial | Humerus | | Radius | | Ulna | | Femur | | Tibia | | Equation used | Stature |
|---|---|---|---|---|---|---|---|---|---|---|---|---|
| | Left | Right | Left | Right | Left | Right | Left | Right | Left | Right | | |
| This study | | | | | | | | | | | | |
| 38 | 307 | 306 | 241 | 239 | | 263 | | 426 | | 348 | F+T | 159.5 |
| 39 | 278 | 279 | | 213 | 237 | 237 | 391 | 394 | 316 | 315 | F+T | 148.2 |
| 64 | 289 | 291 | 238 | 242 | | 267 | 417 | 411 | 363 | 363 | F+T | 158.2 |
| 72 | 316 | 314 | | 267 | | 290 | 438 | 442 | 390 | 391 | F+T | 167.9 |
| 93 | 307 | 309 | 261 | 262 | 280 | 281 | 439 | 444 | | 385 | F+T | 167.3 |
| Tayles 1992 | | | | | | | | | | | | |
| 38 | 307 | | 241 | 238 | | 263 | | | | 345 | T | 158.1 |
| 39 | 277 | 279 | 212 | 213 | 237 | 237 | 391 | 394 | 316 | 315 | F+T | 148.1 |
| 64 | 291 | 289 | 238 | 237 | 260 | 266 | 417 | 411 | 361 | 362 | F+T | 157.6 |
| 72 | 313 | 316 | | 266 | | 288 | 440 | 441 | 387 | 390 | F+T | 167.4 |
| 93 | 308 | 309 | 261 | 262 | 280 | 281 | 440 | 443 | 381 | 383 | F+T | 166.6 |

F= femur, T= tibia

### Adult long bone lengths and stature

Measurements of the long bone lengths of a selection of adults are presented in Table A.3, comparing the two sets of measurements. It is evident that there are no large differences between these measurements. The largest difference totalled 5 mm in the measurement of the right radius of Burial 64. This difference is likely to have been due to the presence of a fresh break that has occurred after the first measurement by Tayles was taken. The majority of measurements are either the same or are one or two millimetres different. A paired t-test in fact showed a statistical difference existed between the two sets of measurements at the 2% level (t-value = 2.5, p-value = 0.02). However, this type of statistical test hypothesises that no difference exists between paired measurements. However, in the field of biological anthropology, a difference of one or two millimetres between observers is considered acceptable. Of the forty paired measurements, 90% differed by no more than 2 mm.

As with the subadult measurements, a few cases arose where discrepancies of opinion concerning whether the bone was complete enough to be measured. Tayles (1992) presented measurements for the left ulna of Burial 64 and the left radius from Burial 39. On the other hand I felt able to obtain a maximum length from the right humerus and right femur from Burial 38 while Tayles (1992) did not.

With regard to stature estimates, the largest difference was in the estimate for Burial 38 totalling 1.4 cm. Other differences were all less than 0.7 cm. A further difference in Burial 38 involved the equation used to estimate stature. Because I measured the femur of this burial, the 'femur + tibia' equation was used to estimate stature.

### Enamel Hypoplasia

The presence of enamel hypoplasia observed by myself and Dr Tayles are compared in Table A.4. In this analysis the largest difference is for Burial 38 and Burial 93 where I recorded three linear enamel hypoplasias and Tayles (1992) did not record any. The teeth from Burial 93 were the most difficult to observe as they were not in a very good condition. There are also a few differences in the total number of teeth observed reflecting personal choices on whether a tooth was complete enough to be accurately observed or not. The observation of linear enamel hypoplasia is obviously not as discrete as one may think. There are variations, such as the depth of the line, that can produce interobserver error.

**Table A.4:** Khok Phanom Di linear enamel hypoplasia compared

| Burial | This study A/N | Tayles 1992 $A^{\#}/N$ |
|---|---|---|
| Subadults | | |
| 12.1 | 2/25 | 1/27 |
| 96 | 0/6 | 0/9 |
| Adults | | |
| 38 | 3/12 | 0/14 |
| 39 | 6/18 | 4/18 |
| 64 | 4/20 | 4/21 |
| 72 | 6/23 | 6/22 |
| 93* | 2/12 | 5/11 |
| Total | 23/116 | 20/122 |
| FET p-value | 0.51 (n.s) | |

A= affected, N = number teeth observed, *many broken teeth, postmortem damage, #only linear enamel hypoplasia (grades 1, 6, 7, and 9 of Tayles 1992 scale)

**Table A.5:** Khok Phanom Di osteoarthritis compared

| | This Study | | | | Tayles 1992 | | | |
|---|---|---|---|---|---|---|---|---|
| | 38 | 39 | 72 | 93 | 38 | 39 | 72 | 93 |
| Shoulders | A | A | P | P | A | A | P | P |
| Elbows | A | A | P | P | P | A | P | P |
| Wrists | A | A | P | A | A | A | P | P |
| Hands | A | A | A | P | A | A | A | P |
| Hips | A | A | P | P | A | A | P | A |
| Knees | | A | A | P | | A | P | P |
| Ankles | A | A | P | A | A | A | P | A |
| Feet | A | A | P | P | / | A | A | A |
| Total | 12/31 | | | | 12/30 | | | |
| FET p-value | >0.99 (n.s.) | | | | | | | |

A= osteoarthritis absent, P= osteoarthritis present. Individual cells shaded slightly darker indicate they are different from the Tayles (1992) analysis.

# Joint Disease

### Osteoarthritis and Vertebral Osteophytosis

The data comparing the final assessment of each joint from my own observations and those of Tayles (1992) are presented in Table A.5 (osteoarthritis) and A.6 (vertebral osteophytosis). Those observations with darker shading than others indicate the differences in the assessment of osteoarthritis. It appears osteoarthritis and vertebral osteophytosis analysis is prone to some degree of interobserver error, with about 20-25% of joints assigned differently. The overall prevalences of the number affected over the number observed are not statistically different for either vertebral osteophytosis or osteoarthritis, as the Fisher's Exact p-values indicate.

When this test for interobserver error was undertaken the joint surfaces were actually graded from 0-4. The grading scheme followed that from Tayles (1992). Grade 0 indicates no joint changes, Grade 1 represents minimal changes, Grade 2 moderate, Grade 3 severe, and Grade 4 joint fusion has occurred. The results of the interobserver error test indicated that the few differences that do exist do not amount to more than one grade. However, it was only later in subsequent analyses of the results that it was decided the grading detail was not directly necessary to obtain a general picture of joint degeneration differences among the samples. Thus, when the grades were divided into osteoarthritis 'present' (grades 2-4) or 'absent' (grades 0-1), the interobserver error became more obvious. However, only four individuals were retested and further tests on more individuals may well reduce this error. Before the results comparing all the samples are taken further than the general prevalences presented in this thesis, more individuals should be retested to obtain a better picture of the interobserver error.

**Table A.6:** Khok Phanom Di vertebral osteophytosis compared

| Burial | This study Cervical | This study Thoracic | This study Lumbar | Tayles 1992 Cervical | Tayles 1992 Thoracic | Tayles 1992 Lumbar |
|---|---|---|---|---|---|---|
| 38 | A | P | A | A | A | P |
| 39 | A | A | A | A | A | A |
| 64 | A | A | P | - | - | A |
| 72 | A | A | P | A | A | A |
| 93 | P | P | P | P | P | P |
| Total | 6/15 | | | 4/13 | | |
| FET p-value | 0.71 (n.s) | | | | | |

A= vertebral osteophytosis absent, P= vertebral osteophytosis present. Those shaded slightly darker indicate they are different from my analysis.

**Table A.7:** Khok Phanom Di caries compared, permanent teeth only

| Burial | This study A/N | Tayles 1992 A/N |
|---|---|---|
| Subadults | | |
| 12.1 | 3/27 | 3/27 |
| Adults | | |
| 38 | 0/12 | 0/16 |
| 39 | 3/18 | 3/19 |
| 64 | 0/25 | 0/25 |
| 72 | 0/26 | 1/28 |
| 93* | 5/20 | 11/23 |
| Total | 11/128 | 18/138 |
| FET p-value | 0.33 (n.s.) | |

A= affected, N = number teeth observed, *many broken teeth, postmortem damage

**Table A.8:** Khok Phanom Di advanced attrition compared, permanent teeth only

| Burial | This study A/N | Tayles 1992 A/N |
|---|---|---|
| 38 | 0/12 | 0/17 |
| 39 | 0/18 | 0/19 |
| 64 | 0/25 | 0/25 |
| 72 | 0/23 | 0/28 |
| 93* | 6/13 | 12/21 |
| Total | 6/91 | 12/110 |
| FET p-value | 0.33 (n.s.) | |

A= affected grade 6 or more (Molnar 1971), N = number teeth observed, *many broken teeth, postmortem damage

# Dental health and disease

The observation of caries, advanced attrition, and antemortem tooth loss are presented in the following tables (Tables A.7 - A.9) comparing my analysis with that of Tayles (1992). There are some differences apparent in the former two analyses only. The largest differences for assessing caries and advanced attrition are both in Burial 93 where six teeth were assessed differently in both analyses. The teeth from Burial 93 were the most difficult to observe as they were not in a very good condition. They and some of the other skeletal material have become more fragile over years of handling. Overall there are a few differences in the total number of teeth observed. This and the other differences reflect personal choices on whether a tooth was complete enough to be accurately observed or not.

**Table A.9:** Khok Phanom Di antemortem tooth loss compared, permanent teeth only

| Burial | This study<br>A/N | Tayles 1992<br>A/N |
|---|---|---|
| 38 | 4/32 | 4/32 |
| 39 | 3/32 | 3/32 |
| 64 | 2/32 | 2/32 |
| 72 | 2/32 | 2/32 |
| 93* | 2/32 | 2/32 |
| Total | 13/160 | 13/160 |

A= lost antemortem , N = number teeth observed, *many broken teeth, postmortem damage

# Appendix B

## Statistics

The statistical analyses in this thesis have been kept simple. Complex statistics on sometimes inadequate data (eg. small sample sizes, missing data) can lead to misleading interpretations. There are basically two types of data used in this thesis, continuous data and nominal scale data.

Continuous data describes that of all the metric variables, such as stature and bone mass measurements. In order to statistically compare metric variables between only two samples 'student's t-tests' are used. For comparing metric data among more than two samples ANOVA (analysis of variance) tests are used.

Nominal scale data described data that is classified by some attribute. In such cases as osteoarthritis, vertebral osteophytosis, and antemortem tooth loss, for example, two classes are described with attributes of present (affected) or absent (unaffected). In comparing such proportional data (number present/total observed), $Chi^2$ tests are employed to compare more than two samples, and Fisher's Exact Tests (FET) are used to compare two samples only. FET are more accurate, irrespective of sample size, than $Chi^2$ for comparing two samples (G.Speirs, pers comm). They are particularly useful when sample sizes are small.

All these four types of tests, for both types of data, are testing whether the null hypothesis (no differences between the data) is rejected or accepted. This involves establishing a critical level for acceptance or rejection. A probability of 5% is most commonly used (Zar 1984). This means that if any of the tests performed produce a p-value of less than 0.05, the data are judged to be significantly different.

In this thesis this level of 5% has been deemed acceptable for ANOVA and $Chi^2$ tests, as well as single t-tests and Fisher's Exact Tests. However, errors can arise when repeated t-tests and FETs are carried out using the same variable. For example, this situation may arise when a $Chi^2$ test has indicated a significant difference exists between the proportion of antemortem tooth loss among the four Thai skeletal samples. This requires FETs being calculated for each combination of two samples to determine between which samples there are significant differences. The results of the FET will, however, include a certain percentage of Type I and II errors. That is, a significant result is obtained when it is not true or an insignificant result is obtained that should have been significant, respectively.

The occurrence of these two errors has an inverse relationship and while a 5% level is widely used as it is "usually considered to a "small enough" chance of committing a Type I error, while not being so small as to result in "too large a chance" of a Type II error" (Zar 1984: 44) it may be considered more important to reduce one or other of these errors. In this thesis it was decided that reducing a Type I error, finding a significant result that in reality was insignificant, was more important than reducing the occurrence of Type II errors. Claiming a significant difference occurred between two samples can lead to further investigation, based on erroneous statistics. Therefore, in multiple two sample comparisons, to reduce the chance of these errors occurring, the critical value was reduced to 0.01 (1%) when using multiple t-tests and Fisher's Exact Tests.

# Appendix C

## Skeletal size versus dental age

For the Figure 4.1 indicating the growth rate as graphed by dental age versus the mean humerus diaphyseal length.

Table C.1: Number of children and the number of humeri measurements available from each sample for each age group.

| Age (years) | Khok Phanom Di | | Ban Lum Khao | | Ban Na Di | |
|---|---|---|---|---|---|---|
| | N children | N m/m | N children | N m/m | N children | N m/m |
| 0-0.9 | | | 1 | 1 | 3 | 5 |
| 1-1.9 | | | 2 | 3 | 2 | 2 |
| 2 | | | 2 | 3 | | |
| 3 | | | 2 | 3 | 1 | 1 |
| 4 | | | | | | |
| 5 | 2 | 2 | | | | |
| 6 | | | | | | |
| 7 | | | 3 | 5 | 1 | 1 |
| 8 | 2 | 4 | | | | |
| 9 | 3 | 4 | | | | |
| 10 | | | | | | |
| 11 | 1 | 1 | | | | |
| 12 | 2 | 2 | 1 | 2 | | |
| 13 | | | 1 | 2 | 1 | 1 |
| 14 | | | 1 | 1 | | |
| Total | 10 | 13 | 13 | 20 | 8 | 10 |

N children = number of children aged from their dentition with humeri able to be measured;
N m/m = number of humeri diaphyseal measurements obtained from children aged from their dentition.

# Appendix D

## Thai stature regression equations

Figure D.1: Thai regression equations for estimating stature from long bone lengths

| **Upper Limb Bones (Sangvichien *et al.* n.d.)** | **Lower Limb Bones (Sangvichien *et al.* 1985)** |
|---|---|
| *Females* | |
| S = 2.7501 Humerus max + 74.0862 | S = 2.5815 Femur max + 49.2412 ± 3.0007 |
| S = 2.6743 Radius max + 92.5011 | S = 2.9716 Tibia max + 51.6015 ± 3.4687 |
| S = 2.7025 Ulna max + 87.1950 | |
| S = 1.6192 (Humerus max + Radius max) + 70.4181 | S = 1.4074 (Femur max + Tibia max) + 48.4454 ± 3.1019 |
| *Males* | |
| S = 3.0197 Humerus max + 70.0340 | S = 1.7289 Femur max + 88.1320 ± 5.3885 |
| S = 3.3658 Radius max + 81.4874 | S = 2.7638 Tibia max + 62.6946 ± 4.5232 |
| S = 3.4970 Ulna max + 72.8812 | |
| S = 1.8816 (Humerus max + Radius max) + 59.2273 | S = 1.4210 (Femur max + Tibia max) + 49.6080 ± 4.2087 |

N.B. All measurements are in cm
max = maximum length

# Appendix E

## Observed Caries Prevalences

Table E.1: Summary of the number of teeth with caries in each sample, for each sex and the total sample.

| | Khok Phanom Di | | Nong Nor | | Ban Lum Khao | | Ban Na Di | | $Chi^2$ | p-value |
|---|---|---|---|---|---|---|---|---|---|---|
| | A/N | % | A/N | % | A/N | % | A/N | % | | |
| Total | 139/1281 | 10.9 | 66/1017 | 6.5 | 40/885 | 4.5 | 24/515 | 4.7 | **40.40** | **<0.01** |
| Female | 96/656 | 14.6 | 44/472 | 9.3 | 32/477 | 6.7 | 12/147 | 8.2 | **20.77** | **<0.01** |
| Male | 43/625 | 6.9 | 22/545 | 4.0 | 8/408 | 2.0 | 12/368 | 3.3 | **16.32** | **<0.01** |
| Male vs Female Fisher's Exact p-value | | | | | | | | | | |
| | **<0.01** | | **<0.01** | | **<0.01** | | 0.03 | | | |

A= affected, N= total number of teeth observed; **Bold** indicates statistical significance at 1% level

Table E.2: Fisher's Exact p-values for inter-sample comparisons of the prevalence of teeth with caries.

| Total | Nong Nor | Ban Lum Khao | Ban Na Di |
|---|---|---|---|
| Khok Phanom Di | **<0.01** | **<0.01** | **<0.01** |
| Nong Nor | - | 0.07 | 0.17 |
| Ban Lum Khao | - | - | 0.90 |
| Females | | | |
| Khok Phanom Di | **<0.01** | **<0.01** | 0.04 |
| Nong Nor | - | 0.15 | 0.74 |
| Ban Lum Khao | - | - | 0.51 |
| Males | | | |
| Khok Phanom Di | 0.04 | **<0.01** | 0.02 |
| Nong Nor | - | 0.09 | 0.60 |
| Ban Lum Khao | - | - | 0.27 |

**Bold** indicates statistical significance at 1% level

# References


Acsádi, Gy., and J. Nemeskéri (1970). *History of Human Life Span and Mortality*. Budapest: Akadémiai Kiadó.

Agelarakis, A. (1996). The archaeology of human bones: prehistoric copper producing peoples in the Khao Wong Prachan Valley, central Thailand. *Indo-Pacific Prehistory Association Bulletin* **14**: 133-138.

Aloia, J.F. (1994). The gain and loss of bone in the human life cycle. *Advances in Nutritional Research* **9**: 1-33.

Altman, R.D. (1997). The syndrome of osteoarthritis. *The Journal of Rheumatology* **24**(4): 766-767.

Andermann, S. (1976). The cnemic index: a critique. *American Journal of Physical Anthropology* **44**: 369-370.

Anderson, L.D., and F.N. Meyer (1991). Fractures of the shafts of the radius and ulna. In *Rockwood and Green's Fractures in Adults*. C.A. Rockwood, D.P. Green, and R.W. Bucholz, (eds). pp. 679-737, Vol. 1. (3rd edition). Philadelphia: Lippincott.

Angel, J.L., J.M. Suchey, M.Y. Iscan, and M.R. Zimmerman (1986). Age at death estimated from the skeleton and viscera. In *Dating and Age Determination of Biological Materials*. pp. 179-220. M.R. Zimmerman and J.L. Angel, (eds). London: Croom Helm.

Aufderheide, A.C., and C. Rodriquez-Martin (1998). *The Cambridge Encyclopaedia of Human Paleopatholgy*. Cambridge: Cambridge University Press.

Bailey, S.M., S.N. Gershoff, R.B. McGandy, A. Nondasta, P. Tantiwongse, D. Suttapreyasri, J. Miller, and P. McCree (1984). A longitudinal study of growth and maturation in rural Thailand. *Human Biology* **56**(3): 539-557.

Bannanurag, R. (1989). *Khok Phanom Di and its socio-cultural implications*. PhD thesis, Department of Anthropology, University of Otago.

Barnett, E., and B.E.C. Nordin (1960). The radiological diagnosis of osteoporosis: a new approach. *Clinical Radiology* **11**: 166-174.

Barton, W.L. (1977). Tropical health: a broad survey. In *Human Ecology in the Tropics*. J.P. Garlick and W.J. Keay, (eds). pp. 87-94. Symposia of the Society for the Study of Human Biology, Vol. XVI. London: Taylor & Francis Ltd.

Bass, W.M. (1987). *Human osteology. A laboratory and field manual*. Columbia: Missouri Archaeological Society.

Bayard, D. (1984). Rank and wealth at Non Nok Tha. In *SE Asian Archaeology at the XV Pacific Science Congress: The origins of agriculture, metallurgy, and the state in Mainland SE Asia*. D. Bayard, (ed.). pp. 87-122. University of Otago Studies in Prehistoric Anthropology, Vol. 16. Dunedin.

Bedford, M.E., K.F. Russell, C.O. Lovejoy, R.S. Meindl, S.W. Simpson, and P.L. Stuart Macadam (1993). Test of the multifactorial aging method using skeletons with known ages-at-death from the Grant Collection. *American Journal of Physical Anthropology* **91**(3): 287-297.

Bergman, R.A.M., and T.H. The (1955). The length of the body and long bones of the Javanese. *Documenta de Medicina Geographia et Tropica* **7**: 197-214.

Bielicki, T. (1998). Growth as an indicator of social inequalities. In *The Cambridge Encyclopedia of Human Growth and Development*. S.J. Ulijaszek, F.E. Johnston, and M.A. Preece, (eds). pp. 54-56. Cambridge: Cambridge University Press.

Black, T.K. (1978). A new method for assessing the sex of fragmentary skeletal remains: femoral shaft circumference. *American Journal of Physical Anthropology* **48**: 227-232.

Bocquet-Appel, J.P., and C. Masset (1982). Farewell to paleodemography. *Journal of Human Evolution* **11**: 321-333.

Bocquet-Appel, J.P., and C. Masset (1985). Paleodemography: Resurrection or ghost? *Journal of Human Evolution* **14**: 107-111.

Bocquet-Appel, J.P., and C. Masset (1996). Paleodemography: Expectancy and false hope. *American Journal of Physical Anthropology* **99**: 571-583.

Bogin, B. (1999). *Patterns of Human Growth*. (2nd edition). Cambridge: Cambridge University Press.

Bogin, B., M. Wall, and R.B. MacVean (1992). Longitudinal analysis of adolescent growth of *ladino* and Mayan school children in Guatemala: effects of environment and sex. *American Journal of Physical Anthropology* **89**: 447-457.

Boldsen, J.L. (1998). Body proportions in a medieval village population: effects of early childhood episodes of ill health. *Annals of Human Biology* **25**(4): 309-317.

Boyd, W.E., C.F.W. Higham, and R. Thosarat (1996). The Holocene palaeogeography of the southeast margin of the Bangkok Plain, Thailand, and its archaeological implications. *Asian Perspectives* **35**(2): 193-207.

Bridges, P.S. (1989). Changes in activities with the shift to agriculture in the southeastern United States. *Current Anthropology* **30**(3): 385-394.

Bridges, P.S. (1991). Degenerative joint disease in hunter-gatherers and agriculturalists from the southeastern United States. *American Journal of Physical Anthropology* **85**: 379-391.

Bridges, P.S. (1992). Prehistoric arthritis in the Americas. *Annual Review of Anthropology* **21**: 67-91.

Brooks, S.T. (1955). Skeletal age at death: the reliability of cranial and pubic age indicators. *American Journal of Physical Anthropology* 13: 567-97.

Brothwell, D. (1981). *Digging Up Bones*. Oxford: British Museum (Natural History) Oxford University Press.

Brothwell, D.R. (1991). On zoonoses and their relevance to paleopathology. In *Human Paleopathology. Current Synthesis and Future Options*. D.J. Ortner and A.C. Aufderheide, (eds). pp. 18-22. Washington: Smithsonian Institution Press.

Buckwalter, J.A., and R.L. Cruess (1991). Healing of the musculoskeletal tissues. In *Rockwood and Green's Fractures in Adults*. C.A. Rockwood, D.P. Green, and R.W. Bucholz, (eds). pp. 181-222, Vol. 1. (3rd edition). New York: Lipincott.

Buikstra, J.E., and D.C. Cook (1980). Paleopathology: an American account. *Annual Review of Anthropology* **9**: 433-470.

Buikstra, J.E., and L.W. Konigsberg (1985). Paleodemography: critiques and controversies. *American Anthropologist* **85**: 316-333.

Buikstra, J.E., and J.H. Mielke (1985). Demography, diet, and health. In *The Analysis of Prehistoric Diets*. R.I.J. Gilbert and J.H. Mielke, (eds). pp. 359-422. Orlando, Florida: Academic Press.

Buikstra, J.E., and D.H. Ubelaker (eds) (1994). *Standards for Data Collection from Human Skeletal Remains*. Fayetteville, Arkansas: Arkansas Archaeological Survey.

Bush, H. (1991). Concepts of health and stress. In *Health in Past Societies. Biocultural interpretations of human skeletal remains in archaeological contexts*. H. Bush and M. Zvelebil, (eds). pp. 11-22. British Archaeological Reports International Series 567. Oxford: Tempvs Reparatvm.

Bush, H. and M. Zvelebil (1991). Pathology and health in past societies: an introduction. In *Health in Past Societies. Biocultural interpretations of human skeletal remains in archaeological contexts*. H. Bush and M. Zvelebil, (eds). pp. 3-9. British Archaeological Reports International Series 567. Oxford: Tempvs Reparatvm.

Byers, S. (1991). Technical note: Calculation of age at formation of radiopaque transverse lines. *American Journal of Physical Anthropology* **85**: 339-343.

Cameron, N. (1998). Adulthood and developmental maturity. In *The Cambridge Encyclopedia of Human Growth and Development*. S.J. Ulijaszek, F.E. Johnston, and M.A. Preece, (eds). pp. 233-235. Cambridge: Cambridge University Press.

Carloni, A.S. (1981). Sex disparities in the distribution of food within rural households. *Food and Nutrition* **7**: 3-12.

Chaiamnuay, P., J. Darmawan, K.D. Muirden, and P. Assawatanabodee (1998). Epidemiology of rheumatic disease in rural Thailand: a WHO-ILAR COPCORD study. *The Journal of Rheumatology* **25**(7): 1382-1387.

Chandrapanond, A., A. Ratchatasilpin, and S. Tansuphasiri (1972). Dietary survey of preschool children and expectant women in Soongnern District, Nakorn Rajisma Province, Thailand. *American Journal of Clinical Nutrition* **25**: 730-735.

Chandrapanond, A., D. Suttapryasri, S. Tansuphasiri, and P. Harinasata (1973). Nutrition, growth and development of Thai village children. *Journal of the Medical Association of Thailand* **56**: 722-731.

Chang, N. (1998). The personal ornaments. In *The Excavation of Nong Nor: a Prehistoric Site in Central Thailand*. C.F.W. Higham and R. Thosarat, (eds). pp. 265-304. Dunedin: Department of Anthropology, University of Otago.

Chong, W.Y. (1988). *Geohydrology of the Town of Khon Kaen, Northeastern Thailand.* Thailand: Asian Institute of Technology.

Chongsuvivatwong, V., S. Pas-Ong, D. McNeil, A. Geater, and M. Duerawee (1996). Predictors for the risk of hookworm infection: experience from endemic villages in southern Thailand. *Transactions of the*

*Royal Society of Tropical Medicine and Hygiene* **90**: 630-633.

Choosiri, P. (1988). *An Analysis of Human Remains from Khok Phanom Di*. Master of Arts thesis, University of Otago.

Clark, G.A., and M. Mack (1988). Reliability assessment of transverse lines. *Human Biology* **60**(2): 283-291.

Cohen, M.N. (1989). *Health and the Rise of Civilisation.* New Haven: Yale University Press.

Cohen, M.N., and G.J. Armelagos (eds) (1984a). *Paleopathology at the Origins of Agriculture.* Orlando: Academic Press.

Cohen, M.N., and G.J. Armelagos (1984b). Paleopathology at the origins of agriculture: editor's summation. In *Paleopathology at the Origins of Agriculture.* M.N. Cohen and G.J. Armelagos, (eds). pp. 585-601. Orlando: Academic Press.

Costa, R.L. (1986). Determination of age at death: dental analysis. In *Dating and Age Determination of Biological Materials*. M.R. Zimmerman and J.L. Angel, (eds). pp. 248-269. London: Croom Helm.

Cox, M., and A. Scott (1992). Evaluation of the obstetric significance of some pelvic characters in an 18th century British sample of known parity status. *American Journal of Physical Anthropology* **89**(4): 431-40.

Cox, T. (1978). *Stress*. London: Macmillan.

Cox, T. and C.J. Mackay (1981). A transactional approach to occupational stress. In *Stress, Work Design and Productivity*. N. Corlett and P. Richardson, (eds). Chichester: Wiley and Sons. (as cited in Cox *et al.* 1983)

Cox, T., S. Cox, and M. Thirlaway (1983). The psychological and physiological response to stress. In *Physiological Correlates of Human Behaviour. Volume I: Basic Issues.* A. Gale and J.A. Edwards, (eds). pp. 255-276. London: Academic Press.

Craig, E.V. (1991). Fractures of the shoulder. Part II: Fractures of the clavicle. In *Rockwood and Green's Fractures in Adults*. C.A. Rockwood, D.P. Green, and R.W. Bucholz, (eds). pp. 928-990, Vol. 1. (3rd edition). Philadelphia: Lippincott.

Cushnagen, J. and P. Dieppe (1991). Study of 500 patients with limb joint osteoarthritis. I. Analysis by age, sex, and distribution of symptomatic joint sites. *Annals of the Rheumatic Diseases* **50**: 8-13.

Dandy, D.J. (1993). *Essential Orthopaedics and Trauma.* (2nd edition). London: Churchill Livingston.

Davis, A. (1996). Schistosomiasis. In *Manson's Tropical Diseases.* G.C. Cook, (ed.). pp. 1413-1456. (20th edition). London: W.B.Saunders.

Debreceny, J. (1998). The ceramic assemblage from phase 2. In *The Excavation of Nong Nor: a Prehistoric Site in Central Thailand*. C.F.W. Higham and R. Thosarat, (eds). pp. 227-264. Dunedin: Department of Anthropology, University of Otago.

Debreceny, J., C.F.W. Higham, and B.F.J. Manly (1998). A statistical analysis of the burials. In *The Excavation of Nong Nor: a Prehistoric Site in Central Thailand.* C.F.W. Higham and R. Thosarat, (eds). pp. 491-508. Dunedin: Department of Anthropology, University of Otago.

Devito, C. and S.R. Saunders (1990). A discriminant function analysis of deciduous teeth to determine sex. *Journal of Forensic Science* **35**: 845-857.

Dias, G., and N. Tayles (1997). 'Abscess cavity' - a misnomer. *International Journal of Osteoarchaeology* **7**: 548-554.

DiBennardo, R., and J.V. Taylor (1979). Sex assessment of the femur: a test of a new method. *American Journal of Physical Anthropology* **50**: 635-638.

Dieppe, P. (1990). Osteoarthritis. A review. *Journal of the Royal College of Physicians London* **24**(4): 262-267.

Doherty, M., A. Jones, and T.E. Cawston (1998). Osteoarthritis. In *Oxford Textbook of Rheumatology*. P.J. Maddison, D.A. Isenberg, P. Woo, and D.N. Glass, (eds). pp. 1515-1553, Vol. 2. (2nd edition). New York: Oxford Medical Publications.

Domett, K.M. (1994). *Pottery manufacturing in the prehistoric Khok Phanom Di population.* Unpublished report, Department of Anatomy and Structural Biology, University of Otago.

Domett, K.M. (1998). Association between degenerative joint disease of the hands and pottery manufacturing in the prehistoric population of Khok Phanom Di, Thailand. *American Journal of Physical Anthropology* Supplement **26**: 72-73.

Douglas, M.T. (1996). *Paleopathology in human skeletal remains from the pre-metal, Bronze and Iron Ages, northeastern Thailand*. PhD thesis, Department of Anthropology, University of Hawaii.

Dreizen, S., C.M. Spirakis and E.R. Stone (1964). The influence of age and nutritional status on "bone scar" formation in the distal end of the growing radius. *American Journal of Physical Anthropology* **22**: 295-306.

Dutour, O. (1986). Enthesopathies as indicators of the activities of neolithic Saharan populations. *American Journal of Physical Anthropology* **71**: 221-224.

Ellner, P.D. and H.C. Neu (1992). *Understanding Infectious Disease.* St Louis: Mosby Year Book.

Evans, A.S. (1998). Epidemiological concepts. In *Bacterial Infections of Humans: Epidemiology and Control*. A.S. Evans and P.S. Brachman (eds). pp. 13-64. (3rd edition). New York: Plenum Medical Book Company.

F.A.O. (1954). *Rice and Rice Diets. A Nutritional Survey.* (Revised edition). Rome: Food and Agriculture Organization of the United Nations.

Fergusson, C.M. (1987). The aetiology of osteoarthritis. *Postgraduate Medical Journal* **63**: 439-445.

Flatt, A.E. (1993). Correction of arthritic deformities of the hand. In *Arthritis and Allied Conditions.* D.J. McCarty and W.J. Koopman, (eds). pp. 919-938, Vol. 1. (12th edition). Philadelphia: Lea and Febiger.

Francis, R.M. (1990). The calcium controversy. In *Osteoporosis 1990.* R. Smith, (ed.). pp. 125-133. London: Royal College of Physicians of London.

Garn, S.M. (1970). *The Earlier Gain and the Later Loss of Cortical Bone in Nutritional Perspective.* Springfield: Thomas.

Garn, S.M., T.V. Sullivan, S.A. Decker, and V.M. Hawthorne (1991). Brief communication: on the optimum number of metacarpals for roentgenogrammetric measurement. *American Journal of Physical Anthropology* **85**(2): 229-232.

Geiger, E., and G. Borgstrom (1962). Fish protein - nutritive aspects. In *Fish as Food. Volume II: Nutrition, Sanitation, and Utilization.* G. Borgstrom, (ed.). pp. 29-114. London: Academic Press.

Genovés, S. (1967). Proportionality of the long bones and their relation to stature among Mesoamericans. *American Journal of Physical Anthropology* **26**: 67-78.

Gershoff, S., R. McGandy, D. Suttapreyasri, C. Promkutkao, A. Nondasuta, U. Pisolyabutra, P. Tantiwongse and V. Viravaidhaya (1977). Nutrition studies in Thailand. II. Effects of fortification of rice with lysine, threonine, thiamine, riboflavin, vitamin A, and iron on preschool children. *American Journal of Clinical Nutrition* **30**: 1185-1195.

Gilbert, B.M. (1973). Misapplication to females of the standard for aging the male os pubis. *American Journal of Physical Anthropology* **38**: 39-40.

Gilbert, B.M., and T.W. McKern (1973). A method for aging the female os pubis. *American Journal of Physical Anthropology* **38**: 31-38.

Giles, E. (1970). Discriminant function sexing of the human skeleton. In *Personal Identification in Mass Disasters.* T.D. Stewart, (ed.). pp. 99-109. Washington: National Museum of Natural History, Smithsonian Institution Press.

Gilles, H.M. (1996). Soil-transmitted helminths (geohelminths). In *Manson's Tropical Diseases.* G.C. Cook, (ed.). pp. 1369-1412. (20th edition). London: W.B.Saunders.

Gindhart, P.S. (1969). The frequency of appearance of transverse lines in the tibia in relation to childhood illnesses. *American Journal of Physical Anthropology* **31**: 17-22.

Gittelsohn, J. (1991). Opening the box: intrahousehold food allocation in rural Nepal. *Social Science and Medicine* **33**: 1141-1154.

Goodman, A.H. (1991). Health, adaptation, and maladaptation in past societies. In *Health in Past Societies. Biocultural Interpretations of Human Skeletal Remains in Archaeological Contexts.* H. Bush and M. Zvelebil, (eds). pp. 31-38. B.A.R. International Series 567. Oxford: B.A.R.

Goodman, A.H. (1993). On the interpretation of health from skeletal remains. *Current Anthropology* **34**(3): 281-288.

Goodman, A.H. (1994). Cartesian reductionism and vulgar adaptationism: issues in the interpretation of nutritional status in prehistory. In *Paleonutrition: The Diet and Health of Prehistoric Americans.* K.D. Sobolik (ed.). pp. 163-177. Centre for Archaeological Investigations, Occasional Paper No. 22., Southern Illinois University.

Goodman, A.H., and G.J. Armelagos (1988). Childhood stress and decreased longevity in a prehistoric population. *American Anthropologist* **90**: 936-944.

Goodman, A.H., and G.J. Armelagos (1989). Infant and child morbidity and mortality risks in archaeological populations. *World Archaeology* **21**(2): 225-243.

Goodman, A.H., G.J. Armelagos, and J.C. Rose (1980). Enamel hypoplasia as indicators of stress in three prehistoric populations from Illinois. *Human Biology* **52**(3): 515-528.

Goodman, A.H., J. Lallo, G.J. Armelagos, and J.C. Rose (1984a). Health changes at Dickson Mounds, Illinois (A.D. 950-1300). In *Paleopathology at the Origins of Agriculture.* M.N. Cohen and G.J. Armelagos, (eds). pp. 271-305. Orlando: Academic Press.

Goodman, A.H., D.L. Martin, and G.J. Armelagos (1984b). Indications of stress from bone and teeth. In *Paleopathology at the Origins of Agriculture.* M.N. Cohen and G.J. Armelagos, (eds). pp. 13-19. Orlando: Academic Press.

Goodman, A.H., R.B. Thomas, A.C. Swedlund, and G.J. Armelagos (1988). Biocultural perspectives on stress in prehistoric, historic and contemporary population research. *Yearbook of Physical Anthropology* **31**: 169-202.

Gotherstrom, A., K. Liden, T. Ahlstrom, M. Kallersjo, and T.A. Brown (1997). Osteology, DNA and sex identification: morphological and molecular sex identifications of five Neolithic individuals from Ajvide, Gotland. *International Journal of Osteoarchaeology* **7**: 71-81.

Green, D.P., and S.A. Rowland (1991). Fractures and dislocations in the hand. In *Rockwood and Green's Fractures in Adults*. C.A. Rockwood, D.P. Green, and R.W. Bucholz, (eds). pp. 441-561, Vol. 1. (3rd edition). Philadelphia: Lippincott.

Grolleau-Raoux, J-L., E. Crubezy, D. Rouge, J-F. Brugne, and S. Saunders (1997). Harris lines: A study of age-associated bias in counting and interpretation. *American Journal of Physical Anthropology* **103**: 209-217.

Haavikko, K. (1970). The formation and the alveolar and clinical eruption of the permanent teeth. *Proceedings of the Finnish Dental Society* **66**: 101-170.

Habberfield-Short, J. (1999). The geoarchaeological context of two Iron Age sites in northeast Thailand. *Archaeology in New Zealand* **42**(3): 213-229.

Hadjimarkos, D.M. (1966). Dietary fluoride and fluorosis in Thailand. *American Journal of Public Health and the Nations Health* **56**(3): 391-2.

Hadler, N.M., D.B. Gillings, H.R. Imbus, P.M. Levitin, D. Makuc, P.D. Ustinger, W.J. Yount, D. Slusser, and N. Moskowitz (1978). Hand structure and function in an industrial setting. *Arthritis and Rheumatism* **21**(2): 210-220.

Hall, M.C. (1961). The trabecular patterns of the neck of the femur, with particular reference to changes in osteoporosis. *Canadian Medical Association Journal* **85**: 1141-1144.

Hall, M.D. (1993). Aspects of the pottery decoration. In *The Excavation of Khok Phanom Di: a Prehistoric Site in Central Thailand. Volume III: The Material Culture (Part I)*. C.F.W. Higham and R. Thosarat, (eds). pp. 239-274. London: The Society of Antiquaries of London.

Hanks, L.M. (1972). *Rice and Man. Agricultural Ecology in Southeast Asia*. Chicago: Aldine Atherton.

Harinasuta, C. (1983). Parasitic diseases in the South (developing world). In *Parasitiology - A Global Perspective*. K.S. Waren and J.Z. Bowers, (eds). pp. 19-44. New York: Springer-Verlag.

Harinasuta, C. (1986). Filariasis in Thailand. In *The 25th Anniversary of the Faculty of Tropical Medicine*. S. Sucharit, S. Areejul, P. Sucharit, S. Vanijanonta and V. Kitikoon (eds). pp. 68-72. Bangkok: Mahidol University.

Harinasuta, C. (1987). Endemic tropical disease in Southeast Asia, with special reference to Thailand. *Annals of Tropical Medicine and Parasitology* **81**(5): 657-669.

Harinasuta, C., and S. Vajrasthira (1964). Soil-transmitted helminths in Thailand: the incidence, distribution and factors influencing the transmission. *Journal of the Medical Association of Thailand* **47**(7): 396-401.

Harkess, J.W., W.C. Ramsey, and J.W. Harkess (1991). Principles of fractures and dislocations. In *Rockwood and Green's Fractures in Adults*. C.A. Rockwood, D.P. Green, and R.W. Bucholz, (eds). pp. 1-180, Vol. 1. (3rd edition). Philadelphia: Lipincott.

Harris, H.A. (1926). The growth of long bones in childhood, with special reference to certain bony striations of the metaphysis and the role of vitamins. *Archives of Internal Medicine* **39**: 785-806.

Harrison, M.H.M, F. Schajowicz and J. Trueta (1953). Osteoathritis of the hip: a study of the nature and evolution of the disease. *Journal of Bone and Joint Surgery* **35B**: 598-626.

Hart, C.A. (1996). Introduction to acute infective diarrhoea. In *Manson's Tropical Diseases*. G.C. Cook, (ed.). pp. 817-823. (20th edition). London: W.B.Saunders.

Haswell-Elkins, M.R., and D.B. Elkins (1996). Food-borne trematodes. In *Manson's Tropical Diseases*. G.C. Cook, (ed.). pp. 1457-1476. (20th edition). London: W.B.Saunders.

Hauck, H.M., J.R. Hanks, and S. Sudsaneh (1959a). Food habits in a Siamese village. *Journal of the American Dietetic Association* **35**: 1143-1148.

Hauck, H.M., A. Rajatsilpin, and D. Thorangkul (1959b). *Observations on the growth, health, and nutritional status of elementary school children, Bang Chan, Thailand, 1952-55*. Proceedings of the 9th Pacific Science Congress, Bangkok, 1957. Vol. 15, pp. 13-15.

Hauck, H.M., and S. Sudsaneh (1959). Food intake and nutritional status in a Siamese village. *Journal of the American Dietetic Association* **35**: 1149-1157.

Heckman, J.D. (1991). Fractures and dislocations of the foot. In *Rockwood and Green's Fractures in Adults*. C.A. Rockwood, D.P. Green, and R.W. Bucholz, (eds). pp. 2041-2182, Vol. 2. (3rd edition). Philadelphia: Lippincott.

Heywood, P.F., and G.C. Marks (1993). Nutrition and health in South-East Asia. *The Medical Journal of Australia* **159**: 133-137.

Higgins, V. (1992). *Research developments in the field of non-specific stress and their relevance to the study of human skeletal remains*. IXth European Meeting of the Paleopathology Association, Barcelona, 1992, pp. 151-156. Museu d'Arqueologia de Catalunya.

Higham, C.F.W. (1984). Prehistoric rice cultivation in Southeast Asia. *Scientific American* **250**(4): 138-146.

Higham, C.F.W. (1989). *The Archaeology of Mainland Southeast Asia: from 10000 B.C. to the Fall of Angkor*. Cambridge: Cambridge University Press.

Higham, C.F.W. (1990). The Bang Pakong Valley research project. In *The Excavation of Khok Phanom Di: a Prehistoric Site in Central Thailand. Volume I: The Excavation, Chronology and Human Burials*. C.F.W. Higham and R. Bannanurag, (eds). pp. 1-17. London: The Society of Antiquaries of London.

Higham, C.F.W. (1993a). The bone, antler and turtle carapace technology. In *The Excavation of Khok Phanom Di: a Prehistoric Site in Central Thailand. Volume III: The Material Culture (Part I)*. C.F.W. Higham and R. Thosarat, (eds). pp. 1-44. London: The Society of Antiquaries of London.

Higham, C.F.W. (1996). *The Bronze Age of Southeast Asia*. Cambridge: Cambridge University Press.

Higham, C.F.W. (1998). The faunal remains from mortuary contexts. In *The Excavation of Nong Nor: a Prehistoric Site in Central Thailand*. C.F.W. Higham and R. Thosarat, (eds). pp. 315-320. Dunedin: Department of Anthropology, University of Otago.

Higham, C.F.W., and R. Bannanurag (1990). *The Excavation of Khok Phanom Di: a Prehistoric Site in Central Thailand. Volume I: The Excavation, Chronology and Human Burials*. London: The Society of Antiquaries of London.

Higham, C.F.W., and R. Bannanurag (1991a). *The Excavation of Khok Phanom Di: a Prehistoric Site in Central Thailand. Volume II: The Biological Remains (Part I)*. London: The Society of Antiquaries of London.

Higham, C.F.W., and R. Bannanurag (1991b). Summary. In *The Excavation of Khok Phanom Di: a Prehistoric Site in Central Thailand. Volume II: The Biological Remains (Part I)*. C.F.W. Higham and R. Bannanurag, (eds). pp. 333-336. London: The Society of Antiquaries of London.

Higham, C.F.W., and R. Bannanurag (1992). Prehistoric traders. *National Geographic Research and Exploration* **8**(2): 180-195.

Higham, C.F.W., and A. Kijngam (1984a). Introduction. In *Prehistoric Investigations in Northeastern Thailand*. C. Higham and A. Kijngam, (eds). pp. 1-3, Vol. i. Oxford: BAR International Series 231.

Higham, C.F.W., and A. Kijngam (1984b). Introduction: Previous research in Northeastern Thailand. In *Prehistoric Investigations in Northeastern Thailand*. C. Higham and A. Kijngam, (eds). pp. 4-21, Vol. i. Oxford: BAR International Series 231.

Higham, C.F.W., and A. Kijngam (eds) (1984c). *Prehistoric Investigations in Northeastern Thailand*. Vol. i, ii, iii. Oxford: B.A.R. International Series 231.

Higham, C.F.W, and A. Kijngam (1984d). The excavation of Ban Na Di, Ban Muang Phruk and Non Kao Noi. In *Prehistoric Investigations in Northeastern Thailand*. C.F.W. Higham and A. Kijngam, (eds). pp. 22-56, Vol. i. Oxford: BAR International Series 231.

Higham, C.F.W., and A. Kijngam (1984e). The mortuary ritual at Ban Na Di. In *Prehistoric Investigations in Northeastern Thailand*. C.F.W. Higham and A. Kijngam, (eds). pp. 413-545, Vol. ii. Oxford: BAR International Series 231.

Higham, C.F.W., and R. Thosarat (1993). *The Excavation of Khok Phanom Di: a Prehistoric Site in Central Thailand. Volume III: The Material Culture (Part I)*. London: The Society of Antiquaries of London.

Higham, C.F.W., and R. Thosarat (1997). *Noen U-Loke and the implications for the origins of early states*. Annual Meeting of the Society for American Archaeology, Nashville.

Higham, C.F.W., and R. Thosarat (eds) (1998a). *The Excavation of Nong Nor: a Prehistoric Site in Central Thailand*. Dunedin: Department of Anthropology, University of Otago.

Higham, C.F.W., and R. Thosarat (1998b). The human burials. In *The Excavation of Nong Nor: A Prehistoric Site in Central Thailand*. C.F.W. Higham and R. Thosarat, (eds). pp. 369-490. Dunedin: Anthropology Department, University of Otago.

Higham, C.F.W., and R. Thosarat (1998c). Introduction: Research Objectives. In *The Excavation of Nong Nor: a Prehistoric Site in Central Thailand*. C.F.W. Higham and R. Thosarat, (eds). pp. 1-22. Dunedin: Department of Anthropology, University of Otago.

Higham, C.F.W., and R. Thosarat (1998d). Noen U-Loke and the implications for the origins of early states. *Bulletin of the Indo-Pacific Prehistory Association* **17**(1: Precongress Issue): 44.

Higham, C.F.W., and R. Thosarat (1998e). *Prehistoric Thailand. From Early Settlement to Sukhotai.* Bangkok: River Books.

Higham, T.F.G. (1993b). The shell knives. In *The Excavation of Khok Phanom Di: a Prehistoric Site in Central Thailand. Volume III: The Material Culture (Part I).* C.F.W. Higham and R. Thosarat, (eds). pp. 177-209. London: The Society of Antiquaries of London.

Higham, T.F.G., and A. Hogg (1998). The radiocarbon chronology. In *The Excavation of Nong Nor: A Prehistoric Site in Central Thailand.* C.F.W. Higham and R. Thosarat, (eds). pp. 23-25. Dunedin: Anthropology Department, University of Otago.

Hillson, S. (1990). *Teeth.* Cambridge: Cambridge University Press.

Hillson, S. (1996). *Dental Anthropology.* Cambridge: Cambridge University Press.

Hirsch, R. , M. Lethbridge-Cejku, W.W. Scott, R. Reichle, C.C. Plato, J. Tobin, et al. (1996). Association of hand and knee osteoarthritis: evidence for a polyarticular disease subset. *Annals of the Rheumatic Diseases* **55**: 25-29.

Hoaglund, F.T., A.C.M.C. Yau, and W.L. Wong (1973). Osteoarthritis of the hip and other joints in southern Chinese in Hong Kong. *Journal of Bone and Joint Surgery* **55A**(3): 545-557.

Hochberg, M.C., C.M. Lethbridge, C.C. Plato, F.M. Wigley, and J.D. Tobin (1991). Factors associated with osteoarthritis of the hand in males: data from the Baltimore Longitudinal Study of Aging. *American Journal of Epidemiology* **134**(10): 1121-7.

Hoffman, J.M. (1979). Age estimations from diaphyseal lengths: two months to twelve years. *Journal of Forensic Science* **24**: 461 - 469.

Hough, A.J., and L. Sokoloff (1989). Pathology of osteoarthritis. In *Arthritis and Allied Conditions.* D.J. McCarty, (ed.). pp. 1571-1594 (11th edition). Philadelphia: Lea and Febiger.

Houghton, P. (1974). The relationship of the preauricular groove of the ilium to pregnancy. *American Journal of Physical Anthropology* **41**: 381-390.

Houghton, P. (1975). The bony imprint of pregnancy. *Bulletin of the New York Academy of Medicine* **51**(5): 655-661.

Houghton, P. (1996). *People of the Great Ocean.* Cambridge: Cambridge University Press.

Houghton, P., and P. de Souza (1975). Discriminant function sexing of prehistoric New Zealand skeletal material from lengths of long bones. *American Journal of Physical Anthropology* **84**(2): 225-229.

Houghton, P., B.F. Leach, and D.G. Sutton (1975). The estimation of stature of prehistoric Polynesians in New Zealand. *Journal of the Polynesian Society* **84**(3): 325-336.

Houghton, P., and W. Wiriyaromp (1984). The people of Ban Na Di. In *Prehistoric Investigations in Northeastern Thailand.* C.F.W. Higham and A. Kijngam, (eds), Vol. ii. pp. 391-411. Oxford: B.A.R. International Series 231.

Hummel, S., and B. Herrmann (1991). Y-Chromosome-specific DNA amplified in ancient human bone. *Naturwissenschaften* **78**: 266-267.

Hummert, J.R., and D.P. Van Gerven (1985). Observations on the formation and persistence of radiopaque transverse lines. *American Journal of Physical Anthropology* **66**: 297-306.

Huss-Ashmore, R., A.H. Goodman, and G.J. Armelagos (1982). Nutritional inference from paleopathology. *Advances in Archaeological Method and Theory* **5**: 395-474.

Hutton, C.W. (1987). Generalised osteoarthritis: An evolutionary problem? *Lancet* **June 27**: 1463-1465.

Iscan, M.Y., and S.R. Loth (1989). Osteological manifestations of age in the adult. In *Reconstruction of Life from the Skeleton.* M.Y. Iscan and K.A.R. Kennedy, (eds). pp. 23-40. New York: Alan R.Liss.

Jackes, M. (1992). Paleodemography: problems and techniques. In *Skeletal Biology of Past Peoples: Research Methods.* S.R. Saunders and M.A. Katzenberg, (eds). pp. 189-224. New York: Wiley-Liss.

Jackes, M., D. Lubell, and C. Meiklejohn (1997). Healthy but mortal: human biology and the first farmers of western Europe. *Antiquity* **71**: 639-658.

Jelliffe, D.B. (1970). General Background. In *Diseases of Children in the Subtropics and Tropics.* D.B. Jelliffe, (ed.). pp. 1-27. (2nd edition). London: Edward Arnold (Publishers) Ltd.

Jelliffe, D.B. and Bennet (1962). *Clinical Obstetrics and Gynecology* **5**: 64. (no title given in primary reference).

Johansson, S.R., and S. Horowitz (1986). Estimating mortality in skeletal populations: influence of the growth rate on the interpretation of levels and trends during the transition to agriculture. *American Journal of Physical Anthropology* **71**: 233-250.

Johnston, F.E. (1962). Growth of the long bones of infants and children at Indian Knoll. *American Journal of Physical Anthropology* **20**: 249-254.

Jowsey, J. (1960). Age changes in human bone. *Clinical Orthopaedics* **17**: 210-218.

Jurmain, R. (1990). Paleoepidemiology of a central California prehistoric population from CA-ALA-329: II. Degenerative disease. *American Journal of Physical Anthropology* **83**: 83-94.

Jurmain, R.D. (1991a). Degenerative changes in peripheral joints as indicators of mechanical stress: opportunities and limitations. *International Journal of Osteoarchaeology* **1**: 247-252.

Jurmain, R.D. (1991b). Paleoepidemiology of trauma in a prehistoric central Californian population. In *Human Paleopathology: Current Syntheses and Future Options*. D.J. Ortner and A.C. Aufderheide, (eds). pp. 241-248. Washington, DC: Smithsonian Institution Press.

Kamalanathan, G.S., H.M. Hauck, and C. Kittiveja (1960a). Incidence of caries among school children in Bang Chan, Thailand, 1953. *American Journal of Clinical Nutrition* **8**: 231-246.

Kamalanathan, G.S., H.M. Hauck, and C. Kittiveja (1960b). Dental development of children in a Siamese village, Bang Chan, 1953. *Journal of Dental Research* **39**(3): 455-461.

Karlberg, J. (1998). The human growth curve. In *The Cambridge Encyclopedia of Human Growth and Development*. S.J. Ulijaszek, F.E. Johnston, and M.A. Preece, (eds). pp. 108-114. Cambridge: Cambridge University Press.

Katz, D., and J.M. Suchey (1989). Race differences in pubic symphyseal aging patterns in the male. *American Journal of Physical Anthropology* **80**: 167-172.

Kealhofer, L. (1996). Human-environmental relationships in prehistory: an introduction to current research in South and Southeast Asia. *Asian Perspectives* **25**(2): 111-117.

Kelley, M.A. (1979). Parturition and pelvic changes. *American Journal of Physical Anthropology* **51**: 541-546.

Kennedy, K.A.R. (1984). Growth, nutrition, and pathology in changing paleodemographic settings in South Asia. In *Paleopathology at the Origins of Agriculture*. M.N. Cohen and G.J. Armelagos, (eds). pp. 169-192. Orlando: Academic Press.

Kennedy, K.A.R. (1989). Skeletal markers of occupational stress. In *Reconstruction of Life from the Skeleton*. M.M. Iscan and K.A.R. Kennedy, (eds). pp. 129-160. New York: Alan R.Liss.

Kennedy, K.A.R. (1998). Markers of occupational stress: conspectus and prognosis of research. *International Journal of Osteoarchaeology* **8**: 305-310.

Kerley, E.R. (1970). Estimation of skeletal age: after about age 30. In *Personal Identification in Mass Disasters*. T.D. Stewart, (ed.). pp. 57-70. Washington: Smithsonian Institution Press.

Kieser, J.A., J. Moggi-Cecchi, and H.T. Groeneveld (1992). Sex allocation of skeletal material by analysis of the proximal tibia. *Forensic Science International* **56**: 29-36.

Kieser, J.A., C.B. Preston, and W.G. Evans (1983). Skeletal age at death: an evaluation of the Miles method of ageing. *Journal of Archaeological Science* **10**: 9-12.

Kijngam, A. (1984). Excavations at Ban Na Di, north east Thailand. In *SE Asia Archaeology at the XV Pacific Science Congress: The origins of agriculture, metallurgy and the state in Mainland SE Asia*. D. Bayard, (ed.). pp. 36-41. University of Otago Studies in Prehistoric Anthropology, Vol. 16. Dunedin.

Knüsel, C.J., S. Goggel, and D. Lucy (1997). Comparative degenerative joint disease of the vertebral column in the medieval monastic cemetery of the Gilbertine priory of St.Andrew, Fishergate, York, England. *American Journal of Physical Anthropology* **103**: 481-495.

Krogman, W.M., and M.Y. Iscan (1986). *The Human Skeleton in Forensic Medicine*. (2nd edition). Springfield, Illinois: Charles C. Thomas.

Kunstadter, P. (1985). Rice in a Lua' subsistence economy, northwestern Thailand. In *Food Energy in Tropical Ecosystems*. D.J. Cattle and K.H. Schwerin, (eds). pp. 21-44. Food and Nutrition in History and Anthropology Volume 4. London: Gordon and Breach Science Publishers.

Larsen, C.S. (1982). The anthropology of St. Catherines Island. 3. Prehistoric human biological adaptations. *Anthropological Papers of the American Museum of Natural History* **57**(3): 157-276.

Larsen, C.S. (1987). Bioarchaeological interpretations of subsistence economy and behaviour from human skeletal remains. In *Advances in Archaeological Method and Theory*. M.B. Schiffer (ed.). pp. 339-445. San Diego: Academic Press.

Larsen, C.S. (1995). Biological changes in human populations with agriculture. *Annual Review of Anthropology* **24**: 185-213.

Larsen, C.S. (1997). *Bioarchaeology: Interpreting behaviour from the human skeleton*. Cambridge: Cambridge University Press.

Larsen, C.S., R. Shavit, and M.C. Griffin (1991). Dental caries evidence for dietary change: an archaeological context. In *Advances in Dental Anthropology*. M.A. Kelley and C.S. Larsen, (eds). pp. 179-202. New York: Wiley Liss.

Leatherwood, E.C. (1966). Letter to the editor. *American Journal of Public Health and the Nations Health* **56**(3): 392-3.

Leatherwood, E.C., G.W. Burnett, R. Chandravejjsmarn, and P. Sirikaya (1965). Dental caries and dental fluorosis in Thailand. *American Journal of Public Health* **55**(11): 1792-1799.

Leigh, R.W. (1925). Dental pathology of Indian tribes of varied environmental food conditions. *American Journal of Physical Anthropology* **8**: 179-199.

Lewis, M., and C. Roberts (1997). Growing pains: the interpretation of stress indicators. *International Journal of Osteoarchaeology* **7**: 581-586.

Lindgren, G. (1998). Changing human growth patterns. In *The Cambridge Encyclopedia of Human Growth and Development*. S.J. Ulijaszek, F.E. Johnston, and M.A. Preece, (eds). pp. 381-386. Cambridge: Cambridge University Press.

Lingström, P., D. Birkhed, Y. Granfeldt, and I. Björck (1993). pH measurements of human dental plaque after consumption of starchy foods using the microtouch and the sampling methods. *Caries Research* **27**: 394-401.

Littleton, J. (1999). Palaeopathology of skeletal fluorosis. *American Journal of Physical Anthropology* **109**: 465-483.

Littleton, J., and B. Frohlich (1993). Fish-eaters and farmers: dental pathology in the Arabian Gulf. *American Journal of Physical Anthropology* **92**: 427-447.

Loka, P., V. Nossar, and A.E. Bauman (1994). A comparative study of the growth of South-East Asian children in south-west Sydney born in Australia and overseas. *Journal of Paediatric Child Health* **30**: 436-438.

Lovejoy, C.O. (1985). Dental wear in the Libben population: its functional pattern and role in the determination of adult skeletal age at death. *American Journal of Physical Anthropology* **68**: 47-56.

Lovejoy, C.O., and K.G. Heiple (1981). The analysis of fractures in skeletal populations with an example from the Libben site, Ottowa County, Ohio. *American Journal of Physical Anthropology* **55**: 529-541.

Lovejoy, C.O., R.S. Meindl, R.P. Mensforth, and T.J. Barton (1985). Multifactorial determination of skeletal age at death: a method and blind tests of its accuracy. *American Journal of Physical Anthropology* **68**: 1-14.

Lovell, N.C. (1997). Trauma analysis in paleopathology. *Yearbook of Physical Anthropology* **40**: 139-170.

Lukacs, J.R. (1989). Dental paleopathology: Methods for reconstructing dietary patterns. In *Reconstruction of Life from the Skeleton*. M.Y. Iscan and K.A.R. Kennedy, (eds). pp. 261-286. New York: Alan R.Liss.

Lukacs, J.R. (1992). Dental paleopathology and agricultural intensification in South Asia: new evidence from Bronze Age Harappa. *American Journal of Physical Anthropology* **87**: 133-150.

Lukacs, J.R. (1995). The 'caries correction factor': a new method of calibrating dental caries rates to compensate for antemortem loss of teeth. *International Journal of Osteoarchaeology* **5**: 151-156.

Lunn, P.G. (1991). Nutrition, immunity and infection. In *The Decline of Mortality in Europe*. R. Schfield, D. Reher and A. Bideau (eds). pp. 131-145. Oxford: Clarendon Press.

Maat, G.J.R. (1984). Dating and rating of Harris's lines. *American Journal of Physical Anthropology* **63**: 291-299.

Macchiarelli, R., and L. Bondilioli (1994). Linear densitometry and digital image processing of proximal femur radiographs: implications for archaeological and forensic anthropology. *American Journal of Physical Anthropology* **93**: 109-122.

Macchiarelli, R., L. Bondioli, L. Censi, M. K. Hernaez, L. Salvadei, and A. Sperduti (1994). Intra- and interobserver concordance in scoring Harris lines: a test on bone sections and radiographs. *American Journal of Physical Anthropology* **95**(1): 77-83.

Maetzel, A., M. Makela, G. Hawker, and C. Bombardier (1997). Osteoarthritis of the hip and knee and mechanical occupational exposure - a systematic overview of the evidence. *Journal of Rheumatology* **24**: 1599-1607.

Maloney, B.K. (1991). The pollen, pteridophyte spore and microscopic charcoal: conclusions. In *The Excavation of Khok Phanom Di: a Prehistoric Site in Central Thailand. Volume II: The Biological Remains (Part I)*. C.F.W. Higham and R. Bannanurag, (eds). pp. 91-94. London: The Society of Antiquaries of London.

Martin, D.L., A.H. Goodman, and G.J. Armelagos (1985). Skeletal pathologies as indicators of quality and quantity of diet. In *The Analysis of Prehistoric Diets*. R.I. Gilbert and J.H. Mielke, (eds). pp. 227-279. Studies in Archaeology. Orlando, Florida: Academic Press.

Mason, G.M. (1991). The molluscan remains: conclusions. In *The Excavation of Khok Phanom Di: a Prehistoric Site in Central Thailand. Volume II: The Biological Remains (Part I).* C.F.W. Higham and R. Bannanurag, (eds). pp. 301-319. London: The Society of Antiquaries of London.

Mason, J.W. (1975b). A historical view of the stress field. Part II. *Journal of Human Stress* 22-36. (as cited in Bush 1991)

Mason, J.W., M.S. Wool, F.F. Wherry, I.I. Pennington, J.V. Brady, and B. Beer (1968). Plasma growth, hormone response to avoidance sessions in the monkey. *Psychosomatic Medicine* **30**: 760-773.

May, R.L., A.H. Goodman, and R.S. Meindl (1993). Response of bone and enamel formation to nutritional supplementation and morbidity among malnourished Guatemalan children. *American Journal of Physical Anthropology* **92**: 37-51.

Mazess, R.B. (1982). On aging bone loss. *Clinical Orthopaedics and Related Research* **165**: 239-252.

McHenry, H., and P.D. Schultz (1976). The association between Harris lines and enamel hypoplasia in prehistoric California Indians. *American Journal of Physical Anthropology* **44**: 507-512.

McKern, T.W. (1970). Estimation of skeletal age: from puberty to about 30 years of age. In *Personal Identification in Mass Disasters*. T.D. Stewart, (ed.). pp. 41-56. Washington: Smithsonian Institution Press.

McKern, T.W. and T.D. Stewart (1957). *Skeletal Age Changes in Young American Males*. Technical Report 45, Environmental Protection Research Division, Quatermaster Research and Development Centre, Natick, Massachusetts, U.S. Army.

Meindl, R.S., and C.O. Lovejoy (1985). Ectocranial suture closure: A revised method for the determination of skeletal age at death based on the lateral-anterior sutures. *American Journal of Physical Anthropology* **68**: 57-66.

Meindl, R.S., C.O. Lovejoy, and R.P. Mensforth (1983). Skeletal age at death: accuracy of determination and implications for human demography. *Human Biology* **55**(1): 73-87.

Meindl, R.S., C.O. Lovejoy, R.P. Mensforth, and R.A. Walker (1985). A revised method of age determination using the os pubis, with a review and tests of accuracy of other current methods of pubic symphyseal aging. *American Journal of Physical Anthropology* **68**: 29-45.

Merbs, C.F. (1983). *Patterns of activity-induced pathology in a Canadian Inuit population.* National Museum of Man Mercury Series. Archaeological Survey of Canada Paper No. 119. Ottawa.

Merbs C.F. (1989). Trauma. In *Reconstruction of Life from the Skeleton.* M.Y. Iscan and K.A.R. Kennedy (eds). pp. 161-190. New York: Alan R Liss.

Merbs, C.F. (1995). Incomplete spondylolysis and healing: a study of ancient Canadian Eskimo skeletons. *Spine* **20**(21): 2328-2334.

Merbs, C.F. (1996). Spondylolysis and spondylolisthesis: a cost of being erect biped or a clever adaptation? *Yearbook of Physical Anthropology* **39**: 201-228.

Merchant, V.L., and D.H. Ubelaker (1977). Skeletal growth of the protohistoric Arikara. *American Journal of Physical Anthropology* **46**: 61-72.

Miles, A.E.W. (1963). The dentition in the assessment of individual age in skeletal material. In *Dental Anthropology*. D.R. Brothwell, (ed.). pp. 191-209. Symposia for the Study of Human Biology, Vol. V. London: Pergamon Press.

Molnar, S. (1971). Human tooth wear, tooth function and cultural variability. *American Journal of Physical Anthropology* **34**: 175-190.

Molnar, S. (1992). *Human Variation. Races, Types, and Ethnic Groups*. (3rd edition). New Jersey: Prentice Hall.

Moore, E.H. (1988). *Moated Sites in Early North East Thailand*. Oxford: B.A.R. International Series.

Moore, M.N. (1993). The burnishing stones. In *The Excavation of Khok Phanom Di: a Prehistoric Site in Central Thailand. Volume III: The Material Culture (Part I).* C.F.W. Higham and R. Thosarat, (eds). pp. 105-118. London: The Society of Antiquaries of London.

Moore, W.J., and E. Corbett (1971). The distribution of dental caries in ancient British populations. I. Anglo-saxon period. *Caries Research* **5**: 151-168.

Moorrees, C.F.A., E.A. Fanning, and E.E. Hunt (1963a). Age variation of formation stages for ten permanent teeth. *Journal of Dental Research* **42**(6): 1490-1502.

Moorrees, C.F.A., E.A. Fanning, and E.E. Hunt (1963b). Formation and resorption of three deciduous teeth in children. *American Journal of Physical Anthropology* **21**: 205-213.

Moskowitz, R.W. (1984). Osteoarthritis: a clinical overview. In *Epidemiology of the rheumatic diseases.* R.C.Lawrence and L.E.Shulman, (eds). pp. 267-276. New York: Gower Medical.

Moskowitz, R.W. (1989). Clinical and laboratory findings in osteoarthritis. In *Arthritis and Allied Conditions.*

D.J. McCarty, (ed.) pp. 1605-1630. (11th edition). Philadelphia: Lea and Febiger.

Moskowitz, R.W. (1993). Clinical and laboratory findings in osteoarthritis. In *Arthritis and Allied Conditions*. D.J. McCarty, (ed.). pp. 1735-1760, Vol. 2. (12th edition). Philadelphia: Lea and Febiger.

Murphy, W.A., and P.A. Kaplan (1995). Temporomandibular joint. In *Diagnosis of Bone and Joint Disorders*. D. Resnick, (ed.). pp. 1699-1754, Vol. 3. (3rd edition). Philadelphia: W.B.Saunders.

Nelsen, K.M. (1999). *The dental health of the people from Noen U-Loke*. Master of Science thesis, Department of Anatomy and Structural Biology, University of Otago.

Nontasut, P., S. Changbumrung, C. Muennoo, K. Hongthong, N. Vudhivai, S. Sanguankiat, and S. Yaemput (1996). Vitamin B1, B2 and B6 deficiency in primary school children infected with hookworm. *Southeast Asian Journal of Medicine and Public Health* **27**(1): 47-50.

Nowell, G.W. (1978). An evaluation of the Miles method of ageing using the Tepe Hassar dental sample. *American Journal of Physical Anthropology* **49**: 271-176.

Nutalaya, P., W. Sophonsakulrat, M. Consuk, and N. Wattanachai (1989). Catastrophic flooding - an agent for landform development of the Khorat Plateau: a working hypothesis. In *Proceedings of the Workshop on Correlation of Quaternary Deposits in South, East and Southeast Asia*. N. Thiramongkol (ed.). pp. 95-115. Bangkok: Department of Geology, Chulalongkorn University.

O'Neill, T.W., E.V. McCloskey, J.A. Kanis, A.K. Bhalla, J. Reeve, D.M. Reid, C. Todd, A.D. Woolf, and A.J. Silman (1999). The distribution, determinants, and clinical correlates of vertebral osteophytosis: a population based survey. *The Journal of Rheumatology* **26**(4): 842-848.

O'Reilly, D.J.W. (1999). *A diachronic analysis of social organisation in the Mun River Valley*. PhD thesis, Department of Anthropology, University of Otago.

Ortiz, A. (1970). Acute diarrhoea in young children. In *Diseases of Children in the Subtropics and Tropics*. D.B. Jelliffe, (ed.). pp. 397-412. (2nd edition). London: Edward Arnold (Publishers) Ltd.

Ortner, D.J. (1968). Description and classification of degenerative bone changes in the distal joint surfaces of the humerus. *American Journal of Physical Anthropology* **28**: 139-156.

Ortner, D.J. (1979). Disease and mortality in the early Bronze Age people of Bab edh-Dhra, Jordan. *American Journal of Physical Anthropology* **51**: 589-598.

Ortner, D.J. (1991). Theoretical and methodological issues in paleopathology. In *Human Paleopathology. Current Syntheses and Future Options*. D.J. Ortner and A.C. Aufderheide, (eds). pp. 5-11. Washington: Smithsonian Institution Press.

Ortner, D.J. (1998). Male-female immune reactivity and its implications for interpreting evidence in human skeletal paleopathology. In *Sex and Gender in Paleopathological Perspective*. A.L. Grauer and P. Stuart-Macadam, (eds). pp. 79-92. Cambridge: Cambridge University Press.

Ortner, D.J., and A.C. Auferderheide (1991). Introduction. In *Human Paleopathology. Current Syntheses and Future Options*. D.J. Ortner and A.C. Auferderheide, (eds). pp. 1-2. Washington: Smithsonian Institution Press.

Ortner, D.J., and W. Putschar (1985). *Identification of Pathological Conditions in Human Skeletal Remains*. Washington D.C.: Smithsonian Institution Press.

Parfitt, A.M. (1988). Bone remodeling: relationship to the amount and structure of bone, and the pathogenesis and prevention of fractures. In *Osteoporosis: Etiology, Diagnosis, and Management*. B.L. Riggs and L.J. Melton III, (eds). pp. 45-93. New York: Raven Press.

Pierson, M. (1998). *Ancient DNA and the people of Khok Phanom Di*. Master of Arts thesis. Department of Anthropology, University of Otago.

Pfeiffer, S., and P. King (1983). Cortical bone formation and diet among protohistoric Iroquians. *American Journal of Physical Anthropology* **60**: 23-28.

Pfeiffer, S.K., and R.A. Lazenby (1994). Low bone mass in past and present aboriginal populations. *Advances in Nutritional Research* **9**: 35-51.

Phenice, T.W. (1969). A newly developed visual method of sexing the os pubis. *American Journal of Physical Anthropology* **30**: 297-302.

Plato, C.C., and A.H. Norris (1980). Bone measurements of the second metacarpal and grip strength. *Human Biology* **52**: 131-149.

Powell, M.L. (1985). The analysis of dental wear and caries for dietary reconstruction. In *The Analysis of Prehistoric Diets*. R.I. Gilbert and J.H. Mielke, (eds). pp. 307-338. Orlando, Florida: Academic Press.

Powell, M.L. (1988). *Status and Health in Prehistory: A Case Study of the Moundville Chiefdom*. Washington: Smithsonian Institution Press.

Preece, M. (1998). Sexual dimorphism of physique. In *The Cambridge Encyclopedia of Human Growth and Development.* S.J. Ulijaszek, F.E. Johnston, and M.A. Preece, (eds). pp. 195. Cambridge: Cambridge University Press.

Putschar, W. (1931). *Entwicklung, Wachstum und Pathologie der Beckenverbindungen des Menschen, mit besonderer Berücksichtigung von Schwangershaft, Geburt und ihren Folgen.* Jena: Gustav Fischer.

Radin, E.L. *et al.* (1973). Response of joints to impact loading: III. Relationship between trabecular microfractures and cartilage degeneration. *Journal of Biomechanics* **6**: 51-57.

Ramalingaswami, V. (1973). Endemic goiter in Southeast Asia. New clothes on an old body. *Annals of Internal Medicine* **78**: 277-283.

Reay, A., and N. Chang (1998). The metal artefacts. In *The Excavation of Nong Nor: a Prehistoric Site in Central Thailand*. C.F.W. Higham and R. Thosarat, (eds). pp. 305-314. Dunedin: Department of Anthropology, University of Otago.

Resnick, D., S.C. Manolagas, G. Niwayama, and M.D. Fallon (1995a). Histogenesis, anatomy, and physiology of Bone. In *Diagnosis of Bone and Joint Disorders.* D. Resnick, (ed.). pp. 609-651, Vol. 2. (3rd edition). Philadelphia: W.B.Saunders.

Resnick, D., and G. Niwayama (eds.) (1981). *Diagnosis of Bone and Joint Disorders.* Volume 1. (1st edition). Philadelphia: W.B.Saunders.

Resnick, D., and G. Niwayama (1995a). Degenerative disease of extraspinal locations. In *Diagnosis of Bone and Joint Disorders.* D. Resnick, (ed.). pp. 1263-1371, Vol. 3. (3rd edition). Philadelphia: W.B.Saunders.

Resnick, D., and G. Niwayama (1995b). Degenerative disease of the spine. In *Diagnosis of Bone and Joint Disorders.* D. Resnick, (ed.). pp. 1372-1462, Vol. 3. (3rd edition). Philadelphia: W.B.Saunders.

Resnick, D., and G. Niwayama (1995c). Osteoporosis. In *Diagnosis of Bone and Joint Disorders.* D. Resnick, (ed.). pp. 1783-1853, Vol. 4. (3rd edition). Philadelphia: W.B.Saunders.

Resnick, D., T.G. Goergen, and G. Niwayama (1995b). Physical injury: concepts and terminology. In *Diagnosis of Bone and Joint Disorders.* D. Resnick, (ed.). pp. 2561-2692, Vol. 5. (3rd edition). Philadelphia: W.B.Saunders.

Ribot, I., and C. Roberts (1996). A study of non-specific stress indicators and skeletal growth in two mediaeval subadult populations. *Journal of Archaeological Science* **23**: 67-79.

Roberts, C. (1991). Trauma and treatment in the British Isles in the Historic Period: a design for multidisciplinary research. In *Human Paleopathology: Current Syntheses and Future Options.* D.J. Ortner and A.C. Aufderheide, (eds). pp. 225-240. Washington, DC: Smithsonian Institute Press.

Roberts, C., and K. Manchester (1995). *The Archaeology of Disease.* (2nd edition). Ithaca, New York: Cornell University Press.

Rodriguez Martin, C., and B. Beranger Meteos (1992). *Interpretation of the skeletal remains of "Los Auchones" (Anaga, Santa Cruz de Tenerife). A case of biocultural isolation.* IXth European Meeting of the Paleopathology Association, Barcelona, 1992, pp. 335-343. Museu d'Arqueologia de Catalunya.

Rogers, J. (1998). Paleopathology of osteoarthritis. In *Osteoarthritis*. K.D. Brandt, M. Doherty, and L.S. Lohmander, (eds). pp. 62-69. (2nd edition). Oxford: Oxford University Press.

Rogers, J., L. Shepstone, and P. Dieppe (1997). Bone formers: osteophyte and enthesophyte formation are positively associated. *Annals of the Rheumatic Diseases* **56**: 85-90.

Rogers, J., and T. Waldron (1995). *A Field Guide to Joint Disease in Archaeology.* Chichester: John Wiley.

Rogers, J., T. Waldron, P. Dieppe, and I. Watt (1987). Arthropathies in palaeopathology: the basis of classification according to most probable cause. *Journal of Archaeological Sciences* **14**: 179-193.

Rogers, T., and S. Saunders (1994). Accuracy of sex determination using morphological traits of the human pelvis. *Journal of Forensic Science* **39**(4): 1047-56.

Roosevelt, A.C. (1984). Population, health, and the evolution of subsistence: conclusions from the conference. In *Paleopathology at the Origins of Agriculture.* M.N. Cohen and G.J. Armelagos, (eds). pp. 559-583. Orlando: Academic Press.

Rosing, F.W. (1983). Sexing immature human skeletons. *Journal of Human Evolution* **12**: 149-155.

Rothschild, B.M. (1997). Porosity: A curiosity without diagnostic significance. *American Journal of Physical Anthropology* **104**: 529-533.

Russell, K.F., S.W. Simpson, J. Genovese, M.D. Kinkel, R.S. Meindl, and C.O. Lovejoy (1993). Independent test of the fourth rib aging technique. *American Journal of Physical Anthropology* **92**: 53-62.

Sangvichien, S., V. Srisurin, V. Wattanayinsakul, P. Theerarattakul, and S. Rakvanichpong (n.d.). *Equations for estimation of Thai's stature from the length of long bones*. Unpublished report, Department of Anatomy, Faculty of Medicine, Siriraj Hospital, Mahidol University, Bangkok.

Sangvichien, S.J., V. Srisurin, and V. Watthanayingsakul (1985). Estimation of stature of Thai and Chinese from the length of femur, tibia and fibula. *Siriraj Hospital Gazette* **37**: 215-218.

Sataragsa, P. (1987). *Engineering Geology of Khorat City, Northeastern Thailand*. Bangkok: Asian Institute of Technology.

Sattenspiel, L., and H. Harpending (1983). Stable populations and skeletal age. *American Antiquity* **48**(3): 489-498.

Saul, F.P., and J.M. Saul (1989). Osteobiography: A Maya example. In *Reconstruction of Life from the Skeleton*. M.Y. Iscan and K.A.R. Kennedy, (eds). pp. 287-302. New York: Alan R. Liss.

Saunders, S.R. (1992). Subadult skeletons and growth related studies. In *Skeletal Biology of Past Peoples: Research Methods*. S.R. Saunders and M.A. Katzenberg, (eds). pp. 1-20. New York: Wiley-Liss.

Saunders, S.R., and R.D. Hoppa (1993). Growth deficit in survivors and non-survivors: biological mortality bias in subadult skeletal samples. *Yearbook of Physical Anthropology* **36**: 127-151.

Schamschula, R.G., B.L. Adkins, D.E. Barmes, G. Charlton, and B.G. Davey (1978). *WHO Study of Dental Caries Etiology in Papua New Guinea*. World Health Organisation Offset Publication No. 40. Geneva: World Health Organisation

Schreurs, W.H.P., P. Migasena, P. Pongpaew, N. Vudhivai, and F.P. Scholp (1976). The vitamin B1, B2 and B6 status of school children in two resettlement areas in northeast Thailand. *Southeast Asian Journal of Tropical Medicine and Public Health* **7**: 586-590.

Schutkowski, H. (1993). Sex determination of infant and juvenile skeletons: I. Morphognostic features. *American Journal of Physical Anthropology* **90**: 199-205.

Scott, E.C. (1979). Dental wear scoring technique. *American Journal of Physical Anthropology* **51**: 213-218.

Scrimshaw, N.S. (1981). Significance of the interactions of nutrition and infection in children. In *Textbook of Pediatric Nutrition*. R.M. Suskind, (ed.). pp. 229-240. New York: Raven Press.

Scrimshaw, N.S., C.E. Taylor, and J.E. Gordon (1959). Interactions of nutrition and infection. *American Journal of the Medical Sciences* **237**: 367-403.

Sealy, J.C., M.K. Patrick, A.G. Morris, and D. Alder (1992). Diet and dental caries among later Stone Age inhabitants of the Cape Province, South Africa. *American Journal of Physical Anthropology* **88**: 123-134.

Seckler, D. (1982). Small but healthy? Some basic problems in the concept of malnutrition. In *Newer Concepts in Nutrition and their Implications for Policy*. P.V. Sukhatme (ed.). pp. 139-148. Pune, India: Maharashtra Association for the Cultivation of Science.

Senewiratne, B., S. Thambipillai, J. Hettiarachchi, and K. Senewiratne (1974). Endemic fluorosis in Ceylon. *Transactions of the Royal Society of Tropical Medicine and Hygiene* **68**(2): 105-113.

Shoocongdej, R. (1996). Rethinking the development of sedentary villages in western Thailand. *Indo-Pacific Prehistory Association Bulletin* **14**: 203-215.

Singh, M., A.R. Nagrath, and P.S. Maini (1970). Changes in trabecular pattern of the upper end of the femur as an index of osteoporosis. *Journal of Bone and Joint Surgery* **52-A**: 457-467.

Sjøvold, T. (1990). Estimation of stature from long bones utilizing the line of organic correlation. *Human Evolution* **5**(5): 431-447.

Smith, G. (1941). *Plague on Us*. Commonwealth Fund (as cited in Barton 1977).

Smith, B.H. (1984). Patterns of molar wear in hunter-gatherers and agriculturalists. *American Journal of Physical Anthropology* **63**: 39-56.

Smith, M.O. (1996). 'Parry' fractures and female-directed interpersonal violence: implications from the Late Archaic of west Tennessee. *International Journal of Osteoarchaeology* **6**: 84-91.

Smythe, H.A. (1983). The medical pathogenesis of generalised osteoarthritis. *Journal of Rheumatology* **10** (Suppl.9): 11-12.

Soames, J.V., and J.C. Southam (1998). *Oral Pathology*. (3rd edition). Oxford: Oxford University Press.

Sokoloff, L. (1980). The pathology of osteoarthrosis and the role of ageing. In *The Aetiopathogenesis of Osteoarthrosis*. G. Nuki, (ed.). Tunbridge Wells, England: Pitman Medical.

Sørenson, P. and T. Hatting (1967). *Archaeological Excavations in Thailand Vol. II: Ban Kao*. Copenhagen: Munksgaard.

Sornmani, S., P. Vivatanasesth, P. Impand, W. Phatihatakorn, and P. Sitabutra (1984). Infection and re-infection rates of opisthorchiasis in the Water Resource Development Area of Nam Pong Project, Khon Kaen Province, northeast Thailand. *Annals of Tropical Medicine and Parasitology* **78**(6): 649-656.

Spector, T.D., and G.D. Campion (1989). Generalised osteoarthritis: a hormonally mediated disease. *Annals of the Rheumatic Diseases* **48**: 523-527.

Springfield, D., and C. Jennings (1991). Pathologic fractures. In *Rockwood and Green's Fractures in Adults*. C.A. Rockwood, D.P. Green, and R.W. Bucholz, (eds). pp. 417-440, Vol. 1. (3rd edition). Philadelphia: Lipincott.

Sreebny, L.M. (1983). Cereal availability and dental caries. *Community Dentistry and Oral Epidemiology* **11**(3): 148-155.

Steckel, R.H. (1995). Stature and the standard of living. *Journal of Economic Literature* **33**: 1903-1940.

Steele, D.G. (1970). Estimation of stature from fragments of long limb bones. In *Personal Identification in Mass Disasters*. T.D. Stewart, (ed.). pp. 85-97. Washington: National Museum of Natural History, Smithsonian Institution Press.

Stephenson, L.S., M.C. Lathan, K.M. Kurz, M. Kinoti, M.L. Odouri and D.W.T. Crompton (1985). Relationship of *Schistosoma haematobium*, hookworm and malarial infections and metrifonate treatment to growth of Kenyan school children. *American Journal of Tropical Medicine and Hygiene* **34**: 1109-1118.

Stewart, T.D. (1970). Identification of the scars of parturition in the skeletal remains of females. In *Personal Identification in Mass Disasters*. T.D. Stewart, (ed.). pp. 127-136. Washington: Smithsonian Institution Press.

Stewart, T.D. (1979). *Essentials of forensic anthropology: especially as developed in the United States.* Springfield, Illinois: C.C.Thomas.

Stini, W.A. (1969). Nutritional stress and growth: sex difference in adaptive response. *American Journal of Physical Anthropology* **31**: 417-426.

Stini, W.A. (1985). Growth rates and sexual dimorphism in evolutionary perspective. In *The Analysis of Prehistoric Diets*. R.I. Gilbert and J.H. Mielke, (eds). pp. 191-226. Orlando, Florida: Academic Press.

Stini, W.A. (1995). Osteoporosis in biocultural perspective. *Annual Review of Anthropology* **24**: 397-421.

Stinson, S. (1985). Sex differences in environmental sensitivity during growth and development. *Yearbook of Physical Anthropology* **28**: 123-147.

Stone, A.C., G.R. Milner, S. Paabo, and M. Stoneking (1996). Sex determination of ancient human skeletons using DNA. *American Journal of Physical Anthropology* **99**: 231-238.

Suchey, J.M., D.V. Wisely, R.F. Green, and T.T. Noguchi (1979). Analysis of dorsal pitting in the os pubis in an extensive sample of modern American females. *American Journal of Physical Anthropology* **51**: 517-540.

Sundick, R.I. (1978). Human skeletal growth and age determination. *Homo* **29**: 228-249.

Suttapreyasri, D., S. Charoenlarp, and S. Suwanakul (1975). Better nutritional status of Thai adults. *Journal of the Medical Association of Thailand* **58**: 357-362.

Suvarnakich, K. (1950). The Thai diet. *Nutrition Reviews* **8**(10): 289-291.

Suwatano, O. (1997). Acute diarrhoea in under five-year-old children admitted to King Mongkut Prachomklao Hospital, Phetchaburi Province. *Journal of the Medical Association of Thailand* **80**(1): 26-33.

Swanson, A.B., and G. de Groot-Swanson (1985). Osteoarthritis in the hand. In *Osteoarthritis*. L. Sokoloff, (ed.). Clinics in the Rheumatic Diseases. Vol. 11, no.2. pp.393-420. London: W.B.Saunders.

Symington, K., and J.C. Rankin (1908). *An Atlas of Skiagrams Illustrating the Development of the Teeth.* London: Longmans, Green.

Tainter, J. (1978). Social inference and mortuary practices, an experiment in numerical classification. *World Archaeology* **7**: 1-15.

Tanner, J.M. (1978). *Education and Physical Growth.* (2nd edition). London: Hodder and Stoughton and New York: International Universities Press.

Tayles, N. (1992). *The people of Khok Phanom Di: Health as evidence of adaptation in a prehistoric Southeast Asian population.* PhD thesis, Department of Anatomy and Structural Biology, University of Otago.

Tayles, N. (1996a). Anemia, genetic diseases, and malaria in prehistoric mainland Southeast Asia. *American Journal of Physical Anthropology* **101**: 11-27.

Tayles, N. (1996b). Tooth ablation in prehistoric Southeast Asia. *International Journal of Osteoarchaeology* **6**: 333-345.

Tayles, N. (1999). *The Excavation of Khok Phanom Di, a Prehistoric Site in Central Thailand. Vol. 5: The People.* London: Society of Antiquaries of London, Research Report no. LXI.

Tayles, N., K.M. Domett, and V. Hunt (1998a). The People of Nong Nor. In *The Excavation of Nong Nor: A prehistoric site in central Thailand.* C.F.W. Higham and R. Thosarat, (eds). pp. 321-368. Monographs in Prehistoric Anthropology, Vol. No.18. Dunedin: Department of Anthropology, University of Otago.

Tayles, N., K.M. Nelsen, H. Buckley, and K. Domett (1998b). The Iron-age people of Noen U-Loke, northeast Thailand: A preliminary review of the first two field seasons. *Bulletin of the Indo-Pacific Prehistory Association* **17**(1): 75. (Abstract)

Tayles, N., K. Domett, and K. Nelsen (2000). Agriculture and dental caries? The case of rice in prehistoric Southeast Asia. *World Archaeology* **32**(1): 68-83.

Thompson, G.B. (1996). *The Excavation of Khok Phanom Di: a Prehistoric Site in Central Thailand. Volume IV: Subsistence and Environment: the Botanical Evidence.* London: The Society of Antiquaries of London.

Thylstrup, A., and O. Fejerskov (1994). *Textbook of Clinical Cariology*. Copenhagen: Munksgaard.

Todd, T.W. (1921). Age changes in the pubic bone. *American Journal of Physical Anthropology* **4**: 1-421.

Trotter, M. (1970). Estimation of stature from intact long limb bones. In *Personal Identification in Mass Disasters*. T.D. Stewart, (ed.). pp. 71-97. Washington: Smithsonian Institution.

Trotter, M., and G.C. Gleser (1951). The effects of aging on stature. *American Journal of Physical Anthropology* **9**: 311-324.

Trotter, M., and G.C. Gleser (1952). Estimation of stature from long bones of American Whites and Negroes. *American Journal of Physical Anthropology* **10**(4): 463-514.

Trotter, M., and G.C. Gleser (1958). A re-evaluation of estimation of stature based on measurements of stature taken during life and of long bones after death. *American Journal of Physical Anthropology* **16**: 79-123.

Turner, C.G. (1979). Dental anthropological indication of agriculture among the Jomon people of central Japan. X. Peopling of the Pacific. *American Journal of Physical Anthropology* **51**: 619-636.

Ubelaker, D.H. (1978). *Human Skeletal Remains: Excavation, Analysis, Interpretation.* (1st edition). Chicago: Adline.

Ubelaker, D.H. (1987). Estimating age at death from immature human skeletons: an overview. *Journal of Forensic Sciences* **32**(5): 1254-1263.

Ubelaker, D.H. (1989). *Human Skeletal Remains.* (2nd edition). Washington: Taraxacum.

Udomochoke, V. (1989). Quaternary stratigraphy of the Khorat plateau area of Northeastern Thailand. In *Proceedings of the Workshop on Correlation of Quaternary Successions in South, East and Southeast Asia*. N. Thiramongkol (ed.). pp. 69-94. Bangkok: Department of Geology, Chulalongkorn University.

Ulijaszek, S.J. (1998). The genetics of growth. In *The Cambridge Encyclopedia of Human Growth and Development.* S.J. Ulijaszek, F.E. Johnston, and M.A. Preece, (eds). pp. 121-123. Cambridge: Cambridge University Press.

van der Linden, F.P.G.M., and H.S. Duterloo (1976). *Development of the Human Dentition: An Atlas.* Hagerstown: Harper and Row.

van Saase, J.L.C.M., L.K.J. van Romunde, A. Cats, J.P. Vandenbroucke, and H.A. Valkenburg (1989). Epidemiology of osteoarthritis: Zoetermeer. Comparison of radiological osteoarthritis in a Dutch population with that in 10 other populations. *Annals of the Rheumatic Diseases* **48**: 271-280.

van Vark, G.N., and W. Schaafsma (1992). Advances in the quantitative analysis of skeletal morphology. In *Skeletal Biology of Past Peoples: Research Methods.* S.R. Saunders and M.A. Katzenberg, (eds). pp. 225-257. New York: Wiley-Liss.

Videman, T., M. Nurminen, and J.D.G. Troup (1990). Lumbar spinal pathology in cadaveric material in relation to history of back pain, occupation, and physical loading. *Spine* **15**(8): 728-740.

Waldron, H.A. (1997). Association between osteoarthritis of the hand and knee in a population of skeletons from London. *Annals of the Rheumatic Diseases* **56**: 116-118.

Waldron, H.A., and M. Cox (1989). Occupational arthropathy: evidence from the past. *British Journal of Industrial Medicine* **46**: 420-422.

Waldron, T. (1987). The relative survival of the human skeleton: implications for palaeopathology. In *Death, Decay and Reconstruction.* A. Boddington, A.N. Garland and R.C. Janaway (eds). pp. 54-64. Manchester: Manchester University Press.

Waldron, T. (1994). *Counting the Dead. The Epidemiology of Skeletal Populations.* Chichester: John Wiley and Sons.

Waldron, T. (1995). Change in the distribution of osteoarthritis over historical time. *International Journal of Osteoarchaeology* **5**: 385-389.

Waldron, T. (1998). A note on the estimation of height from long-bone measurements. *International Journal of Osteoarchaeology* **8**: 75-77.

Walker, P.L. (1986). Sex differences in the diet and dental health of prehistoric and modern hunter-gatherers. *European Meeting of the Paleopathological Association*, Madrid.

Walker, R.A., and C.O. Lovejoy (1985). Radiographic changes in the clavicle and proximal femur and their use in the determination of skeletal age at death. *American Journal of Physical Anthropology* **68**: 67-78.

Wall, C.E. (1991). Evidence of weaning stress and catch-up growth in the long bones of a Central California Amerindian sample. *Annals of Human Biology* **18**(1): 9-22.

Weaver, D.S. (1998). Osteoporosis in the bioarchaeology of women. In *Sex and Gender in Paleopathological Perspective*. A.L. Grauer and P. Stuart-Macadam, (eds). pp. 27-44. Cambridge: Cambridge University Press.

Webb, P.A.O., and J.M. Suchey (1985). Epiphyseal union of the anterior iliac crest and medial clavicle in a modern multiracial sample of American males and females. *American Journal of Physical Anthropology* **68**: 457-466.

Weiss, K.M. (1973). Demographic models for anthropology. *American Antiquity* **38**: 1-88.

Welch, D.J. (1985). *Adaptation to environmental unpredictability: Intensive agriculture and regional exchange at Late Prehistoric centres in the Phimai region in Thailand*. PhD thesis, University of Hawaii.

White, J.C. (1995). Modeling the development of early rice agriculture: ethnoecological perspectives from Northeast Thailand. *Asian Perspectives* **34**(1): 37-68.

WHO (1987). World Health Organization. *Preventing the Tragedy of Maternal Deaths*. Report on the International Safe Motherhood Conference, Narobi, Kenya.

Wilson, C.S. (1985). Staples and calories in Southeast Asia: the bulk of consumption. In *Food Energy in Tropical Ecosystems*. D.J. Cattle and K.H. Schwein, (eds). pp. 65-81. Food and Nutrition in History and Anthropology. London: Gordon and Breach Science Publishers.

Wiriyaromp, W. (1984). *The human skeletal remains from Ban Na Di*. Master of Arts, Department of Anthropology and Department of Anatomy, University of Otago.

Wood, C.S. (1983). Early childhood, the critical stage in human interactions with disease and culture. *Social Science and Medicine* **17**(2): 79-85.

Wood, J.W., G.R. Milner, H.C. Harpending, and K.M. Weiss (1992). The osteological paradox. Problems of inferring health from skeletal samples. *Current Anthropology* **33**(4): 343-370.

Zar, J.H. (1984). *Biostatistical Analysis*. (2nd edition). Englewood Cliffs, New Jersey: Prentice-Hall International.

www.ingramcontent.com/pod-product-compliance
Lightning Source LLC
LaVergne TN
LVHW071210210926
848482LV00015B/769